The FitzHugh-Nagumo Model

MATHEMATICAL MODELLING:
Theory and Applications

VOLUME 10

This series is aimed at publishing work dealing with the definition, development and application of fundamental theory and methodology, computational and algorithmic implementations and comprehensive empirical studies in mathematical modelling. Work on new mathematics inspired by the construction of mathematical models, combining theory and experiment and furthering the understanding of the systems being modelled are particularly welcomed.

Manuscripts to be considered for publication lie within the following, non-exhaustive list of areas: mathematical modelling in engineering, industrial mathematics, control theory, operations research, decision theory, economic modelling, mathematical programmering, mathematical system theory, geophysical sciences, climate modelling, environmental processes, mathematical modelling in psychology, political science, sociology and behavioural sciences, mathematical biology, mathematical ecology, image processing, computer vision, artificial intelligence, fuzzy systems, and approximate reasoning, genetic algorithms, neural networks, expert systems, pattern recognition, clustering, chaos and fractals.

Original monographs, comprehensive surveys as well as edited collections will be considered for publication.

The titles published in this series are listed at the end of this volume.

The FitzHugh-Nagumo Model

Bifurcation and Dynamics

by

C. Rocşoreanu
Department of Mathematics,
University of Craiova,
Craiova, Romania

A. Georgescu
Department of Mathematics,
University of Piteşti,
Piteşti, Romania

and

N. Giurgiţeanu
Department of Economics,
University of Craiova,
Craiova, Romania

KLUWER ACADEMIC PUBLISHERS
DORDRECHT / BOSTON / LONDON

A C.I.P. Catalogue record for this book is available from the Library of Congress.

ISBN 0-7923-6427-9

Published by Kluwer Academic Publishers,
P.O. Box 17, 3300 AA Dordrecht, The Netherlands.

Sold and distributed in North, Central and South America
by Kluwer Academic Publishers,
101 Philip Drive, Norwell, MA 02061, U.S.A.

In all other countries, sold and distributed
by Kluwer Academic Publishers,
P.O. Box 322, 3300 AH Dordrecht, The Netherlands.

Printed on acid-free paper

Printed in the Netherlands.

Table of Contents

INTRODUCTION

The present monograph analyses the *FitzHugh–Nagumo (F–N) model* i.e., the Cauchy problem for some generalized Van der Pol equation depending on three real parameters a, b and c. This model, given in (1.1.17), governs the initiation of the cardiac impulse.

The presence of the three parameters leads to a large variety of dynamics, each of them responsible for a specific functioning of the heart. For physiologists it is highly desirable to have a global view of all possible qualitatively distinct responses of the F–N model for all values of the parameters. This reduces to the knowledge of the global bifurcation diagram.

So far, only a few partial results appeared and they were spread throughout the literature. Our work provides a more or less complete theoretical and numerical investigation of the complex phase dynamics and bifurcations associated with the F–N dynamical system. This study includes the static and dynamic bifurcations generated by the variation of a, b and c and the corresponding oscillations, of special interest for applications. It enables one to predict all possible types of initiations of heart beats and the mechanism of transformation of some types of oscillations into others by following the dynamics along transient phase space trajectories. Of course, all these results hold for the F–N model. The global phase space picture enables one to determine the domain of validity of this model.

Biomathematicians, biophysicists, researchers and graduate students in applied mathematics will find in this book one of the major applications of dynamical systems and bifurcation theory. A secondary audience consists of biologists, cardiologists and undergraduate students in mathematics. Biologists will be interested in the types of oscillations and canard phenomenon, whilst the students can use the book as a case study in nonlinear dynamics and bifurcation.

Large values of c (namely $c > 1+\sqrt{3}$) are considered since it is for these values that relaxation oscillations occur. The case $c < 1+\sqrt{3}$ was left aside, in view of its irrelevancy for our purpose, which is to provide physiologists with a rigorous study of the entire body of predictions associated with the F–N model. However, a mathematical study of this case would be of interest for bifurcation theory.

Another limitation of our analysis arises from the assumption that, in most cases, c is fixed. Only rarely have we considered a variable c.

Chapter 1 covers preliminary matters. It concerns the physiological bases of the initiation of the cardiac electric impulse and the place the F–N model occupies in the context of the mathematical models governing the heart functioning (Section 1.1). Elements of nonlinear phase space dynamics (Section 1.2), definitions and results from bifurcation theory (Section 1.3), singular perturbations and methods for deriving the models of asymptotic approximations (Section 1.4) are introduced and studied.

Our original contribution to the investigation of the F–N model starts with Chapter 2. We deduce the nullclines and the locus of the inflection points of the phase trajectories (Section 2.1). Equlibria are investigated in Section 2.2, and the system linearized about some equlibrium point in Section 2.3. The results are used in Section 2.4 and Section 2.5 to study the geometry of the static bifurcation diagrams at moderate and large values of c, respectively. All this information is used in Section 2.6 to derive the types of hyperbolic equilibria. In Section 2.7, using the center manifold method, the type of some non-hyperbolic equilibria is deduced.

Chapter 3 deals with dynamic bifurcations: Hopf bifurcation (Section 3.1), Bogdanov–Takens codimension-two bifurcations (Section 3.2), homoclinic, double homoclinic, saddle–node homoclinic bifurcations (Section 3.3), Bautin and non-hyperbolic limit cycle bifurcation (Section 3.5). The succession of all transformations leading to the normal form associated with some types of phenomena are presented in detail. For most of the manifolds of the parameter space corresponding to a single type of bifurcation, asymptotic representations are deduced as a result of the local study around the Hopf, Bogdanov–Takens and Bautin bifurcation points. Outside the domain of validity of the asymptotic formulae these manifolds are continued by numerical curves.

In Section 3.4 the breaking saddle-connection bifurcation is investigated. The role of the saddles in the phase portrait is shown and their stable and unstable manifolds are derived. Numerically, other types of codimension-one and -two bifurcations were found. We named them saddle–node–saddle connection bifurcation and saddle–node–saddle with separatrix connection bifurcation, respectively.

A lot of results concerning the asymptotic behaviour of the solution of the F–N model as $c \to \infty$ are available in the literature. Since our aim is to study the dynamics rather than the asymptotics associated with this model, in Chapter 4 we discuss the directions of the general asymptotic investigations (Section 4.1), emphasizing the two time scales for the oscillatory dynamics and deriving the models of inner and outer asymptotic approximation of first (Section 4.2) and higher order (Section 4.3) for the

F–N model as $c \to \infty$. The matching of these models is carried out using the $n \times m$ principle. Some asymptotical results on canards are reported (Section 4.5).

All the previous results are ordered in Chapter 5 to obtain the global bifurcation diagram (i.e., the parametric portrait and the types of dynamics) for the F–N model (Section 5.1). It contains local and nonlocal bifurcations. We found that for the region in the parameter space of the greatest physiological interest the attractive limit cycles prevail. One, two or three limit cycles were revealed. For some regions in the parameter space, two or three limit cycles coexist around various branches of equilibria. For some other regions, two limit cycles around the same branch of equilibria coexist. These situations give rise to a large variety of oscillatory dynamics. In Section 5.2 the basins of attraction are deduced. The transient regimes and their corresponding oscillations are studied in Section 5.3, while the oscillatory regimes and their oscillations are studied in Section 5.4. In Section 5.5 an attempt to explain the initiation of heart beats is done. Concluding remarks of interest to physiologists and mathematical open problems are presented in Sections 5.6 and 5.7.

Appendix A provides expressions for the Liapunov coefficients. They were necessary for the investigation of the Hopf and Bautin bifurcation. The cumbersome computations including these coefficients were carried out with MATHEMATICA. The coefficients were obtained [123] for equations with third degree nonlinearities, more general than those from the F–N model.

Appendix B presents the main ideas underlying the numerical methods based on a variable step found by one of the authors (Nicolaie Giurgiţeanu). These methods were incorporated in the DIECBI application. The availability and the performances of DIECBI were decisive for the cases where analytical results were absent.

The topics treated in this book, as well as the extent of the treatment, reflect the state of the art in nonlinear dynamics. We had at our disposal a lot of recent results on dynamic bifurcation. Applied to the F–N model, they resulted in a quite different study if compared with investigations carried out in previous monographs on particular dynamical systems, e.g., Sparrow's famous book on the Lorenz system [132]. For the Lorenz system, of primary interest was the chaotic behaviour. Our study, somewhat complementary to Sparrow's book, represents its continuation by other methods and having in view other goals, namely, bifurcation and dynamics. For a two-dimensional continuous dynamical system as the F–N model, which does not present chaotic dynamics, we revealed an impressive and unexpectedly large number of bifurcations: all types of generic codimension-one bifurcations described in the literature for planar systems, and 18 types of codimension-two bifurcations. On the one hand, this shows the rich dynamics of the F–N model (there are more than 60 types of topologically non-equivalent dynamics) and on the other, the relevance for bifurcation theory of this model. In conclusion, biologists interested in this model must join their efforts with that of mathematicians in order to obtain a reasonable response of the F–N model.

The idea of writing such a book appeared during the courses and seminars on bifurcation and nonlinear dynamics delivered by the second author at the Universities of Bucharest, Paris and Piteşti. Later on, her former PhD students Carmen Rocşoreanu and Nicolaie Giurgiţeanu combined their efforts in carrying out a comprehensive treatment of the F–N model. We hope that our work will inspire researchers in treating many other nonlinear systems of basic interest to applications.

Carmen Rocşoreanu
Dept. of Mathematics, University of Craiova, Romania

Adelina Georgescu
Dept. of Mathematics, University of Piteşti, Romania

Nicolaie Giurgiţeanu
Dept. of Economics, University of Craiova, Romania

CHAPTER 1

MODELS AND DYNAMICS

In Section 1.1 the physiological and mathematical context of the FitzHugh–Nagumo (F–N) model is presented. The first four subsections contain the physiological basis and the models based on two first order ordinary differential equations (ODE) of initiation of cardiac impulse. Thus, in Section 1.1.1 the types and characteristics of the electrical circuits of the heart are described. The famous Van der Pol model is presented in Section 1.1.2, together with its main types of oscillations. The relaxation oscillations are defined. The Van der Pol generalized model and its particular case, the F–N model, are presented in Section 1.1.3. Different forms of the F–N model are given in Section 1.1.4, together with certain general properties of some of them. The study of one of the forms of the F–N model, namely (1.1.17), is the aim of this book. In the last subsection (1.1.5) the connection with the propagation by rotating waves in the whole cardiac muscle of the electric impulse generated in the sinusal node (Figure 1.1.1) is provided. Models of coupled oscillators, based on systems of ordinary differential equations (SODE) of first order and models with forcing terms related to the F–N model are also mentioned.

Elements of nonlinear finite-dimensional dynamics and bifurcation are presented in Sections 1.2 and 1.3 respectively. The Hopf, Bogdanov–Takens, Bautin and saddle–node separatrix loop bifurcation diagrams are shown. In addition, the structural stability is defined and the characterization of the two-dimensional structurally stable systems is given.

The last section (1.4) concerns elements of asymptotic approximation: singular perturbation problems, nonuniformly valid asymptotic expansions, matching principles, models of asymptotic approximation of different orders.

1.1. MODELS OF THE HEART FUNCTIONING

1.1.1. PROCESSES ASSOCIATED WITH THE PERIODICITY OF THE CARDIAC ELECTRIC SIGNAL

Questions such as how the cardiac electric potential is initiated and propagated, what is the cyclic shape of the cardiac phenomenon, or how is the cardiac rhythm preserved, found their answers using methods from physics, chemistry, and mathematics. The main phenomenon which must be explained is the periodicity of the cardiac electric signal (auto-rhythmicity) [105]. The ability of the heart to beat autonomously is the result of some periodic processes, adjusted at molecular level [22], [23], [28]. In mathematical terms this means that the system of equations modeling the electrical behaviour of cardiac cells must have a periodic and stable solution [100], [144], [60].

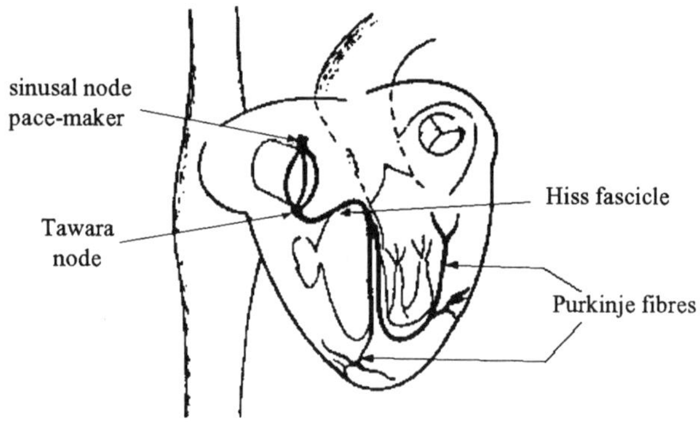

Fig. 1.1.1. Electrical circuits of the heart.

Amongst the electrical circuits of the heart, we quote (Figure 1.1.1):

—the **sinusal node**, situated inside the right atrium, having the fastest rhythm, therefore assumes the role of pace-maker. This node induces about 70 heart beats/minute. If it is hurt, then its function is taken over by the next node;

—the **atrio-ventricle node (Tawara)** represents the connection between the right atrium and ventricle, and produces a delay of about 0.1 seconds. This node induces only 45 beats of heart every minute. This is the reason why, when the sinusal node is hurt, the implantation of a cardiac stimulator is needed;

—the **Hiss fascicle**, consisting of specialized myocardial fibres, whose ramifications develop in both ventricles;

—the **Purkinje fibres** owing to which the excitation wave develops in both ventricles. They contract almost simultaneously.

It follows that the sinusal node and, in exceptional situations, the Tawara node, produces the initiation of the electrical impulse, whilst the other electrical circuits of the heart contribute to its propagation.

At the cell level, during a polarisation-repolarisation cycle, the electrical potential of the cell membrane, x, undergoes four successive stages (Figure 1.1.2) [23]:

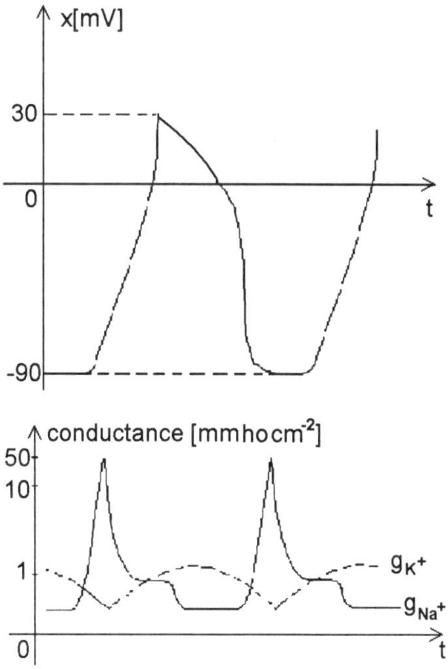

Fig. 1.1.2. The time variation of the electrical potential of a cell membrane and of the conductance g.

— **depolarisation**, associated with the increase of the sodium conduc tance and the sudden penetration of Na^+ ions. The potential increases from -90mV to $+30$mV;

— **slow repolarisation**, when the calcium conductance increases and the Ca^{++} ions penetrate. This phase is absent in the nervous fibre;

— **fast repolarisation**, with the increase of the potassium conductance and the exit of the K^+ ions;

— **the end of repolarisation**, when extra-cellular sodium concentration increases and inter-cellular potassium concentration decreases.

Owing to the topography [145], [88], [89], [101], [75] different shapes of the potential and conductance exist.

The excitation has the following properties [23]:

i) The effect of the stimulus on the polarization is not instantaneous. It takes place with a time delay. In this case we say that the membrane has a capacitive effect. Consequently it is represented as an RC circuit;

ii) The answer of the membrane depends on the stimulus intensity. It follows a law of **all or none** responses: if the stimulus intensity is smaller than a threshold value, the cardiac muscle does not contract, whilst if it is greater than the threshold value, the potential of the cell membrane has the following characteristics:

— the contraction of the cardiac fibre is maximum, irrespective of the stimulus size (**active period**);

— the active period is followed by an **absolute refractory period**, when the cardiac fibre does not respond to the action of another stimulus;

— the absolute refractory period is followed by a **relative refractory period**.

1.1.2. VAN DER POL OSCILLATOR AND RELAXATION OSCILLATIONS

The first attempt to describe, qualitatively, the oscillatory behavior of the heart, was made by Van der Pol, in 1926 [139]. He observed for the first time the relaxation oscillations, by studying an electrical circuit that presents self-entertained oscillations, with an amplitude which does not depend on the initial conditions. One such electrical circuit is depicted in Figure 1.1.3. Denote by R the resistance, C the capacitance, L the inductance, I the current, V the electrical potential, and M the mutual inductance.

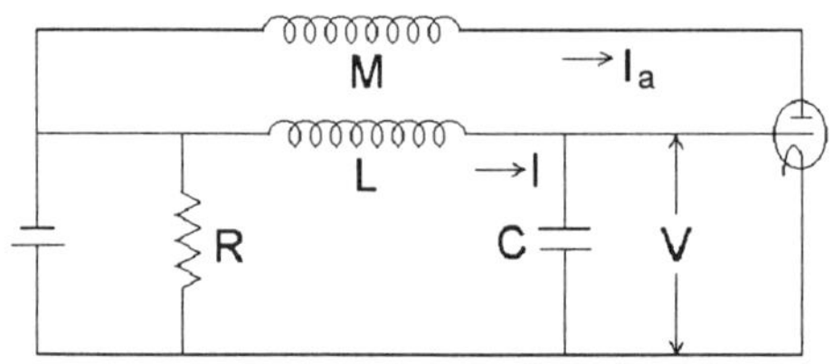

Fig. 1.1.3. Van der Pol electrical circuit.

The equation which characterizes the circuit is

$$L\frac{dI}{d\theta} + RI + V = M\frac{dI_a}{d\theta}$$

with

$$I_a = V - \frac{1}{3}V^3.$$

Since $I = CV'(\theta)$, denoting

$$V = x\sqrt{1 - R\frac{C}{M}}, \qquad \theta = t\sqrt{LC} \quad \text{and} \quad c = \frac{M}{\sqrt{LC}} - R\sqrt{\frac{C}{L}},$$

we have

$$I = \frac{C}{\sqrt{LC}}\sqrt{1 - R\frac{C}{M}\frac{dx}{dt}},$$
$$\frac{dI}{d\theta} = \frac{1}{L}\sqrt{1 - R\frac{C}{M}\frac{d^2x}{dt^2}},$$
$$\frac{dI_a}{d\theta} = \frac{1}{\sqrt{LC}}\sqrt{1 - R\frac{C}{M}}\left(1 - x^2 + \frac{RC}{M}x^2\right)\frac{dx}{dt}$$

and the equation of the electrical circuit becomes the Van der Pol equation

$$\frac{d^2x}{dt^2} + c(x^2 - 1)\frac{dx}{dt} + x = 0. \qquad (1.1.1)$$

For different values of the control parameter c, different types of periodic solutions are obtained (Figure 1.1.4).

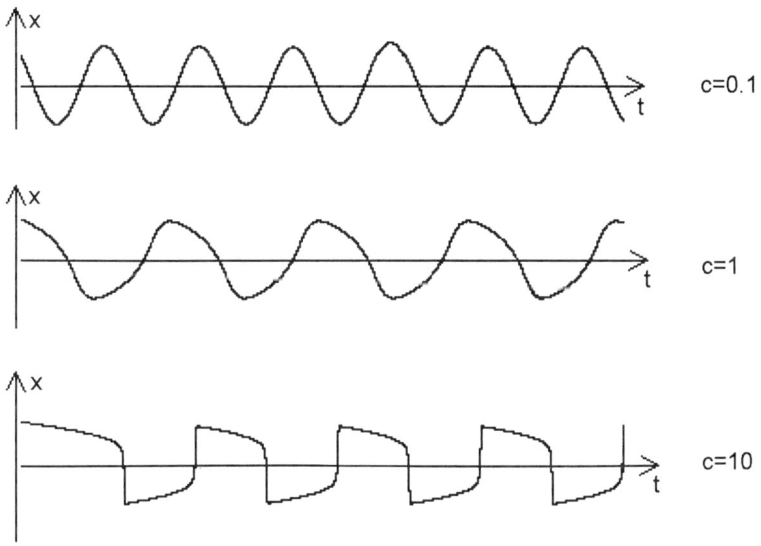

Fig. 1.1.4. Main types of oscillatory solutions of equation (1.1.1).

For large values of c relaxation oscillations occur. They can be found in many material systems.

In particular, in 1933, studying the activity of nerve cells, the English physiologist A.V. Hill stated that relaxation oscillations are the oscillations governing all periodic physiological phenomena. We define these oscillations in a general case [61], [93].

Let us consider the following Cauchy problem for a SODE

$$\begin{aligned} \frac{dx_i}{dt} &= f_i(x_1, \ldots, x_n, c), \\ x_i(0) &= x_i^0, i = \overline{1, n}, \end{aligned} \tag{1.1.2}$$

where c is a large real parameter. By changing the time scale and the parameter according to the transformations

$$t = \tau c, \qquad \varepsilon = \frac{1}{c^2}, \tag{1.1.3}$$

from (1.1.2) it follows that

$$p_i(\varepsilon) \frac{dx_i}{d\tau} = h_i(x_1, \ldots, x_n, \varepsilon). \tag{1.1.4}$$

Assume that p_i and h_i in (1.1.4) are continuous functions which have finite limits as $\varepsilon \to 0$ and there exists a finite number of indices i such as $\lim_{\varepsilon \to 0} p_i(\varepsilon) = 0$.

Suppose that (1.1.4) has a periodic solution corresponding to the closed curve L_ε in the phase space $(x_1, \ldots, x_n)$.

Definition 1.1.1 *We say that a curve C in the space $(x_1, \ldots, x_n)$ is* **strictly transversal** *if it is transversal to any closed curve L_ε, with $0 < \varepsilon \leq \varepsilon_0$, where ε_0 is fixed and there exists $\delta > 0$ independent of ε, such that the acute angle θ_ε between C and L_ε satisfies the inequalities $|\theta_\varepsilon| \geq \delta > 0$.*

Definition 1.1.2 *A periodic solution of system (1.1.4) with period $T(\varepsilon)$ is called a* **relaxation oscillation** *if*

$$\lim_{\varepsilon \to 0} T(\varepsilon)$$

is bounded and nonzero and there exist two strictly transversal curves C_f and C_s such that:

i) there exists i such that

$$\lim_{\varepsilon \to 0} \frac{p_i(\varepsilon)}{h_i(x_\varepsilon, \varepsilon)} = 0, \tag{1.1.5}$$

where $\{x_\varepsilon\} = C_f \cap L_\varepsilon$,

ii) for every i, the limit

$$\lim_{\varepsilon \to 0} \frac{h_i(x_\varepsilon, \varepsilon)}{p_i(\varepsilon)} \tag{1.1.6}$$

is bounded, where $\{x_\varepsilon\} = C_s \cap L_\varepsilon$.

In terms of asymptotic analysis, (1.1.5) and (1.1.6) mean $p_i = o(h_i)$ on $C_f \cap L_\varepsilon$ and $h_i = O(p_i)$ on $C_s \cap L_\varepsilon$ respectively, as $\varepsilon \to 0$. The subscripts f and s stand for *fast* and *slow* respectively.

The systems (1.1.4) can present relaxation oscillations only if they are nonlinear. In addition, the Definition 1.1.2 does not suppose the stability of the oscillation. This is why it is not in perfect agreement with the practicians' idea of relaxation oscillations. The main property of a relaxation oscillation is that it changes fast in a short interval of time.

Using (1.1.3), the Van der Pol equation (1.1.1) reads

$$\varepsilon \frac{d^2x}{d\tau^2} + (x^2 - 1)\frac{dx}{d\tau} + x = 0. \tag{1.1.7}$$

Then, using the Liénard representation [96]

$$y = \varepsilon \frac{dx}{d\tau} + \frac{1}{3}x^3 - x, \tag{1.1.8}$$

equation (1.1.7) can be written in the form of the following system

$$\begin{aligned} \varepsilon \frac{dx}{d\tau} &= y - (\tfrac{1}{3}x^3 - x), \\ \frac{dy}{d\tau} &= -x. \end{aligned} \tag{1.1.9}$$

Therefore, in this particular case, we have $p_1(\varepsilon) \equiv \varepsilon$, $p_2 = 1$ and $h_1 = y - (\frac{1}{3}x^3 - x)$, $h_2 = -x$, where h_1 and h_2 do not depend on ε. It follows that $\frac{p_1}{h_1} \to 0$ along the curve $y = -\frac{2}{3}$ and $\frac{h_1}{p_1} < \infty$ and $\frac{h_2}{p_2} < \infty$ along the curve $y = \frac{x^3}{3} - x$, as $\varepsilon \to 0$. Hence the Van der Pol system (1.1.9) may possess relaxation oscillations [19]. These oscillations require an asymptotic study as $\varepsilon \to 0$, widely illustrated in the literature [4], [61], [22], [76], [111].

1.1.3. GENERALIZED VAN DER POL MODEL. FITZHUGH–NAGUMO MODEL

The generalized Van der Pol type of oscillator is the SODE of the form [61]

$$\begin{aligned} \varepsilon \frac{dx_i}{d\tau} &= F_i(x, y, \varepsilon), \; i = \overline{1, m}, \\ \frac{dy_j}{d\tau} &= G_j(x, y, \varepsilon), \; j = \overline{1, n} \end{aligned} \tag{1.1.10}$$

where $\varepsilon > 0$, $\mathbf{x}(\tau) = (x_1(\tau), \dots, x_m(\tau))$, $\mathbf{y}(\tau) = (y_1(\tau), \dots, y_n(\tau))$ and τ is the time.

The Cauchy problem $x_i(0) = x_i^0$, $y_j(0) = y_j^0$ for (1.1.10) is referred to as the generalized Van der Pol model. A model of this type associated with the variation in time of the electrical potential of the neural membrane, is the Hodgkin–Huxley model for nerve axon [27], [72] [54] and it consists of a Cauchy problem for a system of four first order ODEs.

Starting with this model FitzHugh [40], [39], [41] constructed a two-dimensional model, describing physiological phenomena similar to those corresponding to the Hodgkin–Huxley model [102]. It is the Cauchy problem $x(0) = x^0, y(0) = y^0$ for the following SODE

$$\begin{aligned} \frac{dx}{dt} &= c(x + y + z - \frac{x^3}{3}), \\ \frac{dy}{dt} &= -\frac{1}{c}(x - a + by). \end{aligned} \tag{1.1.11}$$

The parameter z stands for the injected current, x is the electrical potential of the cell membrane and y is an auxiliary variable depending on the refractory period. The parameters a and b are related to the number of channels of the cell membrane which are opened to the Na^+ and K^+ ions and FitzHugh considered $a \geq 0,\ 0 \leq b < 1$.

With the transformation (1.1.3) the system (1.1.11) is written as

$$\begin{aligned} \varepsilon \frac{dx}{d\tau} &= x + y - \frac{x^3}{3} + z, \\ \frac{dy}{d\tau} &= -x - by + a. \end{aligned} \tag{1.1.12}$$

For $a = b = z = 0$, the Van der Pol system (1.1.9) is obtained. Hence, the F–N model generalizes the Van der Pol model.

The elimination of y between the two equations of the system (1.1.11) leads to the F–N equation

$$\frac{d^2x}{dt^2} - (c - c\,x^2 - \frac{b}{c})\frac{dx}{dt} + x(1-b) + b\,\frac{x^3}{3} - a - b\,z = 0. \tag{1.1.13}$$

It depends on four real parameters a, b, c and z, whilst the Van der Pol equation contained a single parameter.

Analyzing the system (1.1.11) for $1 - \frac{2}{3}b < a < 1$, $0 < b < 1$, FitzHugh emphasized the periods of the cardiac cycle. He considered $a = 0.7, b = 0.8, c = 3$ and variable z.

Thus, the stationary state of the physiological system is given by the solution (x, y) of the system

$$\begin{aligned} y + x - \frac{x^3}{3} + z &= 0, \\ x + b\,y - a &= 0. \end{aligned} \tag{1.1.14}$$

Suppose $z = 0$. In this case, (1.1.11) has a stable equilibrium point $P(x_0, y_0)$, with $x_0 \approx 1.19$, $y_0 \approx -0.62$, situated at the intersection of the curve R-H defined by the equation $(1.1.14)_1$ and the straight line d, defined by $(1.1.14)_2$. The phase plane (x, y) and its regions corresponding to the physiological states are given in Figure 1.1.5 [40].

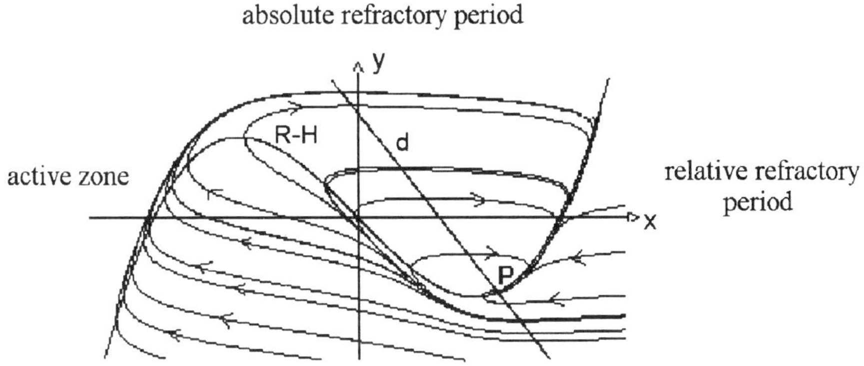

Fig. 1.1.5. The diagram of the physiological states in the F–N model ($a = 0.7$, $b = 0.8$, $c = 3$, $z = 0$).

Consider now the case $z \neq 0$, when the curve $(1.1.14)_1$ is obtained, graphically, by a translation with the quantity z of the R-H curve corresponding to $z = 0$, along the Oy axis. Then, as the solution of (1.1.14) depends also on z, the biological system can evolve either to a stationary state, or to an excitable state (non-stationary).

If the injected current z is negative (cathodic shock) different situations may occur.

For $z = -0.124$, the trajectory of P goes to the new stationary state P_1, without the occurrence of an impulse (Figure 1.1.6a).

For $z = -0.128$, the point P moves into P_2 on a trajectory which corresponds to an impulse, (Figure 1.1.6b) while for $z = -0.4$ a limit cycle occurs and the trajectory of P is attracted by it (Figure 1.1.6c). The occurrence of oscillations when z crosses a critical value and the fact that their shape does not change if z goes on decreasing, emphasizes the law of **all or none** responses of the working heart.

When the injected current, z, is positive (anodic shock) the motion in the (x, y)-plane is sketched in Figure 1.1.6d, emphasizing the phenomenon of breaking anodic excitation.

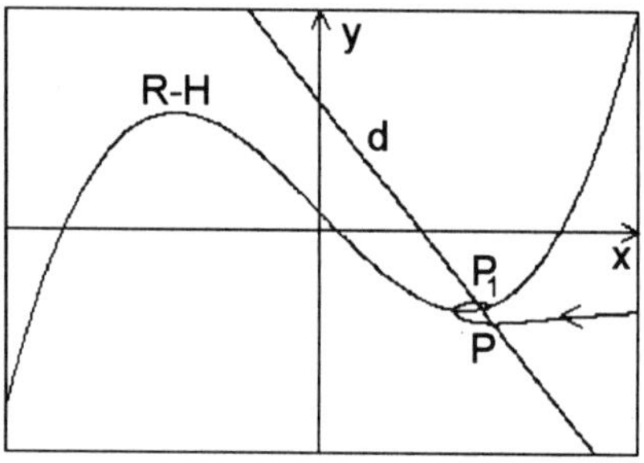

a) $z = -0.124$

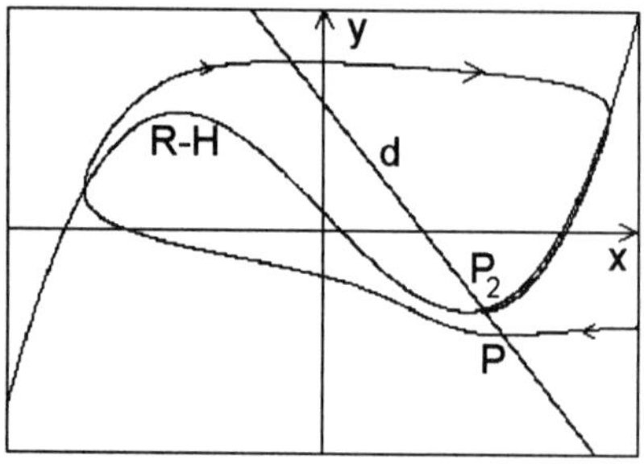

b) $z = -0.128$

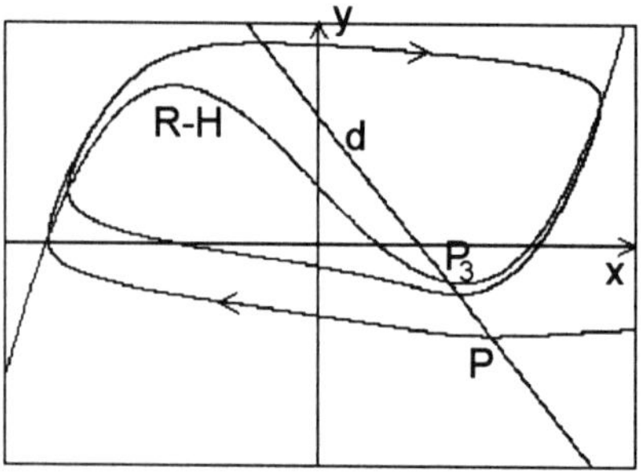

c) $z = -0.4$

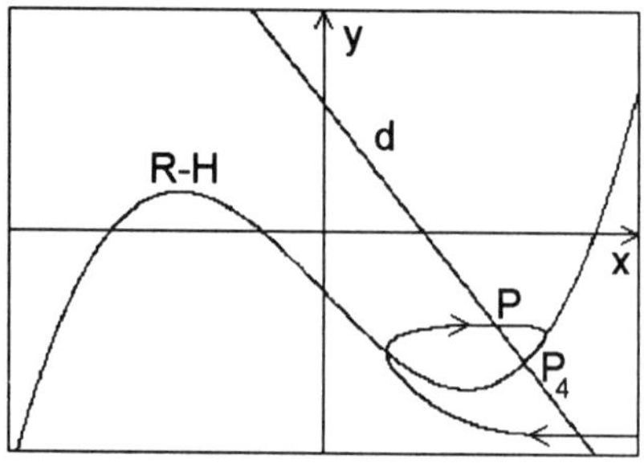

d) $z = 0.4$

Fig. 1.1.6. The diagrams of the dynamics in the (x, y)-plane, for $a = 0.7$, $b = 0.8$, $c = 3$ and $z \neq 0, x \in (-2.2, 2.2)$ and $y \in (-1.5, 1.5)$.

1.1.4. FORMS OF THE FITZHUGH–NAGUMO SYSTEM

In this book we analyze the solution set of the F–N model, for all initial data and all values of the parameters, with the only restriction $c \geq 1+\sqrt{3}$. We recover the cases presented by FitzHugh and other cases occuring in the literature.

We reduce the system (1.1.11) to the form

$$\begin{aligned} \dot{x} &= c\left(x + \overline{y} - \frac{x^3}{3}\right), \\ \dot{\overline{y}} &= -\frac{1}{c}(x - \overline{a} + b\,\overline{y}), \end{aligned} \tag{1.1.15}$$

where

$$\overline{y} = y + z, \quad \overline{a} = a + b\,z. \tag{1.1.16}$$

Here the dot stands for the differentiation with respect to the time t.

We rename the variables $\overline{y}$, $\overline{a}$ as y and a respectively. So, instead of (1.1.15) we write

$$\begin{aligned} \dot{x} &= c\left(x + y - \frac{x^3}{3}\right), \\ \dot{y} &= -\tfrac{1}{c}\,(x - a + b\,y)\,. \end{aligned} \tag{1.1.17}$$

In this book we study the F–N system (1.1.17). In (1.1.17) it is sufficient to consider the case $a \geq 0$. Indeed, with the transformation $x \to -x$, $y \to -y$, (1.1.17) becomes

$$\begin{aligned} \dot{x} &= c\left(x + y - \frac{x^3}{3}\right), \\ \dot{y} &= -\tfrac{1}{c}\,(x + a + b\,y)\,, \end{aligned} \tag{1.1.18}$$

i.e., (1.1.17) is invariant under this transformation if, in addition, we let $a \to -a$.

Therefore, for each fixed t, for a given a, the solutions (x, y) of the system (1.1.17) are symmetric with respect to the origin in the space of a's.

Since (1.1.17) is invariant by the transformation $(t, c) \to (-t, -c)$, in (1.1.17) it is sufficient to consider the case $c \geq 0$. Hence the phase trajectories for $-c$ are the same as for c, but the orientation along the trajectories is changed. Consequently, in order to exhaust the study of the F–N model, in addition to the case $c \geq 1 + \sqrt{3}$ treated in this book it is necessary to study the case $0 \leq c < 1 + \sqrt{3}$.

The change of variable (1.1.3) transforms the system (1.1.17) into the system

$$\begin{aligned} \varepsilon\,\dot{x} &= x + y - \frac{x^3}{3}, \\ \dot{y} &= -x + a - b\,y. \end{aligned} \tag{1.1.19}$$

Here the point stands for the differentiation with respect to the new time variable τ.

Sometimes, instead of system (1.1.19), we use the system

$$\begin{aligned} \dot{x} &= y + x - \frac{x^3}{3}, \\ \dot{y} &= \varepsilon\,(-x + a - b\,y), \end{aligned} \tag{1.1.20}$$

deduced from (1.1.19) using the change of variable $\theta = \tau/\varepsilon$. In (1.1.20), the point stands for the differentiation with respect to θ. Here ε is a small parameter, which corresponds to a large value of c. In addition, $t \ll \tau \ll \theta$. Since the Cauchy problem

$$x(0) = x^0, \quad y(0) = y^0, \tag{1.1.21}$$

for the system (1.1.17) possesses a unique solution, a dynamical system can by associated with this problem.

In a series of papers, Kakiucki and Tchizawa [82], [83], [135] studied the following system, referred to as the F–N system

$$\frac{dv}{d\tau} = -v\,(v-1)\,(v-a_{KT}) - w + I, \qquad \frac{dw}{d\tau} = b_{KT}\,(v - rw)\,, \tag{1.1.22}$$

where $a_{KT} \in (0, 0.5)$ and b_{KT} and r are positive constants. If $I = I_0$ is also constant, the system (1.1.22) reduces to (1.1.17) by the change of variables

$$\begin{aligned} v &= \alpha x + \beta, \\ w &= \delta y + \gamma, \\ \tau &= \eta t, \end{aligned}$$

where

$$\begin{aligned} \alpha^2 &= \sqrt{\frac{P}{9}}, & P &= a_{KT}^2 - a_{KT} + 1, & \beta &= \frac{a_{KT}+1}{3}, \\ \delta &= -3\alpha^3, & \eta^2 &= \frac{1}{b_{KT}}, & \gamma &= I_0 - a_{KT}\beta + (1 + a_{KT})\,\beta^2 - \beta^3, \\ a &= \frac{r\gamma - \beta}{\alpha}, & b &= -\frac{r\delta}{\alpha}, & c &= 3\alpha^2\eta. \end{aligned}$$

If I depends on τ, then (1.1.22) can be associated with a three-dimensional dynamical system. Kakiucki and Tchizawa chosed $I = I_0 + \varepsilon\tau$, $b_{KT} = c_{KT}\varepsilon$, $r > 3/P$, where $\varepsilon > 0$ is a small parameter. By nonstandard analysis methods they investigated certain classes of solutions for (1.1.22) (e.g. ducks, delayed solutions) and their behaviour as $\varepsilon \to 0$ and $r \to 0$.

1.1.5. FITZHUGH–NAGUMO SYSTEM IN THE CONTEXT OF MATHEMATICAL MODELLING IN ELECTROPHYSIOLOGY

Waves generated by the cardiac impulse.
Once initiated, the impulse propagates in the cardiac muscle in wave-like shape. In order to study this wave the electrical potential x is considered not only as a function of time but also of the position of the current point (u, v, w) in the cardiac muscle. The reaction–diffusion equations characterizing the excitable media are used. One of their forms reads

$$\varepsilon\frac{\partial x}{\partial \tau} = \varepsilon^2\,D_1\,\nabla^2\,x + f(x,y), \qquad \frac{\partial y}{\partial \tau} = \varepsilon^2\,D_2\,\nabla^2\,y + g(x,y), \tag{1.1.23}$$

where ∇^2 is the Laplace operator in one, two or three dimensions, f, g are the terms of reaction characterizing the excitable media, D_1, D_2 are transport coefficients and τ is the time.

If x and y do not depend on space variables and choosing

$$f(x, y) = x + y - \frac{x^3}{3}$$

and

$$g(x, y) = -x + a - b\, y,$$

from (1.1.23) the F–N system (1.1.19) is found.

The coupling of excitability with diffusion determines a lot of propagation phenomena [38], [91], [137], [87], [119], [118]. Excitable systems, e.g. the heart, possess travelling waves solutions.

For a one-dimensional excitable media, we have $\nabla^2 = \dfrac{\partial^2}{\partial u^2}$, where u is the space variable.

The simplest motions, corresponding to one-dimensional excitable muscle specific models, are the solitary waves propagating with constant speed [133]. They preserve their shape during propagation, but (contrary to solitons [2]) they annihilate each other as a result of collisions. A solution of (1.1.23) is called a solitary wave which propagates with constant speed α, if it has the self-similar form $x(u, \tau) = x_\alpha(s)$ with $s = u + \alpha\,\tau$. It follows that the solution x_α satisfies an ODE in $\mathbf{R}^2$ which corresponds to a homoclinic orbit (Figure 1.1.7).

In addition, families of periodic waves can exist. Their speed of propagation depends on periods.

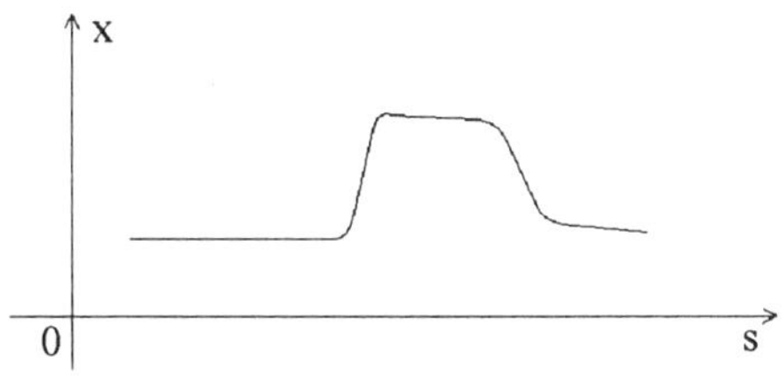

Fig. 1.1.7. Solitary wave

By the F–N model is also meant the Cauchy problem for the following equations

$$\begin{aligned} \frac{\partial^2 x}{\partial u^2} &= \frac{\partial x}{\partial t} - f\,(x) + z, \\ \frac{\partial z}{\partial t} &= b_H x, \end{aligned} \tag{1.1.24}$$

where $f(x) = x(1-x)(x-a_H)$, $0 < a_H < 1$ and b_H is a small positive constant. Here the subscript H stands for Hastings (our constant a is different from a_H). This model possesses periodic travelling wave solutions [70] representing infinite trains of nerve impulses.

This F–N problem also has solutions $x(u,t) = x(s)$, $z(u,t) = z(s)$, where $s = u + c_H t$ and c_H is a constant, corresponding to a single nerve impulse. They are nonconstant travelling waves which satisfy the following system of three first order ODEs

$$\begin{aligned} \frac{dx}{ds} &= y, \\ \frac{dy}{ds} &= c_H y - f(x) + z, \\ \frac{dz}{ds} &= \frac{b_H}{c_H} x. \end{aligned} \tag{1.1.25}$$

In $\mathbf{R}^3$, (the phase space for (1.1.25)), the single equilibrium point is the origin $(x,y,z) = (0,0,0)$. The nonconstant travelling wave solutions of (1.1.24) correspond to doubly asymptotic transient trajectories (i.e., as $t \to \pm\infty$) to the equilibrium point. Therefore, these trajectories are homoclinic orbits of the origin.

The equations (1.1.25) are also referred to as the F–N system.

In the case of a two-dimensional excitable medium, the dynamics is generated by (1.1.23) where $\nabla^2 = \frac{\partial^2}{\partial u^2} + \frac{\partial^2}{\partial v^2}$. In this case circular concentric propagating waves and spiral rotating waves were obtained [1], [84], [88], [108]. They are spatial structures that are formed far from the thermodynamic equilibrium and are studied by Synergetics [53], [29].

The concentric waves are generated by the pace-maker, whose period determines their asymptotic lengths. The spiral waves can be generated by non-excitable non-uniformities of the medium. These *obstacles* destroy the circular front waves. The spiral waves can be also generated by introducing an appropriate perturbation in the relative refractory period. Spiral waves of an excitable medium are characterized by a *core* and a *rotating frequency*, which are unique (Figure 1.1.8) [85], [86].

The existence of a pair of spiral waves generates various types of interactions depending on the distance between their cores. They can be annihilated or new spiral waves can emerge. The interaction between spiral waves with counter-rotation occurs in ventricle fibrillation [62], [77], [79], [110], [142], [143].

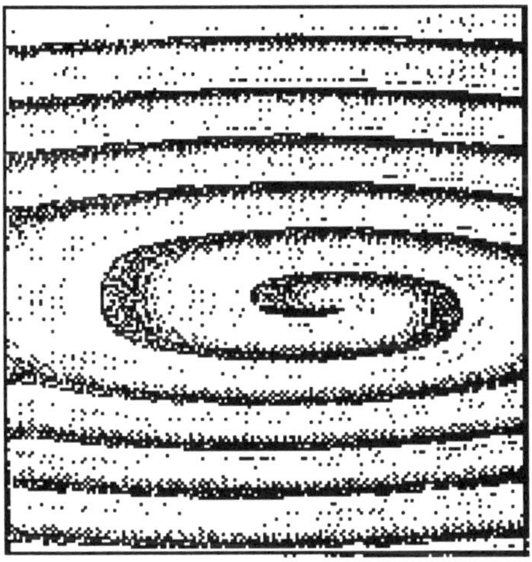

Fig. 1.1.8. Spiral rotating waves in the cardiac muscle.

In the case of a three-dimensional excitable medium $\left(\nabla^2 = \frac{\partial^2}{\partial u^2} + \frac{\partial^2}{\partial v^2} + \frac{\partial^2}{\partial w^2}\right)$ the circular and spiral waves are generalized as spherical and scroll rotating waves [99], [107], [109], [110].

Recent investigations concern the influence of an external electric field on the propagation of spiral waves. Such an intervention in the spiral waves' propagation could offer the possibility of directing them towards regions of cardiac muscle where they can be annihilated [115], [116].

Coupled oscillators.

In various circumstances the heart is considered as a system of coupled nonlinear oscillators describing many phenomena, known from electrocardiograms [74], [103], [127].

The initial oscillators may be of the F–N form (1.1.19) or, more generally, of the form

$$\begin{aligned} \varepsilon\frac{dx_i}{d\tau} &= F_i(x_i, y_i), \\ \frac{dy_i}{d\tau} &= G_i(x_i, y_i), \quad i = \overline{1, n} \end{aligned} \tag{1.1.26}$$

and they may have periodic solutions with different periods.

The coupling of oscillators (1.1.26) is realized as follows

$$\begin{aligned} \varepsilon\frac{dx_i}{d\tau} &= F_i(x_i, y_i), \\ \frac{dy_i}{d\tau} &= G_i(x_i, y_i, \delta) + \delta\, h_i(\mathbf{x}, \mathbf{y}), \end{aligned} \tag{1.1.27}$$

where $i = \overline{1, n}$, $0 < \delta < 1$, $\quad \mathbf{x} = (x_1, ..., x_n)$, $\quad \mathbf{y} = (y_1, ..., y_n)$.

The coupling can also be done with time delay. In this situation, $h_i = h_i(\mathbf{x}, \mathbf{y}, \mathbf{u}, \mathbf{v})$, where

$$\begin{aligned} \mathbf{u} &= (x_1(\tau - \rho_{i1}), ..., x_n(\tau - \rho_{in})), \\ \mathbf{v} &= (y_1(\tau - \rho_{i1}), ..., y_n(\tau - \rho_{in})). \end{aligned} \tag{1.1.28}$$

An example of a representation of the heart by three coupled oscillators S, A, V is presented in [61], where S is the sinusal node, A is the atrium, V is the ventricle and R is a delay, namely the time necessary to cross the Tawara node (Figure 1.1.9).

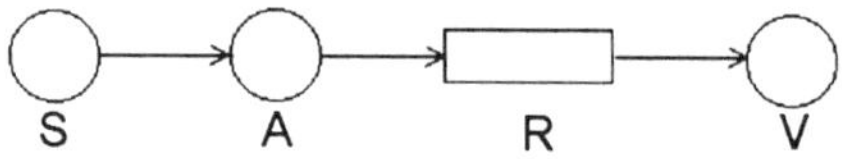

Fig. 1.1.9. The heart as a system of three coupled oscillators.

The coupling between A and V can be variable, generating irregular rhythms of the system, comparable with the cardiac arrhythmia [59].

Self-oscillators.

Non-stationary F–N oscillators with various forcing terms have also been studied. Such an oscillator with a piecewice constant forcing term is [16]

$$\varepsilon \ddot{x} + (x^2 - 1)\dot{x} + x = f(\tau),$$

with

$$f(\tau) = \begin{cases} a, & \tau \in [0, p/2), \\ -a, & \tau \in [p/2, p). \end{cases} \tag{1.1.29}$$

Two-dimensional ODE with periodic forcing terms [94], [95] can be the following [20], [21]

$$\ddot{x} + \nu (x^2 - 1)\dot{x} + x = \nu \lambda \beta \cos(\tau \lambda + \alpha), \tag{1.1.30}$$

$$\ddot{x} + \nu (x^2 - 1)\dot{x} + x = (\alpha \nu + \beta)\cos(k\tau), \tag{1.1.31}$$

with $\nu \gg 1$. These two-dimensional non-autonomous systems can be written as three-dimensional autonomous systems and they can present chaotic dynamics [117], [97].

Certain behaviours of F–N systems with a distributional forcing term were also considered [43]

$$\begin{aligned} \dot{x} &= x - \frac{x^3}{3} - y + I(t), \\ \dot{y} &= c(x + a - by). \end{aligned} \tag{1.1.32}$$

We also mention the stochastic F–N equations [35] or discrete studies [73].

1.2. ELEMENTS OF FINITE-DIMENSIONAL DYNAMICS

1.2.1. DYNAMICAL SYSTEMS AND THE ASSOCIATED SYSTEMS OF AUTONOMOUS ODEs

A process is said **deterministic** if all its future and past evolutions are uniquely defined (i.e., determined) by its present state. The mathematical model of a deterministic process is the dynamical system.

Let M be a metric space and $T=\mathbf{R}$ or $\mathbf{Z}$.

Definition 1.2.1 [6] *The pair* $(M, \{\varphi_t\}_{t\in T})$, *where* $\varphi_t : M \to M$ *are functions satisfying*

i)$\varphi_0 = \mathrm{id}_M$, *where* id *is the identity function on the set* M*;*

ii)$\varphi_{t+s} = \varphi_t \circ \varphi_s, \quad \forall t, s \in T,$

is called a **dynamical system.**

For the sake of simplicity, instead of $(M, \{\varphi_t\}_{t\in T})$ we write $(M, \{\varphi_t\})$.

The space M is called the **phase space** of the system, $T \times M$ is called the **extended phase space** and the functions φ_t are called the **evolution operators (flows)**.

If $T=\mathbf{R}$ the dynamical system is called **continuous,** and if $T=\mathbf{Z}$ it is called **discrete.**

The dynamical system can be defined, equivalently, as a function Φ of time t (continuous or discrete) whose values are the applications φ_t. In this case, instead of $(M, \{\varphi_t\})$ we write (M, Φ), or just Φ.

Definition 1.2.2 *Let* $x^0 \in M$. *The ordered subset of the phase space*

$$\mathrm{Or}(x^0) = \{x \in M, x = \varphi_t(x^0), t \in T\} \tag{1.2.1}$$

is called the **orbit (trajectory, phase trajectory) through** x^0 **of the dynamical system** $(M, \{\varphi_t\})$.

The subset of the extended phase space

$$C\,(x^0) = \{(t, \varphi_t(x^0), t \in T\} \tag{1.2.2}$$

is called the **integral curve through** x^0 **of the dynamical system** $(M, \{\varphi_t\})$.

For $x^0 \in M$ *fixed, the function* $x(., x^0) : T \to M$, $x(t, x^0) = \varphi_t(x^0)$ *is called the* **motion** *or the* **dynamics of** x^0.

It is easy to show that the trajectories are the projections of the integral curves on the phase space. A phase trajectory is also the image of the dynamics of some point.

For continuous dynamical systems the trajectories are curves in the space M, parameterized with respect to the continuous time t and oriented in the direction corresponding to the increase of t. For discrete dynamical systems they are sequences of points in M. The set of all trajectories through x^0 of a dynamical system, as x^0 runs over M, is called the **phase portrait** of the dynamical system.

Definition 1.2.3 *The point $x \in M$ is called an* **equilibrium point** *(fixed point) of the dynamical system $(M, \{\varphi_t\})$ if $\varphi_t(x) = x$ for every $t \in T$. A point of the phase space, which is not an equilibrium point, is called an* **ordinary point**. *A trajectory L is called* **periodic orbit (cycle)** *if it is not an equilibrium point and for every $x \in L$, there exist $T_0 > 0$ such that $\varphi_{t+T_0}(x) = \varphi_t(x)$ for every t in T. The smallest T_0 with this property is called the* **period of the cycle** *L.*

Let (M, φ_t) be a continuous dynamical system.

Definition 1.2.4 *A closed trajectory is called* **isolated** *if there exists a neighborhood of it in the phase space that does not contain any other closed trajectory (or part of a closed trajectory). A phase trajectory which is periodic and isolated (in the phase portrait) is called a* **limit cycle**.

In order to use the theory of dynamical systems in stability problems, it is necessary that the functions φ_t be differentiable, so M must be a differentiable manifold. Although the F–N model corresponds to $M = \mathbf{R}^2$, in the following we give some definitions which are valid for the more general case of differentiable manifolds M.

Definition 1.2.5 [104] *The continuous dynamical system $(M, \{\varphi_t\})$ is called a* **differentiable dynamical system** *if the function*

$$\varphi : R \times M \to M, \qquad \varphi(t, x) = \varphi_t(x), \tag{1.2.3}$$

is differentiable.

Sometimes φ is also referred to as a flow.

As a consequence, the evolution operators φ_t of a differentiable dynamical system are diffeomorphisms.

According to the dimension of M, the dynamical system is called **finite** or **infinite dimensional**. If the dimension of M is n then the dynamical system is called n**-dimensional**.

In the following we shall consider $M = \mathbf{R}^n$, so from now on the elements of M are vectors.

We quote an existence and uniqueness theorem for the Cauchy problem of a SODE in $\mathbf{R}^n$ [71].

Theorem 1.2.1 *Consider the SODE*

$$\dot{\mathbf{x}} = \mathbf{f}(\mathbf{x}), \qquad \mathbf{f} : \mathbf{R}^n \to \mathbf{R}^n, \tag{1.2.4}$$

and let $U \subset \mathbf{R}^n$ *be an open set and* $\mathbf{f}$ *smooth in* U. *Then there exists and is unique the function* $\mathbf{x} : \mathbf{R} \times \mathbf{R}^n \to \mathbf{R}^n$, $\mathbf{x} = \mathbf{x}(\mathbf{t}, \mathbf{x^0})$, *smooth with respect to* $(t, \mathbf{x})$, *such as for every* $\mathbf{x}^0 \in U$ *the following conditions are satisfied:*

i) $\mathbf{x}(0, \mathbf{x}^0) = \mathbf{x}^0$;

ii) there exists $c > 0$ *such that for every* $t \in (-c, c)$, *denoting* $\mathbf{y}(t) = \mathbf{x}(t, \mathbf{x}^0)$, *we have* $\mathbf{y}(t) \subset U$ *and* $\dot{\mathbf{y}}(t) = \mathbf{f}(\mathbf{y}(t))$.

For $\mathbf{x}^0$ fixed, the function $\mathbf{x} = \mathbf{x}(t, \mathbf{x}^0)$, looked at as a function of time, is called **the solution through** $\mathbf{x}^0$ of the system (1.2.4). For every $\mathbf{x}^0 \in U$ it defines two objects, namely

$$C_{\mathbf{f}}(\mathbf{x}^0) = \{(t, \mathbf{x}) | \mathbf{x} = \mathbf{x}(t, \mathbf{x}^0), \quad t \in (-c, c)\} \subset \mathbf{R} \times \mathbf{R}^n,$$

$$\mathrm{Or}_{\mathbf{f}}(\mathbf{x}^0) = \{\mathbf{x} | \mathbf{x} = \mathbf{x}(t, \mathbf{x}^0), \quad t \in (-c, c)\} \subset \mathbf{R}^n,$$

called the **solution curve** and the **orbit through** $\mathbf{x}^0$ **of the vector field** $\mathbf{f}$ (or of the autonomous SODE) respectively.

The following two properties relate differentiable dynamical systems and autonomous SODE.

Property 1.2.1 *Let* $(M, \{\varphi_t\})$ *be a differentiable dynamical system. Then it induces on* M *a vector field defined by*

$$\mathbf{V}(\mathbf{x}) = \left.\frac{d}{dt}\right|_{t=0} \varphi_t(\mathbf{x}) \tag{1.2.5}$$

for every $\mathbf{x} \in M$.

The vector $\mathbf{V}(\mathbf{x})$ is called the **speed of the current** φ_t **at the point** $\mathbf{x}$ and the right-hand side of (1.2.5.) is denoted by $\dot{\mathbf{x}}$. It can be proved that the trajectory through $\mathbf{x}^0$ of the dynamical system is the orbit through $\mathbf{x}^0$ of the induced vector field.

Property 1.2.2 *Let* $\mathbf{f}$ *be a vector field on* $\mathbf{R}^n$. *Then for every* $\mathbf{x}^0 \in \mathbf{R}^n$, *there exists a neighborhood* U *of* $\mathbf{x}^0$ *and there exists* $c > 0$ *and* $\varphi : (-c, c) \times U \to \mathbf{R}^n$, *such that:*

i) φ *is differentiable;*

ii) $\varphi_0 = \mathrm{id}_U$;

iii) $\varphi_t \circ \varphi_s = \varphi_{t+s}$ *for every* $t, s \in (-c, c)$ *such as* $t + s \in (-c, c)$.

The dynamical system $(U, \{\varphi_t\}_{t\in(-c,\,c)})$ is called the **local dynamical system generated by f** and $\mathbf{f}(\mathbf{x})$ is the speed of the current φ_t in $\mathbf{x}$.

The trajectories of $(U, \{\varphi_t\})$ are the same with the orbits of the field $\mathbf{f}$, restricted to U.

In Property 1.2.2. we have

$$\varphi_t(\mathbf{x}^0) = \mathbf{x}(t, \mathbf{x}^0) \qquad \text{and} \qquad \varphi(t, \mathbf{x}^0) = \varphi_t(\mathbf{x}^0). \tag{1.2.6}$$

The next property ensures the uniqueness of the local dynamical system defined by a vector field.

Property 1.2.3 *Let $\varphi_i : (-c_i, c_i) \times U_i \to \mathbf{R}^n$, $i = 1, 2$, be the local dynamical systems corresponding to $\mathbf{x}_1$ and $\mathbf{x}_2$ given by Property 1.2.2 with $U = U_1 \cap U_2 \neq \Phi$ and let $c = \min\{c_1, c_2\}$. Then $\varphi_1|_{(-c,\,c)\times U} = \varphi_2|_{(-c,\,c)\times U}$.*

Properties 1.2.1–1.2.3 allow to identify a dynamical system to its associated vector field. The connection with SODE is evident if $\mathbf{x}(t, \mathbf{x}^0)$ from (1.2.6) is the solution of (1.2.4). This is the reason why we shall not make any distinction between an autonomous SODE, the associated vector field, and the associated dynamical system.

According to these equivalences, the equilibrium points are solutions of the algebraic equation

$$\mathbf{f}(\mathbf{x}) = \mathbf{0}. \tag{1.2.7}$$

1.2.2. INVARIANT SETS IN THE PHASE SPACE. ATTRACTORS

Definition 1.2.6 *A subset S of the set M is* **invariant** *through the dynamical system $(M, \{\varphi_t\})$ if $\varphi_t(\mathbf{x})$ is in S for every $\mathbf{x}$ of S and for every t of $\mathbf{R}$ (i.e., $\varphi_t(S) \subset S$).*

As an example, $\mathrm{Or}(\mathbf{x}^0)$ is an invariant set.

Definition 1.2.7 *The invariant set S is called* **stable** *if for every neighborhood U of S, sufficiently small, there exist a neighborhood V of S, such that for every $\mathbf{x} \in V$ it follows $\varphi_t(\mathbf{x}) \in U$, for every $t > 0$.*

Definition 1.2.8 *The invariant set $S \subset M$ is called an* **attractor** *if there exists its neighborhood U_0 such that for every $\mathbf{x} \in U_0$ we have:*

$$\lim_{t\to\infty} \varphi_t(\mathbf{x}) \in S.$$

This last relationship defines the attractivity of S. If $t \to \infty$ is replaced by $t \to -\infty$, attractivity is replaced by repulsivity and the attractor by the **repulsor**.

Definition 1.2.9 *The set $S \subset M$ is called* **asymptotically stable** *if it is stable and attractive.*

The simplest attractors are equilibrium points and limit cycles (when they are attractive). A phase trajectory which does not belong to an attractor is called a **transient trajectory**.

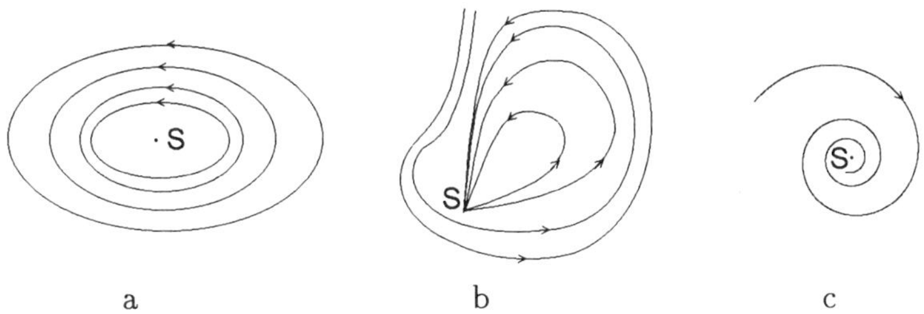

Fig. 1.2.1. The set S consisting of an equilibrium point is: a) stable, but not asymptotically stable; b) unstable, but attractive; c) asymptotically stable [67].

Definition 1.2.10 *Let $\mathbf{x}_0$ be an equilibrium point for SODE* (1.2.4) *with $\mathbf{f} \in \mathbf{C}^1$ and $D\mathbf{f}(\mathbf{x}_0)$ the Jacobi matrix of $\mathbf{f}$ at the point $\mathbf{x}_0$. The system*

$$\dot{\mathbf{x}} = D\mathbf{f}(\mathbf{x}_0)\,\mathbf{x} \tag{1.2.8}$$

is called the **linearized system of the SODE** (1.2.4) *with $\mathbf{f} \in \mathbf{C}^1$ at the point $\mathbf{x}_0$.*

Theorem 1.2.2 *Let $\mathbf{f}$ be a $\mathbf{C}^1$ map. If all the eigenvalues of the matrix $D\mathbf{f}(\mathbf{x}_0)$ has negative real parts, then the equilibrium point $\mathbf{x}_0$ is asymptotically stable. If at least one eigenvalue has positive real part, then $\mathbf{x}_0$ is unstable.*

Definition 1.2.11 *A point $\mathbf{p}$ is an* **ω-limit point** *of the orbit of $\mathbf{x}$ if there exists the sequence $\{\varphi_{t_i}(\mathbf{x})\}$ on $\mathrm{Or}(\mathbf{x})$ such that*

$$\lim_{t_i \to \infty} \varphi_{t_i}(\mathbf{x}) = \mathbf{p}.$$

A point $\mathbf{q}$ is called an **α-limit point** *of the orbit of $\mathbf{x}$ if there exists the sequence $\{\varphi_{t_i}(\mathbf{x})\}$ on $Or(\mathbf{x})$ such that*

$$\lim_{t_i \to -\infty} \varphi_{t_i}(\mathbf{x}) = \mathbf{q}.$$

The set of all ω-limit points of $\mathrm{Or}(\mathbf{x})$ is called the **ω-limit set** of $\mathrm{Or}(\mathbf{x})$ and will be denoted by $\omega(\mathbf{x})$. It is said that the orbit of $\mathbf{x}$ tends asymptotically to $\omega(\mathbf{x})$ as $t \to \infty$. The set of all α-limit points of the orbit of $\mathbf{x}$ is called the **α-limit set** of $\mathrm{Or}(\mathbf{x})$ and is denoted by $\alpha(\mathbf{x})$. It is said that $\mathrm{Or}(\mathbf{x})$ tends asymptotically to $\alpha(\mathbf{x})$ as $t \to -\infty$.

Definition 1.2.12 [67], [69] *An orbit whose α-limit set is an equilibrium point and whose ω-limit set is another equilibrium point is called a* **heteroclinic orbit**. *An orbit whose α-limit set is an equilibrium point, $\mathbf{x}_0$, the same as its ω-limit set, is called a* **homoclinic orbit**. *In this case we say that the orbit is* **doubly asymptotic** *(for $t \to \pm\infty$) to $\mathbf{x}_0$.*

Definition 1.2.13 [67], [69] *The orbit through $\mathbf{x}^0$ is called a* **separatrix** *if in every neighborhood of $\mathbf{x}^0$ there exists a point $\mathbf{x}$ such that $\omega(\mathbf{x}) \neq \omega(\mathbf{x}^0)$ or $\alpha(\mathbf{x}) \neq \alpha(\mathbf{x}^0)$.*

Definition 1.2.14 *The dynamical system $(\mathbf{R}^n, \{\varphi_t\})$ is called* **topologically equivalent** *in a region $U \subset \mathbf{R}^n$ to a dynamical system $(\mathbf{R}^n, \{\psi_t\})$ in a region $V \subset \mathbf{R}^n$ if there exists a homeomorphism $h : \mathbf{R}^n \to \mathbf{R}^n$ with $h(U) = V$, such as the orbits of the first dynamical system from U are transformed into the orbits of the second dynamical system from V, preserving the sense of the dynamics on the trajectories.*

Here **region** *is supposed to be a closed set of $\mathbf{R}^n$, bounded by $(n-1)$-dimensional piecewise smooth submanifolds.*

In certain particular cases the topological equivalence can be expressed analytically. For instance, if

$$\dot{\mathbf{x}} = \mathbf{f}(\mathbf{x}), \qquad \mathbf{x} \in \mathbf{R}^n, \qquad \dot{\mathbf{y}} = \mathbf{g}(\mathbf{y}), \qquad \mathbf{y} \in \mathbf{R}^n$$

and there exist $U, V \subset \mathbf{R}^n$ and $\mathbf{h} : U \to V$ a diffeomorphism such as

$$\mathbf{f}(\mathbf{x}) = D[\mathbf{h}(\mathbf{x})]^{-1}\mathbf{g}(\mathbf{h}(\mathbf{x})),$$

then the two systems are topologically equivalent. In this case, the systems are called **diffeomorphic**.

Another example is the case when there exists a scalar positive map $\mu : \mathbf{R}^n \to \mathbf{R}_+$ such as $\mathbf{f}(\mathbf{x}) = \mu(\mathbf{x})\,\mathbf{g}(\mathbf{x})$ for every $\mathbf{x} \in U$. Then the systems are topologically equivalent, since their orbits are the same and only the speed of movement along the orbits is different. In this case the systems are called **orbitally equivalent**.

Let $\mathbf{x}_0$ be an equilibrium point of the system $\dot{\mathbf{x}} = \mathbf{f}(\mathbf{x})$ and let n_-, n_0, n_+ be the number of eigenvalues with negative, zero or positive real parts of the matrix $D\mathbf{f}(\mathbf{x}_0)$, respectively.

Definition 1.2.15 *The point* $\mathbf{x}_0$ *is called a* **hyperbolic equilibrium** *if* $n_0 = 0$. *The hyperbolic equilibrium is called a* **saddle** *if* $n_- \, n_+ \neq 0$.

Definition 1.2.16 *Let* U *be a neighborhood of the equilibrium* $\mathbf{x}_0$ *and consider the sets*

$$W^s_{\text{loc}}(\mathbf{x}_0) \equiv W^s(\mathbf{x}_0, U) = \{\mathbf{x} \in U | \varphi_t(\mathbf{x}) \in U, \, t \geq 0, \, \lim_{t \to \infty} \varphi_t(\mathbf{x}) = \mathbf{x}_0\}$$

and

$$W^u_{\text{loc}}(\mathbf{x}_0) \equiv W^u(\mathbf{x}_0, U) = \{\mathbf{x} \in U | \varphi_t(\mathbf{x}) \in U, \, t \leq 0, \, \lim_{t \to -\infty} \varphi_t(\mathbf{x}) = \mathbf{x}_0\}.$$

The sets $W^s_{\text{loc}}(\mathbf{x}_0)$ *and* $W^u_{\text{loc}}(\mathbf{x}_0)$ *are called the* **local stable manifold** *of* $\mathbf{x}_0$ *and the* **local unstable manifold** *of* $\mathbf{x}_0$ *respectively.*

Definition 1.2.17 *Let* $\mathbf{x}_0$ *be an equilibrium point. The invariant sets*

$$\begin{aligned} W^s(\mathbf{x}_0) &= \{\mathbf{x} | \lim\nolimits_{t \to \infty} \varphi_t(\mathbf{x}) = \mathbf{x}_0\}, \\ W^u(\mathbf{x}_0) &= \{\mathbf{x} | \lim\nolimits_{t \to -\infty} \varphi_t(\mathbf{x}) = \mathbf{x}_0\} \end{aligned} \tag{1.2.9}$$

are called the **stable set** *and the* **unstable set** *of* $\mathbf{x}_0$, *respectively.*

We have

$$\begin{aligned} W^s(\mathbf{x}_0) &= \bigcup_{t \leq 0} \varphi_t(W^s_{\text{loc}}(\mathbf{x}_0)); \\ W^u(\mathbf{x}_0) &= \bigcup_{t \geq 0} \varphi_t(W^u_{\text{loc}}(\mathbf{x}_0)). \end{aligned}$$

The stable and unstable manifolds of saddle points and the limit cycles are examples of separatices.

Theorem 1.2.3 *Let* $\mathbf{x}_0$ *be a hyperbolic equilibrium. Then the intersections of* $W^s(\mathbf{x}_0)$ *and* $W^u(\mathbf{x}_0)$ *with a sufficiently small neighborhood of* $\mathbf{x}_0$ *contains the submanifolds* $W^s_{\text{loc}}(\mathbf{x}_0)$ *and* $W^u_{\text{loc}}(\mathbf{x}_0)$ *of dimensions* n_- *and* n_+ *respectively. In addition,* $W^s_{\text{loc}}(\mathbf{x}_0)$ *is tangent at* $\mathbf{x}_0$ *to the subspace* E^s *generated by the (generalized) eigenvectors, corresponding to the eigenvalues with strictly negative real part, while* $W^u_{\text{loc}}(\mathbf{x}_0)$ *is tangent at* $\mathbf{x}_0$ *to the subspace* E^u *generated by the (generalized) eigenvectors corresponding to eigenvalues with strictly positive real parts.*

Theorem 1.2.4 (Hartman–Grobman). *If* $\mathbf{x}_0$ *is a hyperbolic equilibrium of the system* $\dot{\mathbf{x}} = \mathbf{f}(\mathbf{x})$, *there exists a neighborhood of* $\mathbf{x}_0$ *in which the system is topologically equivalent to the linearized system* $\dot{\mathbf{x}} = D\mathbf{f}(\mathbf{x}_0)\,\mathbf{x}$.

Theorems 1.2.3 and 1.2.4 are linearization principles showing which properties of linearized associated systems and in what conditions these properties hold also for the nonlinear systems.

Definition 1.2.18 *An orbit is called a* **saddle connection** *if its* α*-limit set and* ω*-limit set are saddle points.*

Some examples of connections between saddles are given in Figure 1.2.2.

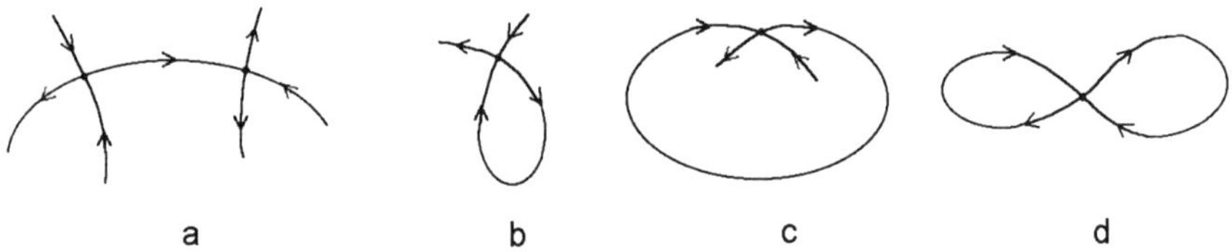

Fig. 1.2.2. Connections between saddles: a) heteroclinic; b) homoclinic; c) 'big' homoclinic; d) double homoclinic.

1.2.3. PLANAR DYNAMICAL SYSTEMS

Consider the two-dimensional dynamical system

$$\dot{\mathbf{x}} = \mathbf{f}(\mathbf{x}), \quad \mathbf{x} \in \mathbf{R}^2, \quad \mathbf{f} \in \mathbf{C}^1. \tag{1.2.10}$$

It is also called a **planar dynamical system.** The results from this Section concern such a system.

Theorem 1.2.5 *In a sufficiently small neighborhood of an ordinary point, the system* (1.2.10) *is topologically equivalent to the product system*

$$\dot{x}_1 = 1, \qquad \dot{x}_2 = 0.$$

This theorem shows the topological structure of the phase space in the neighborhood of ordinary points. However, the global dynamical behaviours cannot be characterized in this way (e.g. the points of the limit cycles or those of homoclinic orbits are ordinary points).

Owing to the Hartman–Grobman theorem the phase portrait of nonlinear dynamical systems in the neighborhood of hyperbolic equilibrium points is topologically equivalent to that of the linearized systems.

If $\mathbf{x}_0$ is an equilibrium point and $A = D\mathbf{f}(\mathbf{x}_0)$, then the eigenvalues $\lambda_{1,2}$ of A are solutions of the secular equation $\lambda^2 -$ trA $\lambda +$ detA$= 0$, where trA and detA are the trace and the determinant of the matrix A, respectively.

There exist three topological equivalence classes of hyperbolic equilibria on the plane: stable foci or nodes (attractors), unstable foci or nodes repulsors) and saddles. Their classification according to the type of the eigenvalues is given in Table 1.2.1 [92].

(n_+, n_-)	Eigenvalues	Phase portrait	Name and stability type
$(0,2)$	$\lambda_{1,2} \in \mathbf{R},\ \lambda_{1,2} < 0$	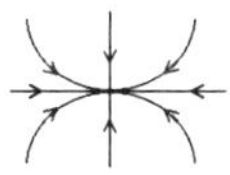	stable node (attractive)
	$\lambda_{1,2} \notin \mathbf{R}, \text{Re } \lambda_{1,2} < 0$	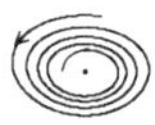	stable focus (attractive)
$(1,1)$	$\lambda_{1,2} \in \mathbf{R}$ $\lambda_1 > 0, \lambda_2 < 0$	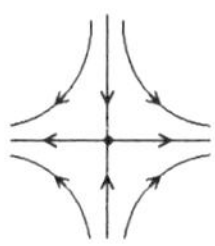	unstable saddle $\lambda_1 + \lambda_2 = 0$ (neutral saddle)
$(2,0)$	$\lambda_{1,2} \in \mathbf{R},\ \lambda_{1,2} > 0$	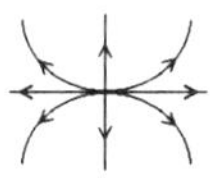	unstable node (repulsive)
	$\lambda_{1,2} \notin \mathbf{R},\ \text{Re } \lambda_{1,2} > 0$		unstable focus (repulsive)

Table 1.2.1. Topological classification of hyperbolic equilibria on the plane.

The stable manifolds of attractive equilibria are two-dimensional. Attractive equilibria do not possess unstable manifolds. Repulsive equilibria have two-dimensional unstable manifolds and do not possess stable manifolds. The saddle points possess both stable and unstable manifolds, which are one-dimensional. The phase portrait of a nonlinear dynamical system in the neighborhood of a non-hyperbolic equilibrium point cannot be deduced from that of the linearized system. It can be characterized only as a result of a nonlinear study. Among the non-hyperbolic equilibrium points of nonlinear systems we mention the saddle–nodes, when $\lambda_1 = 0$, $\lambda_2 \in \mathbf{R}^*$ (Figure 1.2.3).

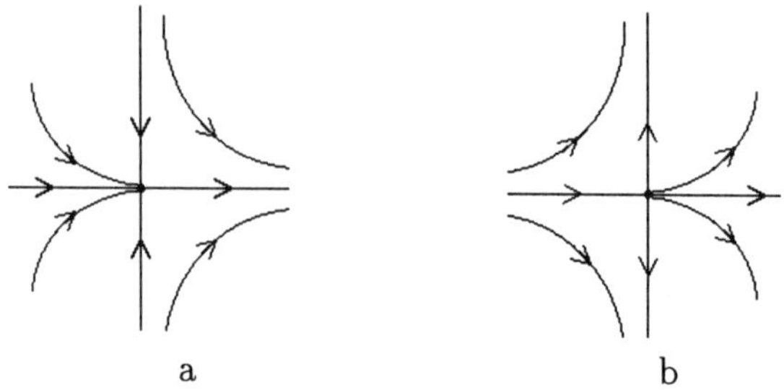

Fig. 1.2.3. Saddle–node equilibria: a) partially attractive; b) partially repulsive.

Other non-hyperbolic equilibria of nonlinear systems will be defined as bifurcation points in Section 1.3.

Homoclinic orbits (doubly asymptotic orbits as $t \to \pm\infty$) of saddle points or saddle–nodes may exist (Figure 1.2.4).

Fig. 1.2.4. Types of homoclinic orbits: a) of a saddle; b) of a saddle–node.

Definition 1.2.19 *A periodic orbit is called* **hyperbolic** *if it is asymptotically stable or unstable, so attractive or repulsive. Otherwise, it is called* **non-hyperbolic** *(Figure 1.2.5).*

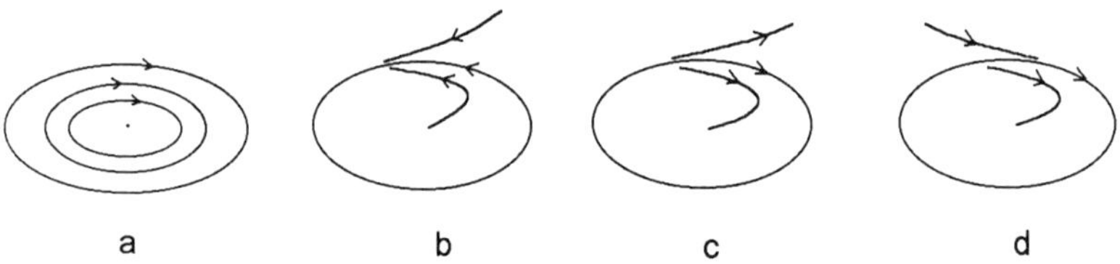

Fig. 1.2.5. a) Periodic orbits which are not limit cycles; b) (hyperbolic) attractive limit cycle; c) (hyperbolic) repulsive limit cycle; d) non-hyperbolic (partially attractive, partially repulsive) limit cycle.

In general, the existence of limit cycles is an open problem and few results exist. Among them we quote:

Theorem 1.2.6 (Bendixson) *Consider the system* (1.2.10) *and let* $D \subset \mathbf{R}^2$ *be an open and simply conex set. If* $\operatorname{div} f \equiv \dfrac{\partial f_1}{\partial x_1} + \dfrac{\partial f_2}{\partial x_2}$ *has a constant sign in D and is not identically zero in D, then the system* (1.2.10) *has no periodic orbits entirely situated in D.*

Theorem 1.2.7 (Poincaré–Bendixson). *Consider the system* (1.2.10) *and* $\mathbf{x}^0 \in \mathbf{R}^2$. *If the* ω*-limit set* $\omega(\mathbf{x}^0)$ *is bounded and does not contain any equilibrium point, then it is a periodic orbit.*

Usually it is very difficult to apply Theorem 1.2.7. This is why we shall avoid applying Bendixson type of theorems, and instead use normal form methods.

1.3. BIFURCATION

1.3.1. STATIC AND DYNAMIC BIFURCATION

Consider the equation

$$\mathbf{f}(\mathbf{x}, \boldsymbol{\alpha}) = 0,\ \mathbf{x} \in \mathbf{R}^n,\ \boldsymbol{\alpha} \in \mathbf{R}^m, \tag{1.3.1}$$

where $\mathbf{x}$ is the unknown vector, $\boldsymbol{\alpha}$ is a parameter vector and $\mathbf{f}$ is a differentiable map, in general nonlinear. Let $(\mathbf{x}^*, \boldsymbol{\alpha}^*) \in \mathbf{R}^n \times \mathbf{R}^m$ be a point such as

$$\begin{aligned} &\mathbf{f}(\mathbf{x}^*, \boldsymbol{\alpha}^*) = 0, \\ &D_{\mathbf{x}}\mathbf{f}(\mathbf{x}^*, \boldsymbol{\alpha}^*) \neq 0. \end{aligned} \tag{1.3.2}$$

Then, the implicit function theorem states the existence, in a neighborhood of $\mathbf{x}^*$, of a map

$$\mathbf{x} = g(\boldsymbol{\alpha}) \tag{1.3.3}$$

with the properties $\mathbf{g}(\boldsymbol{\alpha}^*) = \mathbf{x}^*$ and $\mathbf{f}(\mathbf{g}(\boldsymbol{\alpha}), \boldsymbol{\alpha}) = 0$ for every $\boldsymbol{\alpha}$ in a neighborhood $V_{\boldsymbol{\alpha}^*}$ of $\boldsymbol{\alpha}^*$. In this case $\mathbf{g}(\boldsymbol{\alpha})$ is called the **solution** of (1.3.1). It is unique and depends on $\boldsymbol{\alpha}$. When the conditions of the implicit function theorem are not satisfied, equation (1.3.1) may have no solution for some values of $\boldsymbol{\alpha}$ or may have multiple solutions (even an infinity) for some other values. So, in general, the dependence $\mathbf{x}(\boldsymbol{\alpha})$ is not a function, but a **multiform function**.

Let $S_{\boldsymbol{\alpha}} = \{\mathbf{x}(\boldsymbol{\alpha}) \in R^n |\, \mathbf{f}(\mathbf{x}(\boldsymbol{\alpha}), \boldsymbol{\alpha}) = 0\}$ be the solution set of (1.3.1) for $\boldsymbol{\alpha} \in R^m$ fixed and let $\{S_{\boldsymbol{\alpha}}\}_{\boldsymbol{\alpha} \in R^m}$ be the solution set of (1.3.1). For the sake of brevity, let us denote it by $\{S_{\boldsymbol{\alpha}}\}$. For smooth functions $\mathbf{f}$, $\{S_{\boldsymbol{\alpha}}\}$ consists of continua which intersect themselves at bifurcation points.

When $m = 1$, i.e., the parameter α is one-dimensional, these continua are called **branches of solutions**. This name is also extended to the case $m > 1$.

Definition 1.3.1 *The value* $\boldsymbol{\alpha} = \boldsymbol{\alpha}^*$ *is a* **static bifurcation value** *for* $\{S_{\boldsymbol{\alpha}}\}$ *if by crossing* $\boldsymbol{\alpha}^*$, *the cardinality of* $S_{\boldsymbol{\alpha}}$ *undergoes a jump.*

Definition 1.3.2 *The point* $(\mathbf{x}^*, \boldsymbol{\alpha}^*) \in \mathbf{R}^n \times \mathbf{R}^m$ *is a* **static bifurcation point** *for* $S_{\boldsymbol{\alpha}}$ *if for any neighborhood* $U_{\mathbf{x}^*}$ *of* $\mathbf{x}^*$ *and for every neighborhood* $V_{\boldsymbol{\alpha}^*}$ *of* $\boldsymbol{\alpha}^*$*, there exists* $\boldsymbol{\alpha} \in V^*_{\boldsymbol{\alpha}}$ *and at least two points* $\mathbf{x}'$ *and* $\mathbf{x}''$ *in* $U_{\mathbf{x}^*}$ *such that* $\mathbf{x}'(\boldsymbol{\alpha}), \mathbf{x}''(\boldsymbol{\alpha}) \in S_{\boldsymbol{\alpha}}$.

In this case $\boldsymbol{\alpha}^*$ is a bifurcation value. Often $\mathbf{x}^*$ and $\boldsymbol{\alpha}^*$ are also referred to as bifurcation points, but from the context one can understand whether the bifurcation point is $(\mathbf{x}^*, \boldsymbol{\alpha}^*)$, $\mathbf{x}^*$ or $\boldsymbol{\alpha}^*$.

The bifurcation from Definition 1.3.2 is also called a **bifurcation from a point**. The **bifurcation from a curve of solutions** $\mathbf{x}(\boldsymbol{\alpha})$ of (1.3.1) can also be defined. For the sake of simplicity, let us consider the case $m = 1$.

Definition 1.3.3 *We say that a branch of solutions,* $\mathbf{x}'(\alpha)$*, of* (1.3.1) *is* **bifurcating supercritically (subcritically)** *from the branch* $\mathbf{x}^*(\alpha)$ *at the point* $\mathbf{x}^*(\alpha^*)$ *if for every neighborhood* V *of* $(\alpha^*, \mathbf{x}^*(\alpha^*))$ *there exists a neighborhood* U *of* α^* *such as for* $\alpha \in U$, $\alpha < \alpha^*$ $(\alpha > \alpha^*)$*,* V *does not contain pairs* $(\alpha, \mathbf{x}(\alpha))$*, with* $\mathbf{x}(\alpha) \in S_\alpha$ *,* $\mathbf{x} \neq \mathbf{x}^*$ *and for* $\alpha > \alpha^* (\alpha < \alpha^*)$ *in* V *there exist points* $(\alpha, \mathbf{x}'(\alpha))$ *with* $\mathbf{x}' \neq \mathbf{x}^*$.

When the set $\{S_\alpha\}$ has a limit point situated at infinity (i.e., $\lim_{\alpha\to\alpha^*} \|\mathbf{x}^*(\alpha)\| = \infty$ for $\alpha^* < \infty$) a **bifurcation from 'large' solutions** takes place. The value $\alpha^* = \infty$ is called the **bifurcation value from infinity** if in every neighborhood of α^* there exists α and $x_1, x_2 \in S_\alpha$. These two types of bifurcations are simply called **bifurcations from infinity**. The same name is used for the case when $(\mathbf{x}^*, \boldsymbol{\alpha}^*) = (\infty, \infty)$.

Let $F : \{S_{\boldsymbol{\alpha}}\}_{\boldsymbol{\alpha} \in \mathbf{R}^m} \to \mathbf{R}$ be an injective map.

Definition 1.3.4 *The set* $(\|\boldsymbol{\alpha}\|, F(\mathbf{x}(\boldsymbol{\alpha})))$ *,* $\boldsymbol{\alpha} \in \mathbf{R}^m$ *with* $\mathbf{x}(\boldsymbol{\alpha}) \in S_{\boldsymbol{\alpha}}$ *is called the* **static bifurcation diagram**.

This planar diagram is only qualitative for $m + n > 3$. Often, instead of $\|\boldsymbol{\alpha}\|$, an injective functional of $\boldsymbol{\alpha}$ is taken into account. In the cases $m = 1$, $n < 3$ or $m < 3$, $n = 1$, instead of $\|\boldsymbol{\alpha}\|$ we can take $\boldsymbol{\alpha}$ and instead of $F(\mathbf{x})$ we can take $\mathbf{x}$, such that the static bifurcation diagram is a synthetic representation of the solutions set of (1.3.1) in $\mathbf{R}^3$ [45].

In the static bifurcation diagram, $S_{\boldsymbol{\alpha}}$ may consist of pieces of continua (branches, sheets, etc.). The points of static bifurcation of interest in current applications are defined according to the number and the shape of the branches of solutions near them in the bifurcation diagram and direction of their emergence with respect to the bifurcation point. This definition is expressed in terms of some derivatives of f [25], [52]. A lot of practical definitions, especially for the case $m = n = 1$, are reduced to the description of the static bifurcation diagram around the static bifurcation points. In Figure 1.3.1 such graphical 'definitions' are given. The static bifurcations from a point at finite distance (Figure 1.3.1a,e) or from infinity (Figure 1.3.1d) or from a branch of solutions (1.3.1b) are presented.

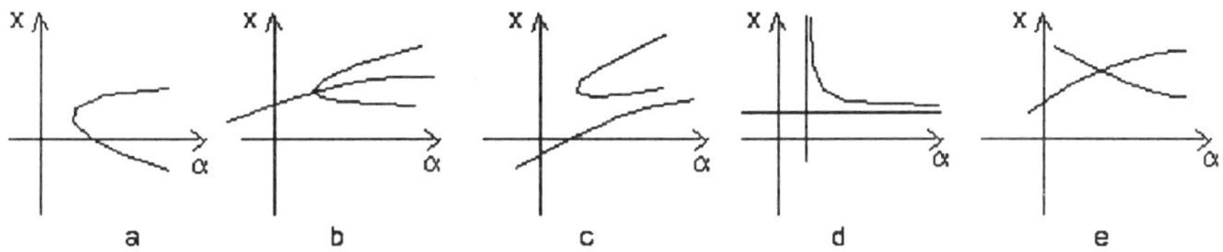

Fig. 1.3.1. Static bifurcation diagrams around a static bifurcation point: a) turning point; b) fork bifurcation point; c) lateral bifurcation; d) bifurcation from infinity; e) transcritical bifurcation.

The branch from which the bifurcation may occur can be isolated (topologically) from the new branches (Figure 1.3.1c). In this case we have a global static bifurcation of lateral type and a local bifurcation of turning type [63], [129].

Let us now introduce the dynamic bifurcation. It characterizes sets of SODEs depending on parameters.

Definition 1.3.5 *The set* $\{\Phi(\boldsymbol{\alpha})\}_{\boldsymbol{\alpha}\in R^m}$ *of dynamical systems* $\Phi(\boldsymbol{\alpha})$, *where* $\boldsymbol{\alpha}$ *is a parameter vector, is called a* **dynamical scheme** *or a* **dynamical system depending on a parameter** *or, even, a* **dynamical system**. *We denote a dynamical scheme by* $\Phi(\boldsymbol{\alpha})$.

The topological equivalence of dynamical schemes takes into account the topological equivalence of the parameter spaces, of the phase spaces and of the phase portraits. In addition, if the phase space is not compact, then the dynamical schemes will be considered in bounded regions of the phase space. More precisely, let us consider two dynamical schemes $\Phi(\boldsymbol{\alpha})$, $\Psi(\boldsymbol{\alpha})$ generated by the following SODEs

$$\dot{\mathbf{x}} = \mathbf{f}(\mathbf{x}, \boldsymbol{\alpha}), \quad \mathbf{x} \in \mathbf{R}^n, \ \boldsymbol{\alpha} \in \mathbf{R}^m, \tag{1.3.4}$$

$$\dot{\mathbf{y}} = \mathbf{g}(\mathbf{y}, \boldsymbol{\beta}), \quad \mathbf{y} \in \mathbf{R}^n, \ \boldsymbol{\beta} \in \mathbf{R}^m, \tag{1.3.5}$$

with **f** and **g** smooth maps.

Definition 1.3.6 *The dynamical scheme associated with* (1.3.4) *is called* **topologically equivalent** *in* $U_{\boldsymbol{\alpha}} \subset \mathbf{R}^n$ *with the dynamical scheme associated with* (1.3.5) *in* $V_{\boldsymbol{\beta}} \subset \mathbf{R}^n$ *if:*

i) there exists a homeomorphism $\mathbf{p} : \mathbf{R}^m \to \mathbf{R}^m$, *with* $\mathbf{p}(\mathbf{0}) = \mathbf{0}$;

ii) there exists a homeomorphism $h_{\boldsymbol{\alpha}} : \mathbf{R}^n \to \mathbf{R}^n$, *with* $h_{\boldsymbol{\alpha}}(U_{\boldsymbol{\alpha}}) = V_{\mathbf{p}(\boldsymbol{\alpha})}, h_{\boldsymbol{\alpha}}(\mathbf{0}) = \mathbf{0}$ *such as for every* $\boldsymbol{\alpha}$, *it transforms the orbits of* (1.3.4) *from* $U_{\boldsymbol{\alpha}}$ *into the orbits of* (1.3.5) *with* $\boldsymbol{\beta} = \mathbf{p}(\boldsymbol{\alpha})$ *from* $V_{\mathbf{p}(\boldsymbol{\alpha})}$, *preserving the sense.*

Here $U_{\boldsymbol{\alpha}}$ *and* $V_{\boldsymbol{\beta}}$ *are regions of* $\mathbf{R}^n$ *depending on parameters.*

Definition 1.3.7 *A dynamical system $\Phi(\boldsymbol{\alpha}^*)$ which is nonequivalent topologically with the dynamical system $\Phi(\boldsymbol{\alpha})$ for all parameters $\boldsymbol{\alpha}$ in a neighborhood of $\boldsymbol{\alpha}^*$ is called a* **bifurcation** *of the dynamical scheme $\Phi(\boldsymbol{\alpha})$.*

Notice that this bifurcation is not the phenomenon. It is a particular dynamical system.

Definition 1.3.8 *The value $\boldsymbol{\alpha}^*$ of the parameter $\boldsymbol{\alpha}$ for which $\Phi(\boldsymbol{\alpha}^*)$ represents a bifurcation is called a* **bifurcation value**.

Definition 1.3.9 *The dynamical system $\Phi(\boldsymbol{\alpha}^*)$ which is a bifurcation is also called a* **bifurcation point** *for the dynamical scheme $\Phi(\boldsymbol{\alpha})$.*

Some bifurcations can be detected by fixing any neighborhood of an equilibrium point. They are called **local bifurcations** or **bifurcations of equilibrium points** (as the Hopf bifurcation, for example). The other bifurcations are called **global bifurcations** (like the homoclinic bifurcation).

In the dynamical context, the solutions of (1.3.1) are the equilibrium points of (1.3.4), static bifurcation being a particular case of dynamic bifurcation. In addition, the dynamical context provides the characterization of the attractivity properties of equilibrium points.

Another particular type of bifurcation is the emergence, from stationary solutions, of nonstationary solutions with atractivity or repulsivity properties. For a long time it was called a **dynamic bifurcation**. We shall call it **classical dynamic bifurcation**. The set of static bifurcation points and that of classical dynamic bifurcation points are intersecting. As an example, the saddle–node homoclinic bifurcation (defined in the following) is both static and classical dynamic bifurcation. A synthetic scheme of bifurcations for (1.3.4) is given by the bifurcation diagram, whose twofold definition is the following.

Definition 1.3.10 *Consider the dynamical scheme $\Phi(\boldsymbol{\alpha})$ and a fixed value, $\boldsymbol{\alpha}^*$, of the parameter. A connected set of the parameter space consisting of $\boldsymbol{\alpha}^*$ and those parameters $\boldsymbol{\alpha}$ for which the system $\Phi(\boldsymbol{\alpha})$ has the phase portrait topologically equivalent to the phase portrait for $\Phi(\boldsymbol{\alpha}^*)$ and which is maximal, is called the* **stratum of $\boldsymbol{\alpha}^*$**. *The boundary of a stratum is called a* **bifurcation boundary**.

The bifurcation boundaries consist of bifurcation values.

Definition 1.3.11 *The set of strata from the parameter space is called the* **parametric portrait**.

Definition 1.3.12 [13], [92] *The parametric portrait together with the phase portraits characterizing each stratum is called the* **bifurcation diagram**.

Definition 1.3.13 *The bifurcation diagrams of topologically equivalent dynamical schemes are called* **equivalent diagrams**.

The definition of the topological equivalence concerned only the dynamical systems defined on compact sets, therefore excluding the dynamical systems defined on $\mathbf{R}^n$. In this way some manifolds of static bifurcation values may not be bifurcation boundaries in the sense of Definition 1.3.10. Even in the planar case ($M = \mathbf{R}^2$), there exist situations when the parametric portrait has an infinity of strata. In the simplest cases the parametric portrait consists of a finite number of regions of $\mathbf{R}^m$ separated by bifurcation boundaries (curves, surfaces). A bifurcation boundary is defined by bifurcation conditions.

Definition 1.3.14 *The difference between the dimension of the parameter space and the dimension of some bifurcation boundary is called the* **codimension** *of the corresponding bifurcations (i.e., dynamical systems) of the system* (1.3.4).

Equivalently, the codimension is the number of independent conditions determining the bifurcation. It is used in classifying the vector fields [92], [64], [66].

In the case $n \geq 3$, the bifurcation boundaries can be dense in some regions of the parameter space and the parametric portrait can have a fractal structure.

1.3.2. STRUCTURAL STABILITY. BIFURCATION. CODIMENSION OF BIFURCATIONS

In Definition 1.3.10 we assumed the existence of some dynamical systems whose phase portrait corresponding to points from certain subsets of the parametric space does not change qualitatively at small perturbations. This allows the connection between bifurcation and structural stability. Consider two dynamical systems

$$\dot{\mathbf{x}} = \mathbf{f}(\mathbf{x}), \quad \mathbf{x} \in \mathbf{R}^n, \tag{1.3.6}$$

$$\dot{\mathbf{x}} = \mathbf{g}(\mathbf{x}), \quad \mathbf{x} \in \mathbf{R}^n, \tag{1.3.7}$$

with $\mathbf{f}$ and $\mathbf{g}$ smooth maps. In particular, $\mathbf{f}$ may be a nonlinear function and $\mathbf{g}$ its linearization: $\mathbf{f}$ occurs as a perturbation of $\mathbf{g}$. In addition, $\mathbf{f}$ may depend on a parameter, while $\mathbf{g}$ may be equal to that value of $\mathbf{f}$ which corresponds to a fixed value of the parameter.

Definition 1.3.15 *The* **distance between the dynamical systems** *(1.3.6) and (1.3.7) in a region $U \subset \mathbf{R}^n$ is defined as the number*

$$d_1 = \sup_{\mathbf{x}\in U} \left\{ \|\mathbf{f}(\mathbf{x}) - \mathbf{g}(\mathbf{x})\| + \left\| \frac{d\mathbf{f}(\mathbf{x})}{d\mathbf{x}} - \frac{d\mathbf{g}(\mathbf{x})}{d\mathbf{x}} \right\| \right\}.$$

The systems are called **ε-close** *(or C^1-close) in U if $d_1 < \varepsilon$, where $\varepsilon > 0$ is a small number.*

Among two ε-close systems, the more complicate system is viewed as obtained by a perturbation of the simpler system.

The norm $\| \bullet \|$ is the vectorial and matriceal norm in $\mathbf{R}^n$ and $M_n(\mathbf{R})$.

Definition 1.3.16 *The system (1.3.6) is called* **strictly structurally stable** *in the region U if every system (1.3.7) sufficiently C^1-close in U is topologically equivalent in U to (1.3.6).*

As a remark, the systems having hyperbolic equilibrium points on the boundary of U or hyperbolic limit cycles that touch this boundary are structurally unstable according to Definition 1.3.16, because small perturbations can move the equilibrium or the limit cycle outside U. In order to avoid this fact, we can consider the dynamical system in the whole space phase. This kind of approach is convenient for the dynamical systems on compact and smooth manifolds. But for systems in $\mathbf{R}^n$, this approach may give $d_1 = \infty$, even for dynamical systems that seems to be very close (for example that have close parameter values). This is why in the case of $\mathbf{R}^n$ another definition of structural stability is necessary. It is given in the following.

Definition 1.3.17 *The system (1.3.6) defined in a region $D \subset \mathbf{R}^n$ is called* **structurally stable** *in region $D_0 \subset D$, if for every system (1.3.7) sufficiently C^1-close in D to (1.3.6) there exist regions U and V contained in D, with $D_0 \subset U$ such that (1.3.6) is topologically equivalent in U to (1.3.7) in V.*

The next theorem gives the complete description of the structural stability of planar dynamical systems.

Theorem 1.3.1 (Andronov–Pontriagin) [3] *Consider the system $\dot{\mathbf{x}} = \mathbf{f}(\mathbf{x})$, $\mathbf{x} \in \mathbf{R}^2$ and let $\mathbf{f}$ be a smooth map. The system is structurally stable in a region $D_0 \subset \mathbf{R}^2$ iff the following conditions are satisfied:*

i) there exists a finite number of equilibrium points in D_0 and all of them are hyperbolic;

ii) there exists a finite number of periodic orbits in D_0 and all of them are hyperbolic;

iii) there are no saddle connections.

For dimensions greater than two a similar result does not exist. A generalization of Theorem 1.3.1 for the case when the phase space is a differentiable two-dimensional bounded manifold is the Peixoto [112], [113] theorem.

Definitions equivalent to (1.3.17) and (1.3.9) for the case of planar systems are given in [68] as follows:

Let $D \subset \mathbf{R}^2$ be a domain surrounded by the curve ∂D and let χ_2^r be the set of C^r vector fields that have not contact with ∂D (i.e., all the trajectories through the points of ∂D enter (or come out from) the domain D).

Definition 1.3.18 *The vector field $f \in \chi_2^r$ is called* **structurally stable** *if it is topologically equivalent in $\overline{D}$ to any vector field $g \in \chi_2^r$ from its neighborhood. A smooth two-dimensional vector field is a* **bifurcation point** *(or simply a* **bifurcation***) for the set of such vector fields if it is not structurally stable.*

The codimension of a bifurcation point for two-dimensional dynamical schemes can be also defined as follows:

Definition 1.3.19 *A vector field (and, so, the corresponding dynamical system) is a* **bifurcation point of codimension 0** *if it is structurally stable. A vector field is a* **bifurcation point of codimension n+1** *if it is not of codimension $0, 1, \ldots, n$ and there exist a neighborhood, in the set χ_2^r of smooth two-dimensional vector fields, that contains only bifurcation points of codimension $0, 1, \ldots, n$ or vector fields equivalent to it.*

1.3.3. TOPOLOGICAL NORMAL FORMS FOR BIFURCATIONS

The investigation of many dynamical systems depending on parameters (i.e., dynamical schemes) leads to the conclusion that, for small dimensions of such schemes and of the parameters, generically there exist only a finite number of classes of equivalence of dynamical schemes. In such a class, all dynamical schemes have (locally) topologically equivalent bifurcation diagrams. In addition, in every class there exists a dynamical scheme generated by a particular polynomial SODE which is the simplest, and is called a normal form [5], [65], [24]. Because of this property the study of dynamical schemes is reduced to the determination of all normal forms for different values of the parameters, topologically equivalent to the given dynamical scheme. More precisely, let us consider the dynamical scheme

$$\dot{\xi} = \mathbf{g}(\boldsymbol{\xi}, \boldsymbol{\beta}, \sigma), \quad \boldsymbol{\xi} \in \mathbf{R}^n, \boldsymbol{\beta} \in \mathbf{R}^k, \sigma \in \mathbf{R}, \tag{1.3.8}$$

with $\mathbf{g}_i$ polynomials in $\boldsymbol{\xi}_i$ with parameters σ, for every $i = 1, 2, .., n$. Assume that for $\boldsymbol{\beta} = \mathbf{0}$ it has an equilibrium point $\boldsymbol{\xi} = \mathbf{0}$, satisfying k bifurcation conditions that determine in $\boldsymbol{\xi} = \mathbf{0}$ a codimension k bifurcation. As the normal forms are polynomial SODEs they will have the form (1.3.8).

Suppose that for (1.3.8) the space of the coefficients of the normal forms has only a finite number of regions corresponding to nontopologically equivalent bifurcation diagrams. Let us also consider the dynamical scheme

$$\dot{\mathbf{x}} = \mathbf{f}(\mathbf{x}, \boldsymbol{\alpha}), \quad \mathbf{x} \in \mathbf{R}^n, \quad \boldsymbol{\alpha} \in \mathbf{R}^k, \tag{1.3.9}$$

which, for $\boldsymbol{\alpha} = \mathbf{0}$, has the equilibrium point $\mathbf{x} = \mathbf{0}$.

Definition 1.3.20 *The dynamical scheme* (1.3.9) *is called* **generic** *if it satisfies a finite number of nondegeneracy conditions of the form*

$$N_i[\mathbf{f}] \neq \mathbf{0}, \quad i = 1, 2, ..., s,$$

where N_i are algebraic functions of some partial derivatives of $\mathbf{f}$ *with respect to* $\mathbf{x}$ *and* $\boldsymbol{\alpha}$, *evaluated at the point* $(\mathbf{x}, \boldsymbol{\alpha}) = (\mathbf{0}, \mathbf{0})$.

Definition 1.3.21 *The dynamical scheme* (1.3.8) *is called a* **topological normal form** *of a bifurcation if every generic dynamical scheme* (1.3.9) *with the equilibrium satisfying the same bifurcation conditions at* $\boldsymbol{\alpha} = \mathbf{0}$, *is locally topologically equivalent, near the origin, to* (1.3.8) *for some values of* σ.

In the above definition, by local topological equivalence we mean that the systems (1.3.8) and (1.3.9) are topologically equivalent in a neighborhood of the origin of the phase space for values of the parameters close to the origin of the parameter space.

Let us remark that in the definition of the normal forms the study is done around the point $(\mathbf{x}, \boldsymbol{\alpha}) = (\mathbf{0}, \mathbf{0})$, where $\mathbf{x} = \mathbf{0}$ is an equilibrium point for (1.3.9) and $\boldsymbol{\alpha} = \mathbf{0}$ is the bifurcation value. Thus, the bifurcations that can be studied by these normal forms take place at equilibrium points. The study based on normal forms is local, being appropriate only for local bifurcations (as Hopf, Bogdanov–Takens or Bautin type). As a consequence, only near such bifurcation points shall we deduce, approximately, the bifurcation boundaries which emerge from these points. The continuation of the bifurcation boundaries far from the local bifurcation values can be carried out only numerically.

The limitation that the normal forms are defined only near the origin is owed to their property of being unfoldings connected with equivalence classes of germs that are, themselves, equivalence classes of functions defined near the origin and transforming the origin of $\mathbf{R}^n$ into the origin of $\mathbf{R}^m$. These equivalence classes are basic notions from the theory of singularities, a theory which includes catastrophe theory and bifurcation theory [5], [34]. The generic dynamical schemes are equivalence classes of dynamical systems that correspond to a particular number and type of non-degenerate conditions and has as their simplest representative a normal form. So, given a SODE (1.3.9) with $n = 2$ and $k = 3$, we must see if, through certain diffeomorfisms, this system can be reduced to one of the normal forms corresponding to codimension $1, 2$ or 3.

We mention that for some types of bifurcation points, algorithms for obtaining normal forms already exist in the literature. Once a topological normal form is found, its bifurcation diagram is universal because it appears as a part of the bifurcation diagrams of generic dynamical systems which present that bifurcation.

The problem of the classification of all possible bifurcation diagrams of the generic schemes locally (near the bifurcation boundaries) has been solved for continuous two-dimensional dynamical schemes only for the case of bifurcations of equilibrium points of codimension smaller than or equal to three [92].

Assume that the dynamical system associated with (1.3.9), where $\boldsymbol{\alpha} = \mathbf{0}$, has the non-hyperbolic equilibrium $\mathbf{x} = \mathbf{0}$. The assumption that the equilibrium is at the origin is done only for sake of simplicity. Suppose also that around the origin the vector field $\mathbf{f}(\mathbf{x}, \mathbf{0})$, denoted, simply, by $\mathbf{f}(\mathbf{x})$, possesses the asymptotic expansion $\mathbf{f} \sim \mathbf{f}_1 + \mathbf{f}_2 + \ldots + \mathbf{f}_k$, as $\|\mathbf{x}\| \to 0$ and its remainder is of order $O\left(\|\mathbf{x}\|^{k+1}\right)$ as $\|\mathbf{x}\| \to 0$. This behaviour indeed is an important restriction. Fortunately it holds for the F–N model.

In spite of the terms $\mathbf{f}_i$ being smaller and smaller as i is larger and larger, their relevance for the topological equivalence, normal forms, and, consequently, bifurcation, is not the same as their order of magnitude. Indeed, the normal form of a system (1.3.9) usually does not contain terms of all orders of the $\mathbf{f}$ expansion. Hence the process of obtaining the normal form eliminated certain terms of orders greater than some of the retained terms. In addition, it is only beyond a certain order that the terms have no influence on the topological equivalence.

Suppose that $\mathbf{f}_r \in H^r$, where H^r is the real vector space of vector fields whose components are homogenous polynomials of degree r, i.e.,

$$\mathbf{f}_r = \sum_{m_1=0}^{r} \dots \sum_{m_n=0}^{r} \sum_{j=1}^{n} a_{\mathbf{m}}^{(j)} \mathbf{x}^{\mathbf{m}} \mathbf{e}_j,$$

where $\mathbf{m} = (m_1, m_2, ..., m_n)$, $\mathbf{x}^{\mathbf{m}} = x_1^{m_1} x_2^{m_2} ... x_n^{m_n}$, $\mathbf{x} = (x_1, x_2, ..., x_n)^{\mathrm{T}}$ and $\mathbf{e}_1$, $\mathbf{e}_2$, ...,$\mathbf{e}_n$ is the natural basis of $\mathbf{R}^n$. Of course $\mathbf{f}_1 = D\mathbf{f}(\mathbf{0})\mathbf{x} = \mathbf{A}\mathbf{x}$ is the linear part of $\mathbf{f}$.

In order to remove some terms $\mathbf{f}_i$, which are not determinative for the normal form, we use the transformation $\mathbf{x} = \mathbf{y} + \mathbf{h}_r(\mathbf{y})$, where $\mathbf{h}_r \in H^r$, $r \geq 2$. If (1.3.9) is of the form $\dot{\mathbf{x}} = \mathbf{A}\mathbf{x} + \mathbf{f}_r(\mathbf{x})$, then we obtain the new system of equations $\dot{\mathbf{y}} = \mathbf{A}\mathbf{y} - L_{\mathbf{A}}\mathbf{h}_r(\mathbf{y}) + \mathbf{f}_r(\mathbf{y}) + \mathrm{O}\left(\|\mathbf{y}\|^{r+1}\right)$, where $L_{\mathbf{A}}$ is a linear operator on H^r, defined by the relation $L_{\mathbf{A}}\mathbf{h}_r(\mathbf{y}) = D\mathbf{h}_r(\mathbf{y})\mathbf{A}\mathbf{y} - \mathbf{A}\mathbf{h}_r(\mathbf{y})$. Hence the transformation $\mathbf{x} \longleftrightarrow \mathbf{y}$ does not introduce smaller order terms. If $L_{\mathbf{A}}$ is invertible then [9] all terms of order $\|\mathbf{y}\|^r$ can be removed if we choose $\mathbf{h}_r(\mathbf{y}) = L_{\mathbf{A}}^{-1}\mathbf{f}_r(\mathbf{y})$. If choosing $\mathbf{h}_r$ with r larger and larger we always obtain invertible corresponding operators $L_{\mathbf{A}}$, then the dynamical system generated by the particular form of (1.3.9) is topologically equivalent to the linear system.

Suppose that $L_{\mathbf{A}}$ is invertible for the system $\dot{\mathbf{x}} = \mathbf{A}\mathbf{x} + \mathbf{f}_r(\mathbf{x})$. In this case $\mathbf{f}_r$ can be removed and the n-tuple of eigenvalues $\boldsymbol{\lambda} = (\lambda_1, ..., \lambda_n)$ of $\mathbf{A}$ is called **non-resonant** of order r. This happens if there exists no $m_j \in \mathrm{N}$, with $\sum_{j=1}^{n} m_j = r$, such that $\lambda_i = \sum_{j=1}^{n} m_j \lambda_j$. If this is not the case, then we say that $\boldsymbol{\lambda}$ is **resonant**, and $\mathbf{f}_r$ cannot be removed. The transformation $\mathbf{x} \longrightarrow \mathbf{y}$ will remove only the terms belonging to the space $B^r = L_{\mathbf{A}}(H^r)$. They are part of $\mathbf{f}_r$. The remaining part cannot be removed and form the so called **resonant terms**. Therefore the normal forms, apart from linear terms, contain resonant terms, of course, up to higher order terms.

In the two-dimensional case, i.e., $n = 2$ in (1.3.9), if the origin is a non-hyperbolic equilibrium, the following situations can occur:

1. $\det\mathbf{A} = 0$, $\mathrm{tr}\mathbf{A} \neq 0$;
2. $\mathrm{tr}\mathbf{A} = 0$, $\det\mathbf{A} \neq 0$;
3. $\mathrm{tr}\mathbf{A} = \det\mathbf{A} = 0$ (and, hence, $\lambda_1 = \lambda_2 = 0$), but $\mathbf{A} \neq \mathbf{0}$;
4. $\mathbf{A} = \mathbf{0}$.

The equalities (inequalities) in these relations are referred to as **degeneracy (non-degeneracy) conditions**. The degeneracy conditions will influence the type of the eigenvalues λ_1 and λ_2 and, hence, the resonance conditions. This is why, for each such case we expect to have a different normal form. This is indeed the case for situations 1–3 [9].

1.3.4. BIFURCATIONS OF PLANAR DYNAMICAL SYSTEMS

In the following some types of dynamical bifurcations and theorems for planar dynamical systems, that will be used in the study of bifurcations in Chapter 3, are presented. Except for Definitions 1.3.7–1.3.9 that are very general and are used in theoretical studies, there exists a lot of other definitions of particular bifurcations, of great interest in applications. In fact, they describe the bifurcation diagrams. We give a few of these definitions.

Definition 1.3.22 *Let* $(\mathbf{x}^*, \alpha^*) = (\mathbf{0}, 0)$ *be the point corresponding to the null solution of* (1.3.4) *with* $n = 2$ *and* $m = 1$. *The point* $(\mathbf{0}, 0)$ *is called a* **Hopf bifurcation point** *if the bifurcation diagram is topologically equivalent to that from Figure* 1.3.2a *or* 1.3.2b.

For the case $n = m = 2$, Definition 1.3.22 is the same, except for the fact that now, the two-dimensional parameter, which must cross $\boldsymbol{\alpha}^*$, can follow any curve in the parameter space but one. This exceptional curve consists of Hopf bifurcation values.

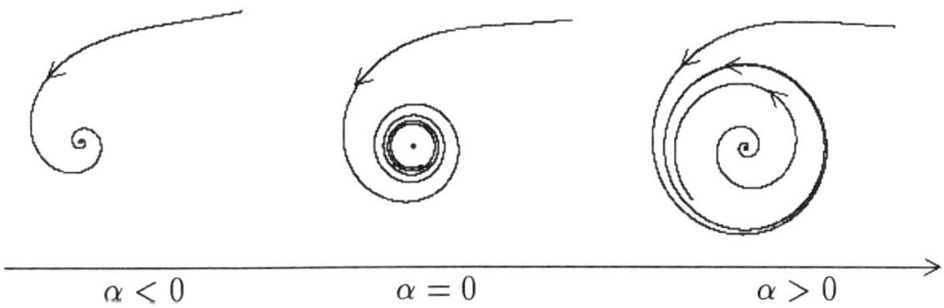

a) Supercritical Hopf bifurcation

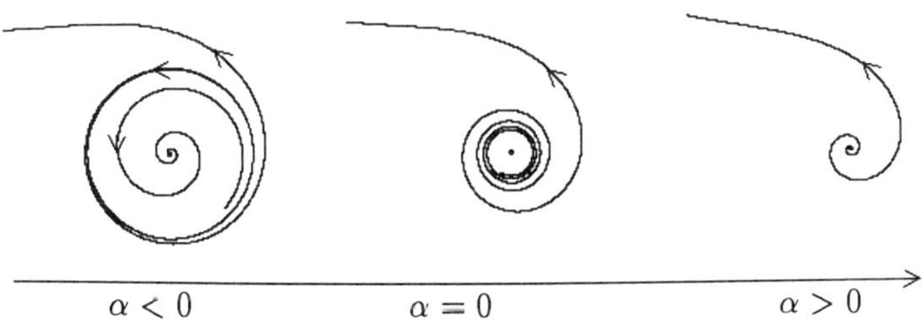

b) Subcritical Hopf bifurcation

Fig. 1.3.2. The Hopf bifurcation.

Diagrams equivalent to those from Figures 1.3.2a and 1.3.2b are given in Figures 1.3.3a and 1.3.3b respectively. Considering that, for Figure 1.3.3, the solution of (1.3.4) is $(\mathbf{x}, \alpha)$ and not $\mathbf{x}$, it follows that these diagrams shaw that the Hopf bifurcation takes place from the branch of solutions $(\mathbf{0}, \alpha)$ situated on the α-axis of the parameter, at a point that, for the linearized dynamical system, corresponds to a centre. The solutions of this branch are attractive foci for $\alpha < 0$ and repulsive foci for $\alpha > 0$ (Figure 1.3.3a). At the Hopf bifurcation point $(\mathbf{0}, 0)$, attractive limit cycles emerge. Their diameter increases with α, for $\alpha > 0$. Since the bifurcation of periodic solutions takes place for α greater than the critical value $\alpha^* = 0$ (of bifurcation), this Hopf bifurcation is called **supercritical**. Similarly, in Figures 1.3.2b and 1.3.3b, the Hopf bifurcation is **subcritical** [25].

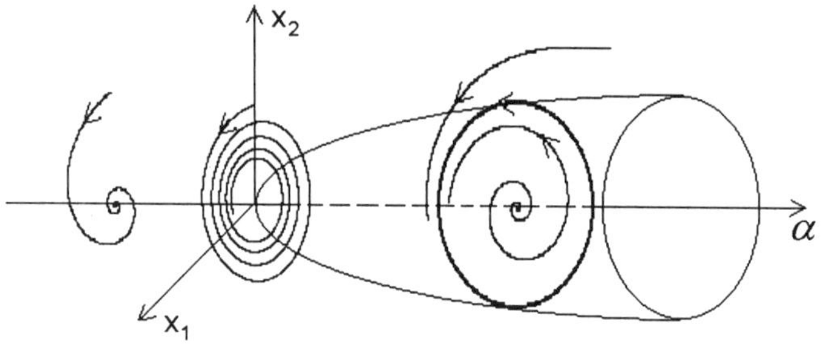

a) Supercritical Hopf bifurcation diagram

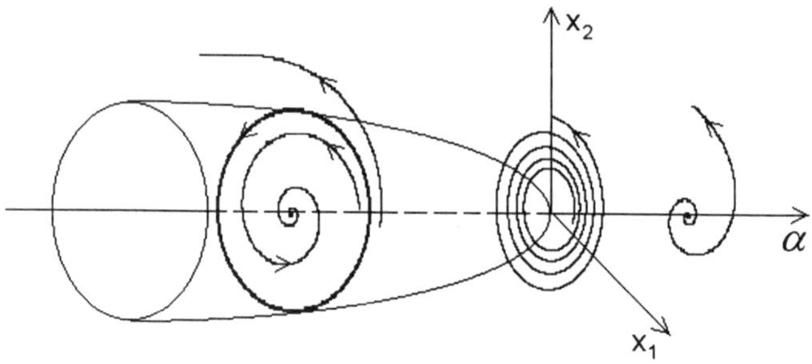

b) Subcritical Hopf bifurcation diagram

Fig. 1.3.3. Uniparametric Hopf bifurcation diagrams.

The systems

$$\begin{aligned} \dot{y}_1 &= \alpha\, y_1 - y_2 \pm\, y_1\, (y_1^2 + y_2^2), \\ \dot{y}_2 &= y_1 + \alpha\, y_2 \pm\, y_2\, (y_1^2 + y_2^2), \end{aligned} \tag{1.3.10}$$

present Hopf bifurcations. They are normal forms for this kind of bifurcations.

Using polar coordinates ρ and θ, (1.3.10) becomes

$$\begin{aligned} \dot{\rho} &= \rho\, (\alpha \pm\, \rho^2), \\ \dot{\theta} &= 1 \end{aligned} \tag{1.3.11}$$

and can be integrated explicitly. Equations $(1.3.11)_1$ and $(1.3.11)_2$ are uncoupled. For $(1.3.11)_1$ a static bifurcation diagram can be done, in which the trivial solution corresponds to the branch of trivial solutions, whilst the nontrivial solutions correspond to the limit cycles in Figure 1.3.3. More precisely, the system (1.3.11) with the minus sign has, for every α, the equilibrium $\rho = 0$ which is attractive for $\alpha < 0$, weakly attractive (centre for the linearized system and attractive with a slowly vanishing speed as $t \to \infty$ for the nonlinear system) for $\alpha = 0$ and repulsive for $\alpha > 0$. In addition, for $\alpha > 0$, (1.3.11) has the attractive equilibrium $\rho = \sqrt{\alpha}$. The static bifurcation diagram of (1.3.11) corresponds to the diagram in Figure 1.3.2a of (1.3.10), having for $\alpha > 0$, in the neighborhood of $\mathbf{y} = \mathbf{0}$, a unique limit cycle. It is an attractive circle of radius $\sqrt{\alpha}$. In the case of (1.3.10) with the plus sign the situations of Figures 1.3.2b and 1.3.3b take place.

For systems (1.3.4) more complicated than (1.3.10), sufficient conditions for the topological equivalence to (1.3.10) are found. Sometimes, these theorems are taken as definitions for the Hopf bifurcation. They are important because they give sufficient conditions for the existence of limit cycles. In general this problem is difficult to be solved in the theory of dynamical systems. Sometimes these sufficient conditions are expressed by using the eigenvalues of the linearized system corresponding to (1.3.4). Indeed, the first transformation used to reduce (1.3.4) to its normal form is constructed with the aid of the eigenvectors of the linearized system.

Theorem 1.3.2 [67] **(Hopf)** *Consider the system*

$$\dot{\mathbf{x}} = A(\alpha)\,\mathbf{x} + F(\mathbf{x}, \alpha), \tag{1.3.12}$$

with $F \in C^k$, $k > 2$, $\mathbf{x} \in \mathbf{R}^2$, $\alpha \in \mathbf{R}$, *such that* $F(\mathbf{0}, \alpha) = 0$ *and* $D_{\mathbf{x}} F(\mathbf{0}, \alpha) = \mathbf{0}$ *for every* α *with sufficientely small* $|\alpha|$. *Let* $\lambda(\alpha) = \mu(\alpha) \pm i\,\omega(\alpha)$, *with* $\mu(0) = 0$ *and* $\omega(0) \neq 0$, *be the eigenvalues of the linear part* $A(\alpha)$ *at the origin. If* $\frac{d\mu}{d\alpha}(0) \neq 0$, *then, in every neighborhood* U *of the origin of* $\mathbf{R}^2$ *and for every* $\alpha_0 > 0$ *there exist* α^* *with* $0 < |\alpha^*| < \alpha_0$ *such that for* $\alpha = \alpha^*$ *the system* (1.3.12) *has a periodic nontrivial orbit in* U *(that is* (1.3.12) *presents a Hopf bifurcation).*

The next theorem gives sufficient conditions to reduce (1.3.4) to the normal form (1.3.10).

Theorem 1.3.3 [92] **(The normal form for the Hopf bifurcation).** *Consider the system* (1.3.4) *with* $\mathbf{f} \in C^1$, $n = 2$, $m = 1$, *having for* $\alpha = 0$ *the equilibrium point* $\mathbf{x} = \mathbf{0}$. *If, for small* α, *the eigenvalues are* $\lambda_{1,2}(\alpha) = \mu(\alpha) \pm i\,\omega(\alpha)$ *with* $\mu(0) = 0$, $\omega(0) > 0$ *and the following nondegeneracy conditions are satisfied:*

i) $\mu'(0) \neq 0$;

ii) $\ell_1(0) \neq 0$, *with* ℓ_1 *the first Liapunov coefficient,*
then the system is locally topologically equivalent, near the origin, to one of the normal forms (1.3.10).

The Hopf bifurcation is one of the generic codimension-one bifurcations. The types of generic bifurcations of codimension-one for (1.3.4) with $n = 2$ are schematically presented in Figure 1.3.4, as given in [67].

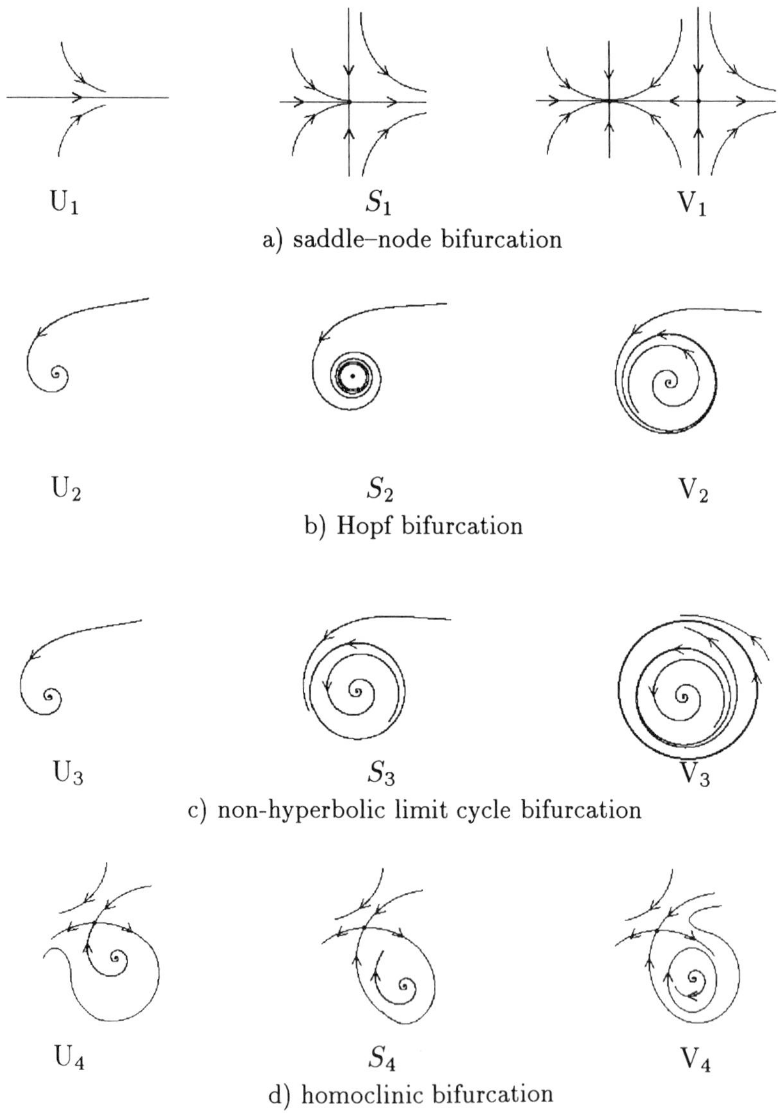

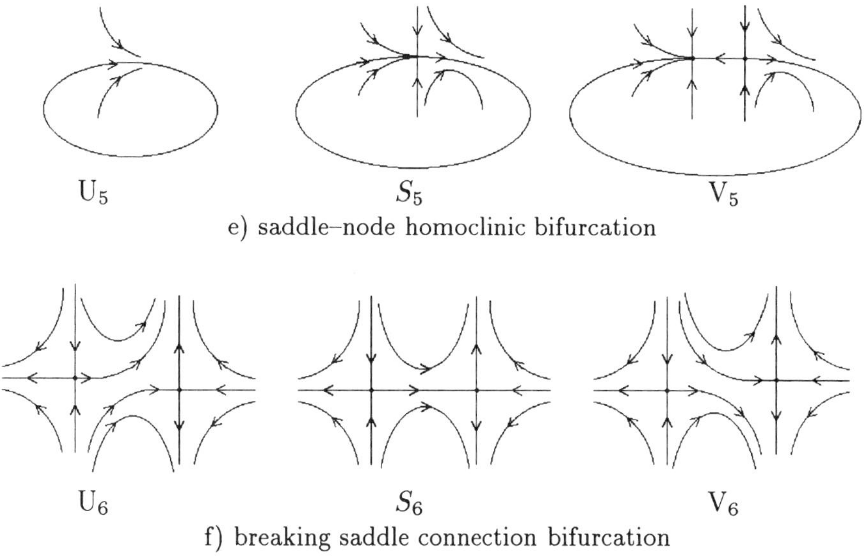

Fig. 1.3.4. Generic codimention–one bifurcations for planar systems.

In Figure 1.3.4, U_i, S_i and V_i , $i = \overline{1,6}$ are strata of the parameter space around the bifurcation value α^*. The set $U_i \cup S_i \cup V_i$ is a neighborhood of α^*. The dynamical systems corresponding to points from U_i and V_i are structurally stable, whilst those corresponding to points from S_i are topologically equivalent to the dynamical system obtained for α^*. If $\alpha \in \mathbf{R}$, U_i and V_i are portions of curves, whilst S_i is α^*. If $\boldsymbol{\alpha} \in \mathbf{R}^2$, U_i and V_i are portions of the plane, whilst S_i is a curve containing $\boldsymbol{\alpha}^*$. If $\boldsymbol{\alpha} \in \mathbf{R}^3$, U_i and V_i are three-dimensional domains, whilst S_i is a surface containing the point $\boldsymbol{\alpha}^*$.

In the following some codimension-two bifurcations are defined and sufficient conditions for the occurence of these bifurcations are given. They will be used in the study of the system (1.1.17).

Definition 1.3.23 *Consider the point* $(\boldsymbol{\alpha}^*, \mathbf{x}^*) = (\mathbf{0}, \mathbf{0})$ *corresponding to the zero solution of* (1.3.4) *with* $n = m = 2$. *The point* $(\mathbf{0}, \mathbf{0})$ *is called a* **Bautin bifurcation point** *if its bifurcation diagram around the bifurcation value is topologically equivalent to that from Figure* 1.3.5.

The Bautin bifurcation is a degenerated Hopf bifurcation [11], [12]. The $O\beta_2$-axis corresponds to the Hopf bifurcation values, but for $\beta_1 = 0$ the Hopf bifurcation turns from subcritical to supercritical. The equilibrium is attractive for $\beta_1 < 0$, repulsive for $\beta_1 > 0$, weak attractive for $\beta_1 = 0$, $\beta_2 < 0$ and weak repulsive for $\beta_1 = 0$, $\beta_2 > 0$. The curve Ba which emerges from the origin is tangent to $O\beta_2$ at the origin. The points of the curve Ba correspond to non-hyperbolic limit cycle bifurcation.

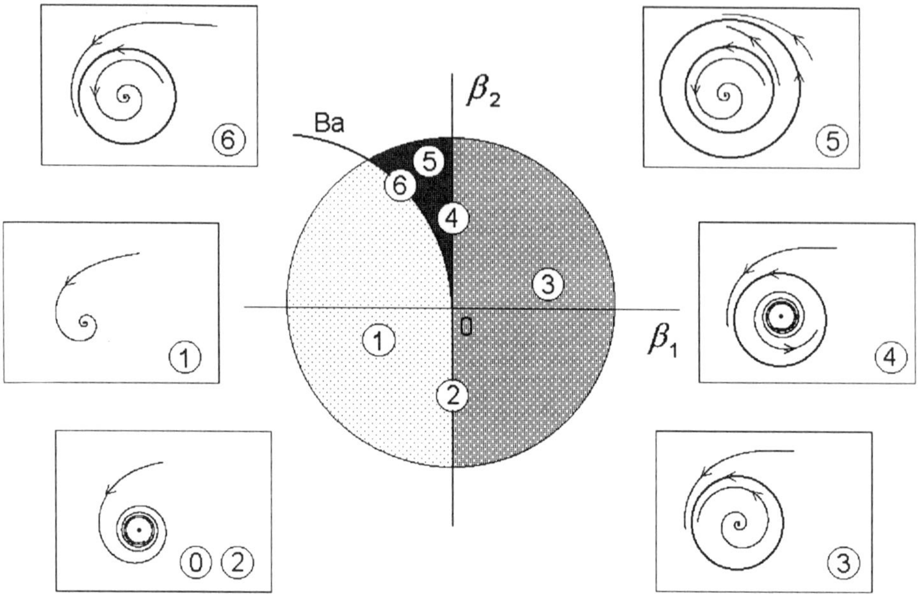

Fig. 1.3.5. Bifurcation diagram near a Bautin bifurcation value.

Consider the system

$$\dot{\mathbf{z}} = (\beta_1 + i)\,\mathbf{z} + \beta_2\,\mathbf{z}|\mathbf{z}|^2 + s\,\mathbf{z}|\mathbf{z}|^4,\ \mathbf{z} \in \mathbf{C},\ s = \pm 1. \tag{1.3.13}$$

Equation (1.3.13) represents the normal form for the Bautin bifurcation. For $s = -1$, using the polar coordonates $\mathbf{z} = \rho\, e^{i\theta}$ (1.3.13) becomes

$$\begin{aligned} \dot{\rho} &= \rho\,(\beta_1 + \beta_2 \rho^2 - \rho^4), \\ \dot{\theta} &= 1. \end{aligned} \tag{1.3.14}$$

Its bifurcation diagram corresponds to that from Figure 1.3.5 for the system (1.3.13). If $s = 1$ the bifurcation diagram remains the same, but the flow on the trajectories of the system (1.3.13) is reversed. In first asymptotic approximation, as $\beta_1 \to 0$, the curve Ba is defined by the equation

$$Ba: \qquad \beta_2^2 + 4\,\beta_1 = 0, \qquad \beta_2 > 0. \tag{1.3.15}$$

A sufficient condition for the existence of the Bautin bifurcation is given by the following theorem. This theorem also states that (1.3.13) is the corresponding normal form.

Theorem 1.3.4 [92] **(The normal form for the Bautin bifurcation)** *Consider the system* (1.3.4) *with* $n = m = 2$ *and* $\mathbf{f} \in C^1$ *having the equilibrium* $\mathbf{x} = \mathbf{0}$ *for* $\boldsymbol{\alpha} = \mathbf{0}$. *Assume that, for small* $\boldsymbol{\alpha}$, *the eigenvalues* $\lambda_{1,2}(\boldsymbol{\alpha}) = \mu(\boldsymbol{\alpha}) \pm i\,\omega(\boldsymbol{\alpha})$ *have* $\mu(\mathbf{0}) = \mathbf{0}$, $\omega(\mathbf{0}) > 0$ *and* $\ell_1(\mathbf{0}) = 0$. *If the following nondegeneracy conditions:*

i) $\ell_2(\mathbf{0}) \neq 0$ *where* ℓ_2 *is the second Liapunov coefficient;*

ii) the map $\boldsymbol{\alpha} \to (\mu(\boldsymbol{\alpha}), \ell_1(\boldsymbol{\alpha}))^T$ *is regular at* $\boldsymbol{\alpha} = \mathbf{0}$;

are satisfied, then the system is locally topologically equivalent near the origin to the system (1.3.13), *where* $s = \operatorname{sign}(\ell_2(\mathbf{0}))$.

The Liapunov coefficients ℓ_1 and ℓ_2 are the coefficients of the terms of the third and fifth order, respectively, of an intermediate system occuring during the derivation of the normal form (1.3.13). Their expressions will be given in Section 3.5.1.

Definition 1.3.24 [7], [17], [26], [134] *Assume that the point* $(\boldsymbol{\alpha}^*, \mathbf{x}^*) = (\mathbf{0}, \mathbf{0})$ *is the null solution of* (1.3.4) *with* $n = m = 2$. *The point* $(\mathbf{0}, \mathbf{0})$ *is called a* **Bogdanov–Takens bifurcation point** *if the bifurcation diagram around the bifurcation value is topologically equivalent to that in Figure* 1.3.6.

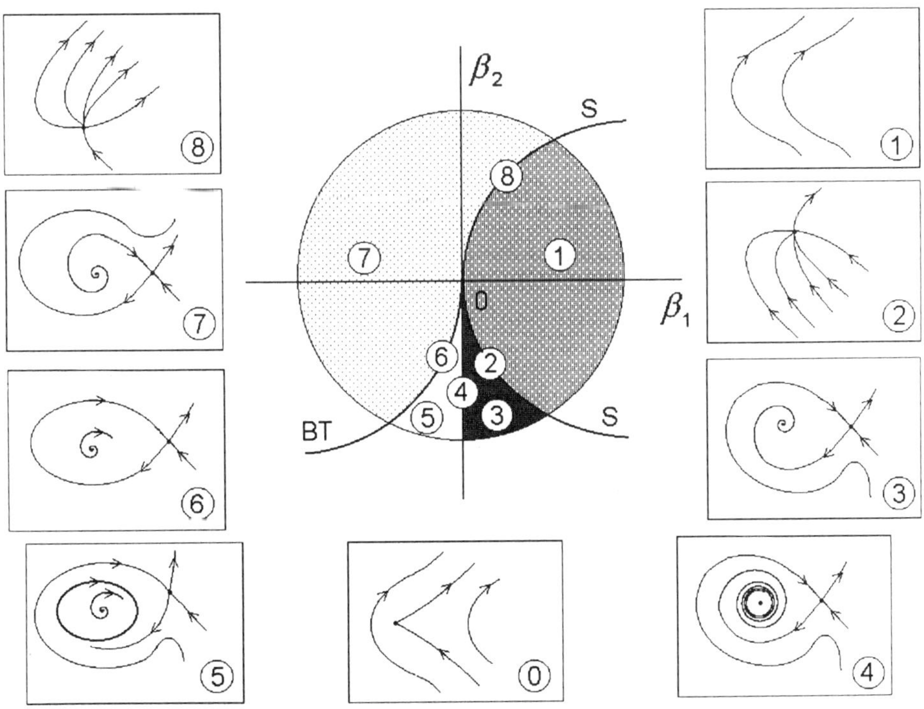

Fig. 1.3.6. Bifurcation diagram near a Bogdanov–Takens value.

The following systems

$$\begin{aligned} \dot{\eta}_1 &= \eta_2, \\ \dot{\eta}_2 &= \beta_1 + \beta_2\,\eta_1 + \eta_1^2 + s\,\eta_1\,\eta_2, \end{aligned} \tag{1.3.16}$$

with $s = \pm 1$, represent the normal form for the Bogdonov-Takens bifurcation. The diagram from Figure 1.3.6 corresponds to $s = -1$. The $O\beta_2$-axis, with $\beta_2 < 0$, contains the Hopf bifurcation values, whilst the curve S contains saddle–node bifurcation values and is tangent to the $O\beta_2$-axis at the origin. The curve BT consists of homoclinic bifurcation values and emerges from the origin. It has the equation

$$BT : \beta_1 = -\frac{6}{25}\,\beta_2^2 + o(\beta_2^2), \quad \beta_2 < 0. \tag{1.3.17}$$

The system (1.3.16) has a single attractive limit cycle in domain 5 and has no limit cycles outside this domain. For $s = 1$ the parametric portrait is the same, but the limit cycle is repulsive and the sense along the trajectories is reversed.

Assume that the system (1.3.4) with $n = m = 2$, $\mathbf{f} \in C^1$ has, for $\boldsymbol{\alpha} = \mathbf{0}$, the point of equilibium $\mathbf{x} = \mathbf{0}$ and both eigenvalues of the linearized system are zero. If $\mathbf{v}_0$, $\mathbf{v}_1$ ($\mathbf{w}_1$, $\mathbf{w}_0$, respectively) are the eigenvector and the generalized eigenvector of $D\mathbf{f}(\mathbf{0})$, (respectively $(D\mathbf{f}(\mathbf{0})^T)$, and they are choosen such as $\langle \mathbf{v}_0, \mathbf{w}_0 \rangle = \langle \mathbf{v}_1, \mathbf{w}_1 \rangle = 1$, then the homeomorphism $\mathbf{x} = y_1\,\mathbf{v}_0 + y_2\,\mathbf{v}_1$ transforms (1.3.4) into the system $\dot{\mathbf{y}} = \mathbf{g}\,(\mathbf{y}, \boldsymbol{\alpha})$.

Let

$$a_{20} = \frac{\partial^2 g_1}{\partial y_1^2}(\mathbf{0},\mathbf{0}), \qquad b_{20} = \frac{\partial^2 g_2}{\partial y_1^2}(\mathbf{0},\mathbf{0}), \qquad b_{11} = \frac{\partial^2 g_2}{\partial y_1 \partial y_2}(\mathbf{0},\mathbf{0}).$$

Theorem 1.3.5 [92] **(The normal form for the Bogdanov–Takens bifurcation)** *Assume that the system* (1.3.4) *with* $n = m = 2$ *and* $\mathbf{f} \in C^1$, *has, for* $\boldsymbol{\alpha} = \mathbf{0}$, *the equilibrium* $\mathbf{x} = \mathbf{0}$ *and both eigenvalues of the linearized system are equal to zero. If the following nondegeneracy conditions:*

i) $D\mathbf{f}\,(\mathbf{0},\mathbf{0}) \neq 0$;
ii) $a_{20} + b_{11} \neq 0$;
iii) $b_{20} \neq 0$;
iv) the application $(\mathbf{x}, \boldsymbol{\alpha}) \to \left(\mathbf{f}\,(\mathbf{x}, \boldsymbol{\alpha}), \mathrm{tr}\frac{\partial \mathbf{f}}{\partial \mathbf{x}}\,(\mathbf{x}, \boldsymbol{\alpha}), \det\frac{\partial \mathbf{f}}{\partial \mathbf{x}}\,(\mathbf{x}, \boldsymbol{\alpha})\right)$ *is regular at* $(\mathbf{x}, \boldsymbol{\alpha}) = (\mathbf{0}, \mathbf{0})$;

are satisfied, then the system is locally topologically equivalent to the normal form (1.3.16), *with* $s = \mathrm{sign}[b_{20}\,(a_{20} + b_{11})]$.

Definition 1.3.25 *Assume that the point* $(\boldsymbol{\alpha}^*, \mathbf{x}^*) = (\mathbf{0}, \mathbf{0})$ *is the trivial solution of* (1.3.4) *with* $n = m = 2$. *The point* $(\mathbf{0}, \mathbf{0})$ *is called a* **saddle–node separatrix loop bifurcation point** *if the bifurcation diagram around the bifurcation value is topologically equivalent to that of Figure* 1.3.7.

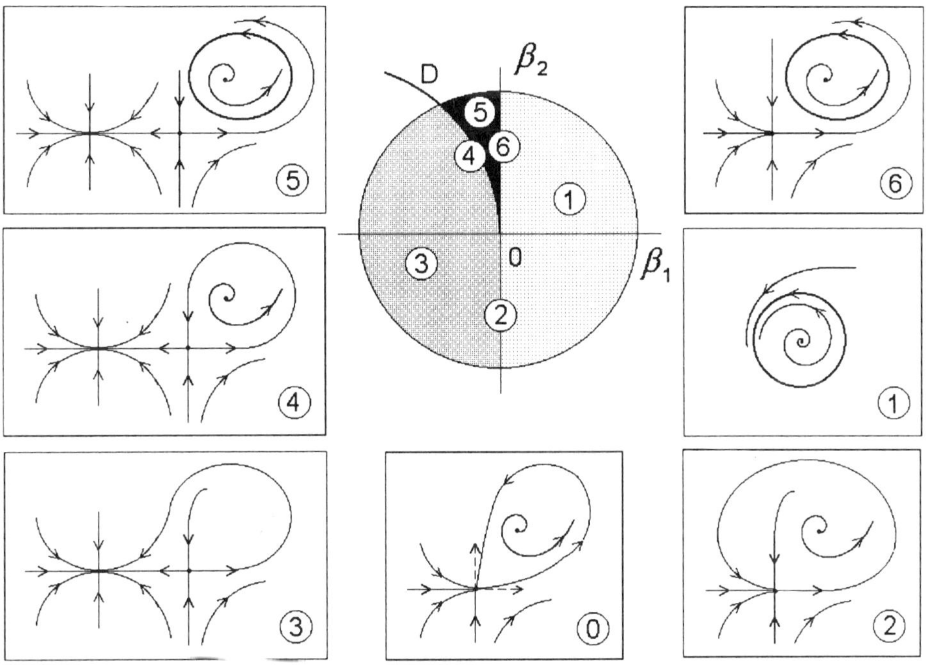

Fig. 1.3.7. Bifurcation diagram near a separatrix loop saddle–node.

For points of the $O\beta_2$-axis, the system has a saddle–node equilibrium point. The curve D consists of homoclinic bifurcation values and it is tangent to $O\beta_2$-axis in $\beta = 0$. As the $O\beta_2$-axis, with $\beta_2 < 0$, is crossed a saddle–node homoclinic bifurcation occurs.

Theorem 1.3.6 [130], [131] *Assume that the system* (1.3.4) *with* $n = m = 2, \mathbf{f} \in C^1$ *has, for* $\boldsymbol{\alpha} = \mathbf{0}$, *an equilibrium point* $\mathbf{x} = \mathbf{0}$ *of saddle–node type, with a negative eigenvalue (the other being zero) and a separatrix loop L. Consider, in the Banach space of all vector fields* $\mathbf{f}(\mathbf{x}, \boldsymbol{\alpha})$, *the codimension-two submanifold* Σ *of all fields* $\mathbf{f}(\mathbf{x}, \mathbf{0})$ *possessing such a point of equilibrium. If the familiy* $\mathbf{f}(\mathbf{x}, \boldsymbol{\alpha})$ *is transversal to* Σ *at* $\boldsymbol{\alpha} = \mathbf{0}$, *then* $(\mathbf{0}, \mathbf{0})$ *is a saddle–node separatrix loop bifurcation point.*

Notice that the normal forms are associated only with the bifurcations of equilibrium points.

From the Bautin, Bogdanov–Takens, and saddle–node with separatrix loop bifurcations points, one or several curves of bifurcation values of homoclinic type, non-hyperbolic limit cycle type, and saddle–node homoclinic type emerge. Some of them may occur even in the absence of the quoted bifurcations. Since for them the normal forms are not known, the numerical computations are very useful.

There exist theorems, such as the Center Manifold Theorem, allowing one to reduce the dimension of a given system near a local bifurcation [141], [92].

Thus, consider the system (1.3.6) with $n = 2$ and $\mathbf{f}$ sufficiently smooth. Suppose that $\mathbf{x} = \mathbf{0}$ is an equilibrium point whose corresponding eigenvalues are $\lambda_1 = 0$ and $\lambda_2 \neq 0$. Let E^c be the linear eigenspace corresponding to $\lambda_1 = 0$ and let φ_t be the flow associated with (1.3.6).

Theorem 1.3.7 (Center Manifold Theorem) *There is a locally defined smooth one-dimensional invariant manifold $W^c_{\text{loc}}(\mathbf{0})$ of* (1.3.6) *that is tangent to E^c at $\mathbf{x} = \mathbf{0}$. Moreover, there is a neighborhood U of $\mathbf{x} = \mathbf{0}$ such that if $\varphi_t(x) \in U$ for all $t \geq 0$ $(t \leq 0)$, then $\varphi_t(x) \to W^c_{\text{loc}}(\mathbf{0})$ for $t \to \infty$ $(t \to -\infty)$.*

The manifold W^c_{loc} is called **the center manifold**.

In a basis formed by the eigenvectors, system (1.3.4) with $n = 2$ is written as

$$\begin{aligned} \dot{u} &= g(u, v), \\ \dot{v} &= \delta v + h(u, v), \end{aligned} \tag{1.3.18}$$

where $\delta \neq 0$ and the functions g and h have Taylor expansions in a neighborhood of (0,0) in $\mathbf{R}^2$, which contains no constant and no linear terms.

According to Theorem 1.3.7, there exists $v = \mathcal{V}(u)$ the center manifold. The following result holds.

Theorem 1.3.8 *Consider system* (1.3.18), *with $\delta \neq 0$. The flow in the center manifold is determined by the equation*

$$\dot{u} = g(u, \mathcal{V}(u)). \tag{1.3.19}$$

If the solution $u = 0$ of (1.3.19) *is stable (unstable), then the solution $(u, v) = (0, 0)$ of* (1.3.18) *is stable (unstable).*

A lot of other types of bifurcation, e.g. codimension-three bifurcations, have not been displayed in our presentation, but they cannot be avoided if, in the model, three parameters occur [81], [92].

1.4. REGULAR AND SINGULAR PERTURBATIONS

1.4.1. ASYMPTOTIC EXPANSIONS AND MATCHING PRINCIPLES

Let $f : D \to \mathbf{C}$, $f = f(z)$ be a map and let $z_0 \in \overline{D}$ be a point of accumulation of the set $D \subset \mathbf{C}$. We are interested in the asymptotic behaviour of the map f as $z \to z_0$, i.e., of the values $f(z)$, as $z \to z_0$. The point z_0 is called an **asymptotic point**.

Two functions f and $g : D \to \mathbf{C}$, $f = f(z)$, $g = g(z)$ stand in the **order relation** O in D, as $z \to z_0$ (and it is written $f = O(g)$), if there exists a neighborhood U_{z_0} of z_0 and a constant $A \geq 0$ such as $|f(z)| \leq A|g(z)|$, for every $z \in U_{z_0} \cap D$.

The functions f and g are in **order relation** o, as $z \to z_0$, in D (and it is written $f = o(g)$), if for every $\varepsilon > 0$, there exists a neighborhood U_ε of z_0 such as $|f(z)| < \varepsilon |g(z)|$ for every $z \in U_\varepsilon \cap D$.

Instead of $f = o(g)$, we can write $f \ll g$, as $z \to z_0$, in D and it is said that f **is much smaller than** g around the asymptotic point z_0.

The functions f and g are **asympotically equivalent** (or equal) as $z \to z_0$ if

$$\lim_{z \to z_0} \left| \frac{f(z)}{g(z)} \right| = 1.$$

We write this as $f \sim g$, as $z \to z_0$, in D. The relation $\sim$ is an equivalence relation, whilst the relations O and o are not. The functions f and g are called of the **same order** and we write $f = \mathrm{Ord}(g)$ if $f = \mathrm{O}(g)$ and $g = \mathrm{O}(f)$ as $z \to z_0$ in D. It can be prooved that $f = \mathrm{Ord}(g)$ iff there exists the constants $0 < A, B < \infty$ such as $A|g| \leq |f| \leq B|g|$ as $z \to z_0$, in D.

Finding the asymptotic behaviour of a function f, as $z \to z_0$, means to find a sequence of (standard) functions, which attached series has the sum asymptotically equivalent to f.

The standard functions $\delta_i : D \to \mathbf{C}$, $i = 1, 2, ..., N \leq \infty$ form an **asymptotic sequence** if $\delta_{i+1} \ll \delta_i$. Let us remark that, in spite of the name of 'sequence', the set of functions δ_i can be finite. Consider the constants a_i, $i = 1, ..., N$. Then the sum $\sum_{i=1}^{N} a_i \delta_i$ is called an **asymptotic series with respect to the asymptotic sequence** $\{\delta_i\}_{i=\overline{1,N}}$. In general, the asymptotic series are divergent.

The asymptotic series $\sum_{i=1}^{N} a_i \delta_i$ with respect to the asymptotic sequence $\{\delta_i\}_{i=\overline{1,N}}$ is an **asymptotic expansion of the function** f if the remainder $r_N(z) \equiv f(z) - \sum_{i=1}^{N} a_i \delta_i$ is much smaller than δ_N, i.e., $|r_N(z)/\delta_N(z)| \to 0$ as $z \to z_0$.

It follows that the asymptotic expansion of the function f with respect to the asymptotic sequence $\{\delta_i\}_{i=1,2,...,N}$, as $z \to z_0$, is unique and has the form

$$f \sim \sum_{i=1}^{N} a_i \delta_i, \tag{1.4.1}$$

where

$$a_1 = \lim_{z \to z_0} \frac{f(z)}{\delta_1(z)},$$

$$a_k = \lim_{z \to z_0} \frac{f(z) - \sum_{i=1}^{k-1} a_i \delta_i}{\delta_k}, \quad k = 2, ..., N.$$

The domain of validity of an asymptotic expansion is that for which $r_N \ll \delta_N$. Let us remark that f has the same order of magnitude with δ_1 and with any other partial sum of the series (1.4.1).

In general, a function has many asymptotic expansions around different asymptotic points z_0. The first difficult problem is to find the order of magnitude of f, and therefore of the standard functions δ_i, especially when the expression of f is unknown, as in the case of the solutions $f(z, p)$ of differential equations, where p is the parameter on which f depends. The presence of the parameter has as a consequence the dependence on that parameter of the order relations between f and other functions. Therefore the parameter induces a nonuniformity of the asymptotic expansions.

1.4.2. PERTURBATION PROBLEMS

The asymptotic theory has two big chapters: asymptotic analysis, and the perturbation theory. Asymptotic analysis studies the order relations, the sequences, the series, and the asymptotic expansions. We gave the definitions above in the set $\mathbf{C}$. Similary, they can be given in $\mathbf{R}^n$ or in Banach spaces (finite- or infinite-dimensional) by replacing the modulus with an apropriate norm. In the asymptotic analysis the variable with respect to which the asymptotic study is carried out is called the **asymptotic variable**, whilst the rest of the variables are called the **asymptotic parameters** [51].

In the perturbation theory the asymptotic variable is a physically small parameter, $\varepsilon > 0$, $\varepsilon \ll 1$, and the independent variables are considered as asymptotic parameters. The problems containing a small parameter are called **perturbation problems**.

Let $x(t, \varepsilon)$ be a solution of a differential equation in which t is the independent variable and $\varepsilon > 0$ a small parameter. An asymptotic expansion of it, as $\varepsilon \to 0$, with respect to the asymptotic sequence $\alpha_1(\varepsilon), \alpha_2(\varepsilon)$

$, ..., \alpha_N(\varepsilon)$, as $\varepsilon \to 0$, is the **Poincaré expansion**

$$x(t,\varepsilon) \sim x_1(t)\,\alpha_1(\varepsilon) + x_2(t)\,\alpha_2(\varepsilon) + ... + x_N(t)\,\alpha_N(\varepsilon) \text{ as } \varepsilon \to 0, \quad (1.4.2)$$

or, more generally, the expansion

$$x(t,\varepsilon) \sim x_1(t,\varepsilon) + x_2(t,\varepsilon) + ... + x_N(t,\varepsilon) \text{ as } \varepsilon \to 0, \quad (1.4.3)$$

where

$$x_N \ll x_{N-1} \ll ... \ll x_1, \text{ as } \varepsilon \to 0 \quad (1.4.4)$$

and

$$\lim_{\varepsilon \to 0} \frac{x(t,\varepsilon) - \sum_{i=1}^{k} x_i(t,\varepsilon)}{\alpha_k(\varepsilon)} = 0. \quad (1.4.5)$$

The sequence $\alpha_i, i = 1, 2, ..., N$ used to write the expansion (1.4.3) does not appear explicitly.

In perturbation problems the nonuniformity of the asymptotic expansions with respect to the independent variable (that is the asymptotic parameter) is very important. Thus, we can have several asymptotic expansions of a given function $x(t,\varepsilon)$, as $\varepsilon \to 0$, in different regions of the domain of the independent variable. The problem of their matching must then be solved. A problem is called a problem of **regular perturbations** if it contains a small parameter ε and its solution has a single asymptotic expansion as $\varepsilon \to 0$, in the whole domain Ω of the independent variable, that is, its expansion is uniformly valid in Ω.

A problem is called a problem of **singular perturbations** if it contains a small parameter ε and its solution has in Ω at least two asymptotic expansions as $\varepsilon \to 0$, that is, that the expansion is not uniformly valid all over the domain of the independent variable [42], [140].

A singular perturbations problem is of **boundary layer** type if in the domain of the independent variable there exists a small region such that when crossed the solution changes its order, and so has a fast variation. Usually the boundary layer is near a frontier, on the boundary of the domain of definition of the solution (this is why it is called a boundary layer). When the independent variable is the time and the boundary layer is near the initial instant, the layer is called **initial layer** . The asymptotic expansion of the solution inside the boundary layer is called the **inner asymptotic expansion**, whilst that from outside is called the **outer asymptotic expansion**. Inside the boundary layer a new independent variable is introduced, in order to magnify it. It is called the **inner variable**. In the case of the initial layer, instead of inner and outer variables we have a **fast** and a **slow** variable.

There exists several rules for matching the asymptotic expansions of a solution, called **matching principles**. Amongst them we mention the **n $\times$ m principle**, for problems of boundary layer type. These problems contain the small parameter as a coefficient of the higher order derivatives (or of some of them). The principle states as follows: the outer asymptotic expansion truncated at n terms of the inner asymptotic expansion, truncated at m terms is equal to the inner asymptotic expansion truncated at m terms of the outer asymptotic expansion truncated at n terms of the solution.

In singular perturbations problems, the specification of the order of magnitude of the independent variable with respect to ε is necessary. As a consequence, inner limits $\lim\limits_{\varepsilon\to 0,\tau \text{ fixed}} x\left(t\left(\tau,\varepsilon\right),\varepsilon\right)$ and outer limits $\lim\limits_{\varepsilon\to 0,t \text{ fixed}} x\left(t,\varepsilon\right)$, where τ is the inner variable and t the outer variable, must be considered. As an exemple, $\tau = t/\varepsilon$ when $t = \mathrm{O}(\varepsilon)$ and $\varepsilon = \mathrm{O}(t)$ as $\varepsilon \to 0$ and $t \to 0$. We recall that the τ variable is called fixed from the asymptotic viewpoint if $\tau = \mathrm{Ord}(1)$ as $\varepsilon \to 0$, that is there exists the constants (bounded) $A, B > 0$ such as $A < \tau < B$ (i.e., τ is neither too big, nor too small).

Let us see, for instance, what the 1×1 principle means, supposing that the appropriate asymptotic sequence is the power sequence

$$1, \varepsilon, \varepsilon^2, ..., \text{ as } \varepsilon \to 0. \tag{1.4.6}$$

As the expansions with respect to (1.4.6) are Taylor series, they are appropriate when the function to be expanded is sufficiently smooth in ε.

Let $x\left(t,\varepsilon\right)$ be a real function, where $t > 0$ is the independent variable and ε $(0 < \varepsilon \ll 1)$ is the small parameter. Suppose that we have an initial layer $[0, \delta]$ along the time-axis, where $\delta \ll 1$, so this layer is also very small. The outer expansion of x as $\varepsilon \to 0$ with respect to the sequence (1.4.6) is

$$x\left(t,\varepsilon\right) \sim X_0\left(t\right) + \varepsilon X_1\left(t\right) + \varepsilon^2 X_2\left(t\right) + ... \quad \text{as } \varepsilon \to 0,\, t = O(1). \tag{1.4.7}$$

If $\delta = \mathrm{Ord}(\alpha\left(\varepsilon\right))$, then a new variable $\tau = t/\alpha\left(\varepsilon\right)$ is introduced in the initial layer, which will be asymptotically fixed, that is $\tau = \mathrm{Ord}\left(1\right)$, as $\varepsilon \to 0$. Owing to this change of variable the initial layer was magnified. For the sake of simplicity, consider $\alpha\left(\varepsilon\right) = \varepsilon$. Then x becomes a function of τ and ε, that is $x\left(t,\varepsilon\right) = x\left(\varepsilon\tau,\varepsilon\right) = \hat{x}\left(\tau,\varepsilon\right)$. Suppose that the asymptotic sequence used in writing the inner asymptotic expansion as $\varepsilon \to 0$ is also (1.4.6), so this expansion reads

$$x\left(t,\varepsilon\right) \sim \hat{x}_0\left(\tau\right) + \varepsilon\, \hat{x}_1\left(\tau\right) + \varepsilon^2\, \hat{x}_2\left(\tau\right) + ... \quad \text{as} \varepsilon \to 0,\, \tau = O\left(1\right). \tag{1.4.8}$$

An important remark is that because the first term of the asymptotic sequence (1.4.6) is equal to 1, the first coefficients of the expansions (1.4.7) and (1.4.8) are the limits of the function to be expanded, more precisely $X_0(t) = \lim_{\varepsilon\to 0} x(t,\varepsilon)$ and $\hat{x}_0(\tau) = \lim_{\varepsilon\to 0} \hat{x}(\tau,\varepsilon)$.

With this remark, the 1×1 matching principle becomes

$$\lim_{\varepsilon\to 0,\tau \text{ fixed}} X_0(\varepsilon\tau) = \lim_{\varepsilon\to 0,t \text{ fixed}} \hat{x}_0(t/\varepsilon) \tag{1.4.9}$$

or, equivalently

$$\lim_{t\to 0} X_0(t) = \lim_{\tau\to\infty} \hat{x}(\tau). \tag{1.4.10}$$

If one of the two limits from (1.4.10) does not exist, then the 1×1 principle can not be applied. In this situation we apply another matching principle, namely the **intermediary matching principle**. In order to do so, we introduce a new independent variable

$$\eta = \frac{t}{\beta(\varepsilon)}, \tag{1.4.11}$$

where $\alpha(\varepsilon) \ll \beta(\varepsilon) \ll 1$, as $\varepsilon \to 0$. So, η =Ord(1) in an intermediary domain situated between the initial layer and its exterior. This is why η is called the **intermediary variable**. In this domain, $t \ll 1$ and $\tau \gg 1$ as $\varepsilon \to 0$, so there exists the possibility to match the expansions (1.4.7) and (1.4.8) using the intermediary matching principle. In order to obtain it, in these expansions we replace $t = \eta\beta(\varepsilon)$ and $\tau = \eta\beta(\varepsilon)/\alpha(\varepsilon)$. The expansions obtained cannot be considered as asymptotic series with respect to the sequence (1.4.6) since ε occurs in the coefficients X_i and $\hat{x}_i$. This is why they will be asymptotic expansions whith respect to two unknown asymptotic sequences, that will be determined during the matching process.

In the $n \times m$ matching process, an intermediary series is obtained, included in (1.4.7) and (1.4.8), in the common domain of validity. Correspondingly, an asymptotic expansion as $\varepsilon \to 0$, uniformely valid for $t \in \mathbf{R}$, will be the sum of (1.4.7) and (1.4.8), minus the intermediary series. As a result, a composite series is obtained, i.e., depending on t and τ.

1.4.3. MODELS OF ASYMPTOTIC APPROXIMATION

Consider a problem

$$B(\varepsilon, x) = 0, \tag{1.4.12}$$

for a perturbation equation

$$T(\varepsilon, x) = 0, \tag{1.4.13}$$

where ε is the small parameter, x is the unknown function, B is the (limit and/or initial) conditions operator and T is a nonlinear differential operator. Formally, for the study of the asymptotic behavior of the solution x as $\varepsilon \to 0$, (1.4.7) is substituted in (1.4.13) and the condition that the coefficients of the powers of ε in the obtained equation vanish is considered. Thus the equations

$$T_k(x) = 0, \ k = 0, 1, ... \tag{1.4.14}$$

are found. They are called the **order k^{th} asymptotic approximation equations** of (1.4.13) as $\varepsilon \to 0$. Suppose that (1.4.12), (1.4.13) is a problem of regular perturbations. Then, substituting (1.4.7) into (1.4.12) it follows

$$B_k(x) = 0, \qquad k = 0, 1, ... \tag{1.4.15}$$

The corresponding problems

$$B_k(x) = 0, \qquad T_k(x) = 0 \tag{1.4.16}$$

are called the **models of asymptotic approximation of order** k for the given model (1.4.12), (1.4.13) as $\varepsilon \to 0$ in Ω, where Ω is the domain of definition of the independent variable t. The sketched method to deduce these models is called the **regular perturbations method**.

In the case of singular perturbations problems this method is applied several times, with the difference that the deduction of the conditions attached to every model is the most difficult part of the method and is carried out in different ways for different types of singular perturbation problems. We shall apply it in a particular case in Sections 4.2 and 4.3.

The first asymptotic approximation problem (for $k = 0$) is called the **reduced problem**. It is usually the only one which is studied, and it is of paramount importance for the other models. It contains equations which are strongly nonlinear, whilst the other asymptotic approximation models are affine (linear and nonomogenous), so they can be solved more easily.

CHAPTER 2

STATIC BIFURCATION AND LINEARIZATION OF THE FITZHUGH–NAGUMO MODEL

In this chapter we start with a study of the system (1.1.17), assuming x and y to be functions of the time t and $a \geq 0$. The nullclines, the curves of points of inflection of phase trajectories (Section 2.1), the surface of equilibrium points, and the curves of static bifurcation values in the parameters plane (b, a) (Section 2.2) are deduced.

The genuine static study contains no information about the dynamics around equilibria. Hence it says nothing about the type of equilibria. Thus, in spite of this chapter being mainly devoted to static bifurcation diagrams, we include sections devoted to the sign and nature of the eigenvalues of the systems linearized around equilibria of the nonlinear system. This will complete the static diagrams with dynamic features.

The study of the eigenvalues, eigenvectors, and eigen-directions of the linearized system is performed in Section 2.3. In the hyperbolic case the connection with the nonlinear case is made. The nonlinear static bifurcation diagrams (Section 2.4) and their asymptotic behaviour as $c \to \infty$ (Section 2.5) are shown. A separation of the (b, a)-plane into regions in which the hyperbolic equilibrium points have the same type is then made (Section 2.6).

In Section 2.7, computing the center manifold and the restriction of the F–N system to this manifold, the type of some non-hyperbolic equilibria is deduced.

2.1. GEOMETRIC PROPERTIES OF PHASE TRAJECTORIES

2.1.1. NULLCLINES

The curves of equations

$$\begin{aligned} y &= -x + \frac{x^3}{3}, \\ y &= \frac{a - x}{b}, \end{aligned} \tag{2.1.1}$$

for $b \neq 0$ and of equations

$$y = -x + \frac{x^3}{3},\qquad x = a, \tag{2.1.2}$$

for $b = 0$ are the nullclines of the system (1.1.17). They are represented in Figure 2.1.1. The equations (2.1.1) and (2.1.2) are obtained from (1.1.17) by taking $\dot{x}= 0$ and $\dot{y}= 0$.

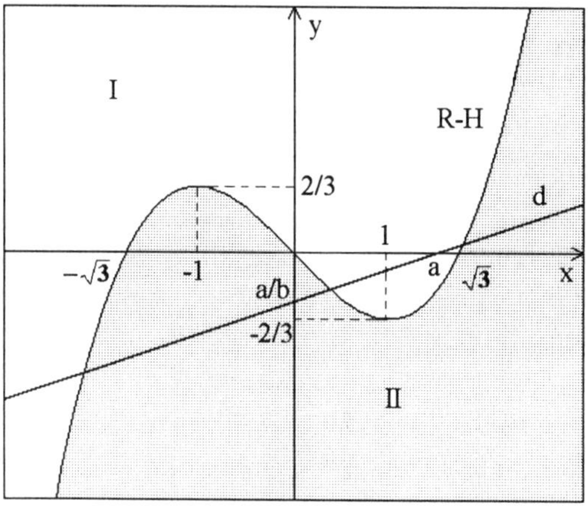

Fig. 2.1.1. Nullclines of the F–N system.

The nullcline $(2.1.1)_1$ is known as the Riemann–Hugoniot (R–H) curve. It passes through the origin and remains unchanged at the variation of the parameters. The nullcline $(2.1.1)_2$ is a straight line, d, of slope $\left(-\frac{1}{b}\right)$, cutting the Oy-axis at $y = a/b$ and the Ox-axis at $x = a$. Hence it depends on a and b, but not on c. For $b = 0$ the second nullcline is vertical.

The equilibria are situated at the intersection of the two nullclines. Consequently, the system may possess one, two, or three equilibria.

2.1.2. PHASE TRAJECTORIES

In general the trajectories are curves defined by multiform functions of x

$$y = y\left(x, a, b, c\right). \tag{2.1.3}$$

The tangent to a trajectory is given by $\left(\dot{x}, \dot{y}\right)$. It follows that the slope of the tangent to the trajectory is

$$y' = \frac{\dot{y}}{\dot{x}},\qquad \dot{x}\neq 0. \tag{2.1.4}$$

Thus, at the points where $\dot{y}= 0$, the tangent to the trajectory is parallel to the Ox-axis, whilst at the points where $\dot{x}= 0$, it is parallel to the Oy-axis. At the first glance, the R–H seems to be a separatrix and even a phase trajectory. This is not the case, as it can be immediately shown. Indeed, by differentiating $(2.1.1)_1$ and taking into account (1.1.17) we do not obtain an identity. However, the influence of R–H on the phase space is notable. Indeed, at any point of R–H the tangent to the trajectory is parallel to the Oy-axis (Figure 2.1.2).

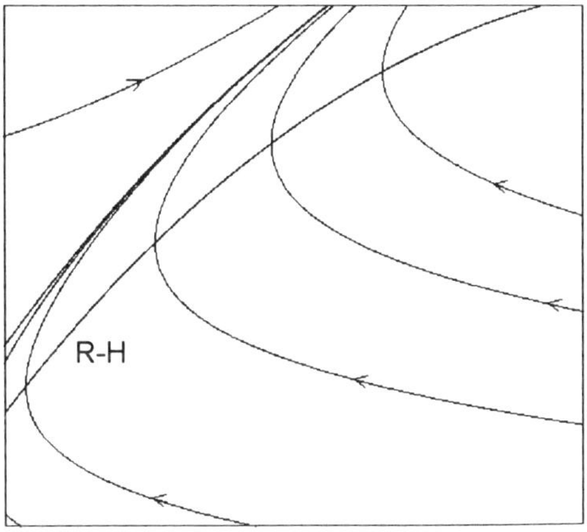

Fig. 2.1.2. Phase trajectories cutting R–H.

The R–H curve separates the phase plane into two regions (Figure 2.1.1). Since for all points in I, $\dot{x}> 0$, then following any phase trajectory x is increasing. Similarly, since for all points in II, $\dot{x}< 0$, then on the phase trajectories in this region x is decreasing.

The investigation of the points of inflection will show that, as the R–H curve is crossed by a trajectory, the trajectory remains in the same half-plane determined by the tangent to the trajectory at the intersection point [49]. The tangent itself is normal to the x-direction. It follows that the presence of the R–H curve looks as an obstacle, returning the trajectories in the half-plane from which they came. That is why it plays an important role in the study of the phase portrait.

2.1.3. INFLECTION POINTS

Taking into account (2.1.4) and (1.1.17), a simple computation shows that

$$y' = \frac{dy}{dx} = \frac{x - a + by}{c^2\left(-x - y + \frac{x^3}{3}\right)}. \tag{2.1.5}$$

The points of inflection (x, y) of the phase trajectories for which $\dot{x}\neq 0$ are solutions of the equation

$$y'' = 0, \tag{2.1.6}$$

where

$$y'' = \frac{d^2y}{dx^2} = \frac{\ddot{y}\,\dot{x} - \ddot{x}\,\dot{y}}{\dot{x}^2}, \qquad \dot{x}\neq 0. \tag{2.1.7}$$

Then (2.1.6), (1.1.17) and (2.1.7) imply that the inflection points are situated on the curves $y = y(x)$ defined by the following equation [121]

$$\begin{aligned}
&y^2c\left(-bx^2+b-1\right)+y\left[x^5\frac{bc}{3}-x^3\frac{c}{3}\left(1+\frac{b^2}{c^2}+4b\right)\right.\\
&\left.+\,x^2ac+x\left(\frac{b^2}{c}+bc-\frac{b}{c}-c\right)-ac+\frac{ab}{c}\right]\\
&+2x^6\frac{c}{9}-x^5\frac{ac}{3}-x^4\frac{1}{3}\left(\frac{b}{c}+2c\right)+x^3\frac{a}{3}\left(\frac{b}{c}+4c\right)\\
&+x^2\frac{b-1}{c}+xa\left(\frac{2}{c}-\frac{b}{c}-c\right)-\frac{a^2}{c}=0
\end{aligned} \tag{2.1.8}$$

For points (x, y) of the phase space with $x \neq \pm\sqrt{1-\frac{1}{b}}$, if the discriminant of (2.1.8) is positive, i.e., $\Delta(x, a, b, c) > 0$, the curve of the inflection points has two branches $y = y_1(x)$ and $y = y_2(x)$. If $\Delta(x, a, b, c) = 0$ a single branch $y = y(x)$ of inflection points exists, whilst if $\Delta(x, a, b, c) < 0$, no branch of inflection points exists. For $x = \pm\sqrt{1-\frac{1}{b}}$ two additional points are obtained. However, the curve $x = x(y)$ deduced from (2.1.8) can have at most six branches. Therefore in the general case complicated forms of the curves of the inflection points of the phase trajectories are expected.

Taking into account that along the trajectories $\ddot{x}= c\left[\dot{x}\,(1-x^2)+\dot{y}\right]$ and $\ddot{y}= -\frac{1}{c}\left[\dot{x}\,+b\,\dot{y}\right]$, it follows the equivalent to (2.1.8) form for $\dot{x}\neq 0$

$$\frac{1}{c}\left(\dot{x}\right)^2+c\left(\dot{y}\right)^2+\dot{x}\dot{y}\left[\frac{b}{c}+c\left(1-x^2\right)\right]=0. \tag{2.1.9}$$

It follows that the points (x, y) where $\dot{x}= 0$ and $\dot{y}\neq 0$ are not inflection points, but turning bifurcation points for the trajectories. The points (x, y) with $\dot{y}= 0$ and $\dot{x}\neq 0$ are extremum points for trajectories.

The points (x, y) for which $\dot{x}= 0$ and $\dot{y}= 0$ are equilibrium points. Therefore they are trajectories degenerated to a point and, consequently, the problem of their inflection points has no meaning. Formally, the equilibria satisfy (2.1.9) and therefore, (2.1.8). This is why, in the graphical representations, these points seem to be situated at the intersection of the R–H curve and the curves of points of inflection. However, this is not the case: the equilibria are limit points of the curves of inflection points and do not belong to them.

Some curves of points of inflection for the phase trajectories, corresponding to (2.1.8), are represented in Figure 2.1.3. The figures are made for $x, y \in (-2.5, 2.5)$.

The form of these curves strongly depends on parameters. They are useful in the representation of the phase portrait.

The asymptotic behaviour of some parts of the curves of inflection points is the same as the asymptotic behaviour of the R–H curve, as $x \to \pm\infty$. Indeed, from (2.1.8) we have immediately the expressions of y as functions of x. Dividing these expressions by x^2 and letting $x \to \pm\infty$ we obtain the following asymptotic expansions for the two branches

$$y_1 \sim \frac{x^3}{3} - x\left(-\frac{1}{3b} + \frac{b}{3c^2} + \frac{4}{3} + \frac{1}{3bc} - \frac{1}{c}\right) \quad \text{as } x \to \pm\infty \tag{2.1.10}$$

$$y_2 \sim -\frac{2}{3b}x + \frac{a}{b} \quad \text{as } x \to \pm\infty \tag{2.1.11}$$

valid for asymptotically fixed a, b and c.

This means that, in the first approximation, the asymptotic behaviour of y_1 and R–H and of y_2 and the nullcline **d**, are the same. This implies a difficulty in studying the phase portrait between the R–H curve and the outer branches of the curves of points of inflection. In some cases, a part of the limit cycle is situated between the R–H curve and these branches.

We also have

$$y_{1,2} \sim A_{1,2} + B_{1,2}x, \text{ as } x \to 0 \tag{2.1.12}$$

where

$$\begin{aligned} A_{1,2} &= \frac{ac - \frac{ab}{c} \pm \sqrt{\left(ac + \frac{ab}{c}\right)^2 - 4a^2}}{2c\,(b-1)}, \\ B_{1,2} &= \frac{a\left(\frac{b}{c} + c - \frac{2}{c}\right) - A_{1,2}\,(b-1)\left(\frac{b}{c} + c\right)}{2c\,(b-1)\,A_{1,2} + \frac{ab}{c} - ac}. \end{aligned}$$

These formulae are particularly important for those cases where the curves of inflection points pass through the origin.

The importance of the inflection curves in the formation and destruction of concave limit cycles will be emphasized in Section 5.4.2.

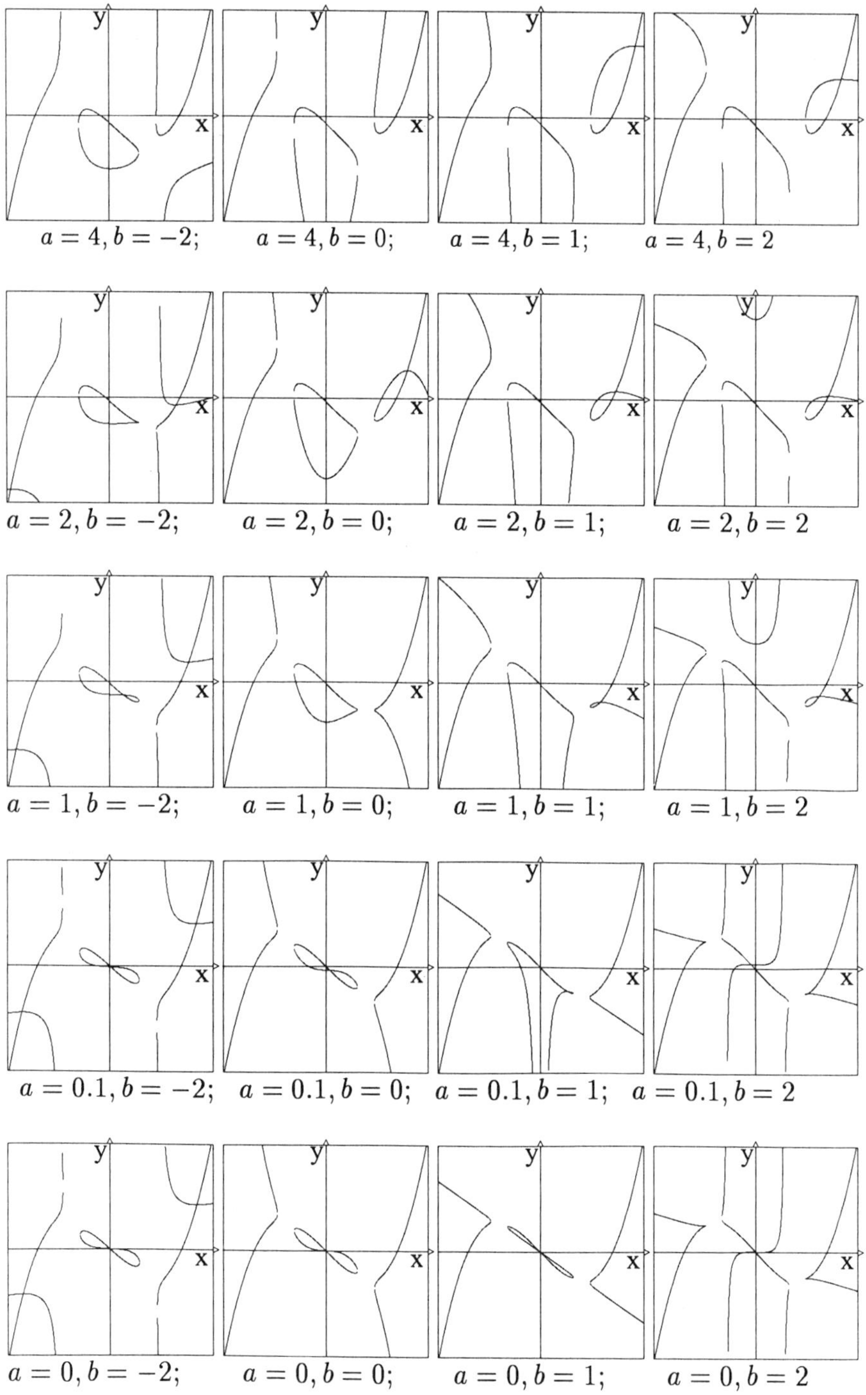

Fig. 2.1.3. Curves of inflection points of the phase trajectories for $c = 5$.

2.2. EQUILIBRIA

The phase space of the dynamical system generated by (1.1.17) is the (x, y)-plane. The equilibrium points of this system satisfy (2.1.1) if $b \neq 0$ and (2.1.2) if $b = 0$, i.e., they are situated at the intersection of the two null-clines, a result already used in the previous section. Eliminating y between $(2.1.1)_1$ and $(2.1.1)_2$, we obtain

$$x^3 - 3x\left(1 - \frac{1}{b}\right) - 3\frac{a}{b} = 0, \quad b \neq 0. \tag{2.2.1}$$

In the (b, a, x)-space, (2.2.1) defines the solution surface, denoted by Σ_1. For $b = 0$, the intersection between the nullclines $(2.1.2)_1$ and $(2.1.2)_2$ is the curve Γ

$$x = a, \quad y = \frac{a^3}{3} - a, \tag{2.2.2}$$

consisting of limit points for Σ_1. Hence $\Sigma = \Sigma_1 \cup \Gamma$ represents the surface of the abscissae of the equilibrium points, for $b, a \in \mathbf{R}$ (Figure 2.2.1). Since (2.2.1) does not depend on c, the geometry of Σ depends on c neither.

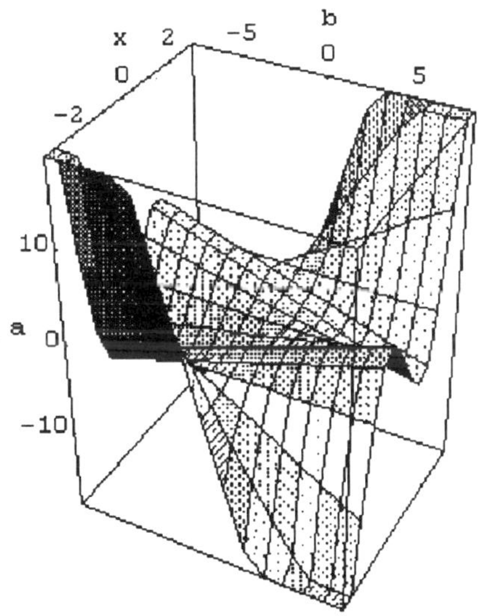

Fig. 2.2.1. The surface Σ.

For $b = 1$ and $a = 0$ the equation (2.2.1) has a triple solution $x = 0$. No other triple solution of (2.2.1) exists.

If

$$D \equiv \left(\frac{1}{b} - 1\right)^3 + 9\frac{a^2}{4b^2} < 0,$$

(which implies either $b > 1$ or $b < 0$), equation (2.2.1) has three real distinct solutions

$$\begin{aligned} x_1 &= 2\sqrt{1-\frac{1}{b}}\sin\Phi, \\ x_2 &= -2\sqrt{1-\frac{1}{b}}\sin\left(\Phi+\frac{\pi}{3}\right), \\ x_3 &= 2\sqrt{1-\frac{1}{b}}\sin\left(-\Phi+\frac{\pi}{3}\right), \end{aligned} \tag{2.2.3}$$

where

$$\Phi = \frac{1}{3}\arcsin\left[-\frac{3a}{2b\sqrt{(1-\frac{1}{b})^3}}\right].$$

As $D < 0$ iff $\Phi \in \left(-\frac{\pi}{6}, \frac{\pi}{6}\right)$, we have $x_2 < x_1 < x_3$.

If $D > 0$, equation (2.2.1) has a single real solution

$$x_0 = \sqrt[3]{\frac{3a}{2b}+\sqrt{D}} + \sqrt[3]{\frac{3a}{2b}-\sqrt{D}}. \tag{2.2.4}$$

We have $D = 0$ iff $\Phi = \pm\pi/6$. In addition, if $D = 0$, (2.2.1) has two distinct real solutions (one of them being double) or the triple solution $x = 0$ for $(b, a) = (1, 0)$. More precisely, $x_1 = x_3 = \sqrt{1-\frac{1}{b}}$, if $\Phi = \pi/6$ and $x_1 = x_2 = -\sqrt{1-\frac{1}{b}}$ if $\Phi = -\pi/6$. Hence, as $(b, a) \neq (1, 0)$ and $D = 0$ two roots coincide, while for $D > 0$ they become complex conjugate such that a single real solution exist.

The intersections of the surface Σ with the planes $a =$constant are the one-parameter static bifurcation diagrams. They represent the variation of x_i with respect to the parameter b [121].

For $a \geq 0$, up to a topological equivalence, there exists four shapes of static bifurcation diagrams. They are obtained for (1) $a = 0$, (2) $a \in (0, \sqrt{3})$, (3) $a = \sqrt{3}$, (4) $a > \sqrt{3}$ (Figure 2.2.2).

For a fixed a, we also have $\lim_{b\to 0, b>0} x_0 = \lim_{b\to 0, b<0} x_1 = a$, so the branches of solutions x_0 and x_1 collide for $b = 0$. The branch x_0 also collides with x_3 for some $b > 0$ and for every fixed $a > 0$. In addition, for a fixed $a > \sqrt{3}$, the branch x_0 collides with x_2 for two negative values of b.

Remark also that for a fixed a, we have

$$\lim_{b\to\infty, (b\to-\infty)} x_3 = \sqrt{3},$$

$$\lim_{b\to 0, b<0} x_3 = \infty,$$

$$\lim_{b\to\infty,(b\to-\infty)} x_2 = -\sqrt{3},$$

$$\lim_{b\to 0, b<0} x_2 = -\infty,$$

$$\lim_{b\to\infty,(b\to-\infty)} x_1 = 0.$$

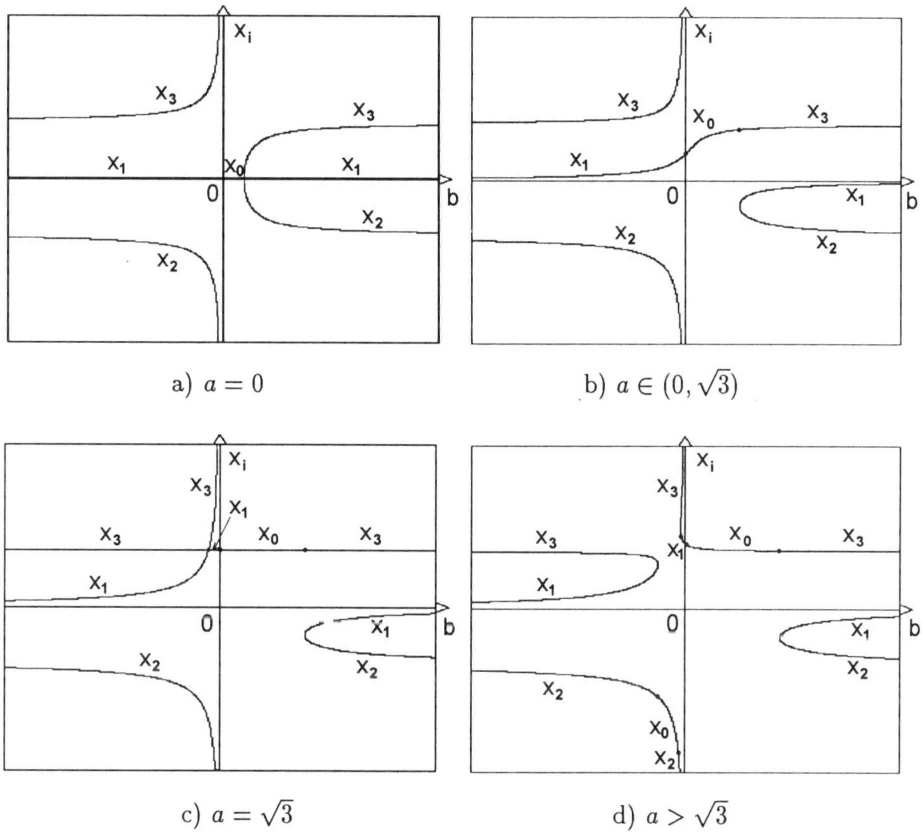

Fig. 2.2.2. Static bifurcation diagrams.

Other expressions of x_1, x_2 and x_3 in terms of a and b can be easily obtained.

Remark that, in spite of the fact that we defined four real solutions x_0, x_1, x_2, x_3, maximum three of them exist.

Consequently, taking into account the values of the parameters a and b, the dynamical system associated with (1.1.17) has three, two or one equilibrium points. Correspondingly, according to the position of the (b, a) point, in the parameter space, the surface Σ has one or three sheets, which coalesce along some spatial curves. The projections of these curves on the (b, a)-plane are two curves $S_{1,2}$ defined by the condition $D = 0$, for $b \neq 0$. Thus, they have the equations

$$S_{1,2}: \qquad a = \pm\frac{2}{3}|b|\sqrt{\left(1-\frac{1}{b}\right)^3}, \quad b \in (-\infty, 0) \cup [1, \infty), \qquad (2.2.5)$$

where the sign + corresponds to S_1. For (b, a) situated on these curves, two equilibria of the F–N system coincide if $(b, a) \neq (1, 0)$ and three equilibria coincide if $(b, a) = (1, 0)$.

As it was to be expected, the equations (2.2.5) do not depend on the parameter c.

The asymptotic behaviour of S_1 for large values of b is the following

$$a \sim \frac{2}{3}b - 1 \quad \text{as } b \to \infty$$
$$a \sim -\frac{2}{3}b + 1 \quad \text{as } b \to -\infty.$$

For S_2 a behaviour symmetric with respect to the b-axis occurs. The graphical representation, for $b \in [-c, c]$, of the curves S_1 and S_2 and their asymptotes is given in Figure 2.2.3.

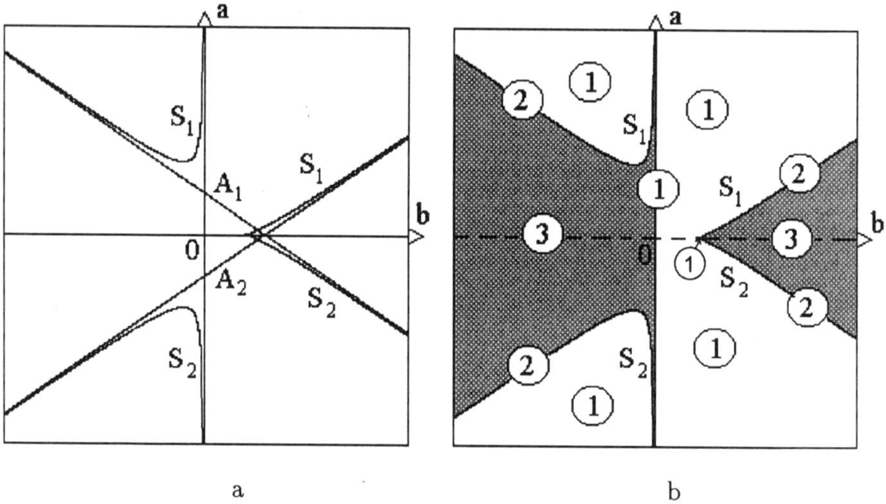

Fig. 2.2.3. a) The curves $S_{1,2}$ and their asymptotes $A_{1,2}$; b) the number of equilibria corresponding to every region of the (b, a)-plane.

The straight lines A_1 of the equation $a = 1 - \frac{2b}{3}$ and A_2 of the equation $a = -1 + \frac{2b}{3}$, the asymptotes to S_1 and S_2, intersect the Oa-axis at the points $(0,1)$, $(0,-1)$, respectively, and intersect each other at the point $\left(\frac{3}{2},0\right)$. The curves S_1 and S_2 are intersecting at the point $(1,0)$. The extremum points (minimum and maximum) of the curves S_1 and S_2 are $\left(-\frac{1}{2},\sqrt{3}\right)$ and $\left(-\frac{1}{2},-\sqrt{3}\right)$ respectively.

The static bifurcation takes place for $D=0$ or $b=0$. This is why, in the (b,a)-plane, the static bifurcation values are situated on the curves S_1, S_2 and on the Oa-axis.

2.3. EIGENVALUES OF THE LINEARIZED SYSTEM. EIGENVECTORS AND EIGEN-DIRECTIONS

2.3.1. THE LINEARIZED SYSTEM

The system obtained by linearization of (1.1.17) around an equilibrium point (x_i, y_i), $i=0,1,2,3$ has the form

$$\begin{aligned} \dot{X} &= c\left(1-x_i^2\right)X + cY, \\ \dot{Y} &= -\frac{X}{c} - \frac{bY}{c}. \end{aligned} \tag{2.3.1}$$

The associated characteristic equation

$$\lambda^2 + \lambda\left(cx_i^2 - c + \frac{b}{c}\right) - b\left(1-x_i^2\right) + 1 = 0 \tag{2.3.2}$$

has the solutions

$$\lambda_{1,2} = \frac{c}{2}\left[1 - x_i^2 - \frac{b}{c^2} \pm \sqrt{\left(1 - x_i^2 + \frac{b}{c^2}\right)^2 - \frac{4}{c^2}}\right], \tag{2.3.3}$$

which are the eigenvalues of (2.3.1).

The signs of the real parts of λ_1 and λ_2 will determine the attractivity properties of the equilibrium points. The nature of λ_1 and λ_2 will be used in the study of Hopf bifurcation, in Section 3.1.

The eigenvalues λ_1 and λ_2 are complex if

$$\left(1 - x_i^2 + \frac{b}{c^2}\right)^2 - \frac{4}{c^2} < 0. \tag{2.3.4}$$

or, equivalently,

$$\frac{1}{c^4}\left(b + c^2 - c^2x_i^2 - 2c\right)\left(b + c^2 - c^2x_i^2 + 2c\right) < 0.$$

Consider, in the (b, a, x)-space, the parabolic cylinders

$$\mathrm{B}_1' : \quad b = c^2x^2 - c^2 - 2c, \tag{2.3.5}$$

$$\mathrm{B}_2' : \quad b = c^2x^2 - c^2 + 2c. \tag{2.3.6}$$

The condition (2.3.4) states that if (x_i, y_i) is an equilibrium point such that (b, a, x_i) is situated between the cylinders B_1', B_2', then the corresponding eigenvalues of (x_i, y_i) are complex, whilst if (b, a, x_i) is situated outside the region delimited by B_1', B_2', the eigenvalues are real. For (b, a, x_i) situated on B_1' or B_2', the eigenvalues are real and equal. Namely, $\lambda_1 = \lambda_2 = c\left(1 - x_i^2\right) + 1$ for B_1' and $\lambda_1 = \lambda_2 = c\left(1 - x_i^2\right) - 1$ for B_2'.

In order to determine the sign of Re λ_1 and Re λ_2 it is necessary to investigate the position, in the (b, a, x)-space, of the points for which at least one of the relations

$$\mathrm{Re}\ \lambda_1 = 0 \text{ or } \mathrm{Re}\ \lambda_2 = 0 \tag{2.3.7}$$

holds. Thus, in the case of complex eigenvalues, the condition (2.3.7) is

$$1 - x_i^2 - \frac{b}{c^2} = 0. \tag{2.3.8}$$

Consider now, in the (b, a, x)-space, the parabolic cylinder

$$\mathrm{B}_3' : \quad b = c^2\left(1 - x^2\right). \tag{2.3.9}$$

Then, the condition (2.3.8) shows that if (x_i, y_i) is an equilibrium such that (b, a, x_i) is situated on the cylinder B_3' and in the space region between the cylinders B_1',B_2', then Re λ_1 = Re $\lambda_2 = 0$ and Im $\lambda_1 = -$ Im $\lambda_2 = \sqrt{c^2\left(1 - x_i^2\right)^2 - 1}$. Therefore $\lambda_{1,2} = \pm i\sqrt{1 - c^2\left(1 - x_i^2\right)^2}$.

In the case of real eigenvalues, the condition (2.3.7) becomes

$$1 - x_i^2 - \frac{b}{c^2} \pm \sqrt{\left(1 - x_i^2 + \frac{b}{c^2}\right)^2 - \frac{4}{c^2}} = 0. \tag{2.3.10}$$

Consider the surfaces

$$\mathrm{B}_{4-7}' : \quad x = \pm\sqrt{1 - \frac{1}{b}}, \quad b \in (-\infty, 0) \cup [1, \infty) \tag{2.3.11}$$

in the (b, a, x)-space, where $\mathrm{B}_{4,5}'$ correspond to the sign plus, $\mathrm{B}_{6,7}'$ to the minus sign, $\mathrm{B}_{4,7}'$ to $b \geq 1$ and $\mathrm{B}_{5,6}'$ to $b < 0$. The condition (2.3.10) shows that if (x_i, y_i) is an equilibrium point such that (b, a, x_i) is situated on one of the surfaces B'_{4-7}, then $\lambda_{1,2} \in \mathbf{R}$, $\lambda_1 = 0$ or/and $\lambda_2 = 0$.

2.3.2. THE NATURE AND SIGN OF THE EIGENVALUES. DISCUSSION IN THE (b, x)-PLANE

Let B_1, B_2, B_3 be the orthogonal projections of the parabolic cylinders B'_1, B'_2, B'_3 on the (b, x)-plane. These are the parabolas of equations (2.3.5), (2.3.6) and (2.3.9), respectively.

For $c > 1 + \sqrt{3}$, the parabolas B_1 and B_2 do not intersect themselves, B_1 intersects the Ox-axis at $\left(0, \pm\sqrt{1 + \frac{2}{c}}\right)$ and the Ob-axis at $\left(-c^2 - 2c, 0\right)$ and B_2 intersects the Ox-axis at $\left(0, \pm\sqrt{1 - \frac{2}{c}}\right)$ and the Ob-axis at $\left(-c^2 + 2c, 0\right)$. The parabola B_3 intersects the Ox-axis at $(0, \pm 1)$ and the Ob-axis at $\left(c^2, 0\right)$. The parabola B_1 intersects B_3 at the points $\overline{Q}_1(-c, \sqrt{1 + \frac{1}{c}})$ and $\overline{Q}_4\left(-c, -\sqrt{1 + \frac{1}{c}}\right)$, while B_2 intersects B_3 at the points $\overline{Q}_2\left(c, \sqrt{1 - \frac{1}{c}}\right)$ and $\overline{Q}_3\left(c, -\sqrt{1 - \frac{1}{c}}\right)$ (Figure 2.3.1).

In the (b, x)-plane, the parabolas B_1 and B_2 determine two domains: a domain between the branches of B_1 and B_2, where $\lambda_1 = \overline{\lambda}_2 \in \mathbf{C}$ and the outer domain, where $\lambda_1, \lambda_2 \in \mathbf{R}$. In addition, for equilibrium points (x_i, y_i) with (b, x_i) situated on B_1, we have $\lambda_1 = \lambda_2 = c + 1 - cx_i^2$, whilst for (b, x_i) situated on B_2 we have $\lambda_1 = \lambda_2 = c - 1 - cx_i^2$. On the portion of B_3 situated between B_1 and B_2 we have Re $\lambda_1 =$ Re $\lambda_2 = 0$. Consequently, if (b, x_i) is one of the points $\overline{Q}_i$, $i = 1, 2, 3, 4$, we have $\lambda_1 = \lambda_2 = 0$.

In domain 1 in Figure 2.3.1, defined by B_1, B_2 and B_3, we have $\lambda_{1,2} \in \mathbf{C}$, Re $\lambda_1 =$ Re $\lambda_2 > 0$. Outside this domain, but still between B_1 and B_2 (the two domains 2 in Figure 2.3.1), we have $\lambda_{1,2} \in \mathbf{C}$, Re $\lambda_1 =$ Re $\lambda_2 < 0$.

Let B_{4-7} be the projections of the cylindrical surfaces $B'_{4,7}$ on the (b, x)-plane. These are the curves of equations (2.3.11), represented also in Figure 2.3.1. The straight line $x = 1$ is asymptotic to B_5 as $b \to -\infty$ and to B_4 as $b \to \infty$, whilst the straight line $x = -1$ is asymptotic to B_6 as $b \to -\infty$ and to B_7 as $b \to \infty$.

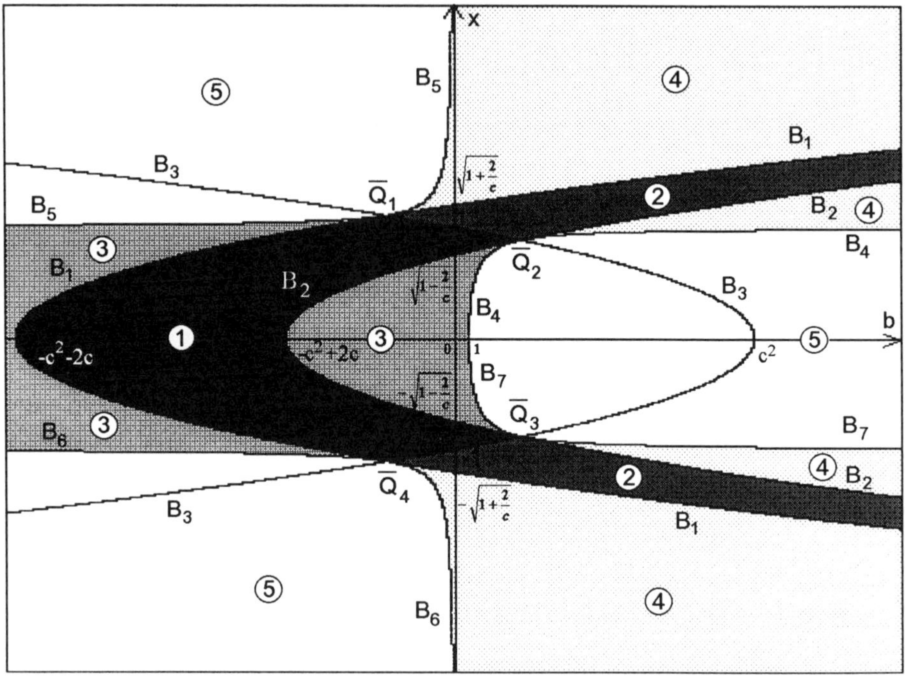

Fig. 2.3.1 The curves B_i, $i = \overline{1,7}$ and the regions of the $(b,\, a)$-plane where the nature of $\lambda_{1,2}$ and the sign of Re $\lambda_{1,2}$ are preserved:
(1) $\lambda_{1,2} \in \mathbf{C}$, Re $\lambda_1 =$ Re $\lambda_2 > 0$, Im λ_1, Im $\lambda_2 \neq 0$;
(2) $\lambda_{1,2} \in \mathbf{C}$, Re $\lambda_1 =$ Re $\lambda_2 < 0$, Im λ_1, Im $\lambda_2 \neq 0$;
(3) $\lambda_{1,2} \in \mathbf{R}$, $\lambda_1, \lambda_2 > 0$;
(4) $\lambda_{1,2} \in \mathbf{R}$, $\lambda_1, \lambda_2 < 0$;
(5) $\lambda_{1,2} \in \mathbf{R}$, $\lambda_1 < 0, \lambda_2 > 0$ or $\lambda_1 > 0$, $\lambda_2 < 0$.

If (x_i, y_i) is an equilibrium point with (b, x_i) situated on one of the curves B_{4-7} and $b \neq \pm c$, then only one of the eigenvalues λ_1, λ_2 is zero. Let us remark that B_5 and B_1 are tangent at $\overline{Q}_1$, B_6 and B_1 are tangent at $\overline{Q}_4$, B_4 and B_2 are tangent at $\overline{Q}_2$ and B_7 and B_2 are tangent at $\overline{Q}_3$.

The curves B_{4-7} separate, in the (b, x)-plane, another domains (3, 4 and 5 in Figure 2.3.1). They have an important role in the determination of the type of the equilibria. Thus, for the equilibrium point (x_i, y_i) such that (b, x_i) is situated in the three domains numbered with 3, we have $\lambda_{1,2} \in \mathbf{R}$, $\lambda_{1,2} > 0$. For (b, x_i) situated in the four domains numbered with 4 we have $\lambda_{1,2} \in \mathbf{R}$, $\lambda_{1,2} < 0$ and for (b, x_i) situated in the three domains 5 we have $\lambda_{1,2} \in \mathbf{R}^*$ and λ_1 and λ_2 are of oposite signs. In addition, for (b, x_i) situated in the domains 5 and on B_3, we have $\lambda_1 = -\lambda_2 \in \mathbf{R}$.

2.3.3. THE NATURE AND SIGN OF THE EIGENVALUES. DISCUSSION IN THE (b, a)-PLANE

The discussion in the (b, x)-plane was the simplest. Now we associate it with the discussion in the (b, a)-plane, since this is the correct place for the bifurcation.

Also, this time we must be very careful because some points of the (b, a)-plane can represent the projection of several equilibria.

Let C_1' and C_2' be the spatial curves obtained intersecting the surface $\sum$ with the cylinders B_1' and B_2' respectively. The projections on the (b, a)-plane of the curves C_1' and C_2' are the curves

$$C_{1,2}: \quad a = \pm\left(-\frac{2b}{3} + 1 + \frac{2b}{3c} + \frac{b^2}{3c^2}\right)\sqrt{1 + \frac{2}{c} + \frac{b}{c^2}}, \tag{2.3.12}$$

$$C_{3,4}: \quad a = \pm\left(-\frac{2b}{3} + 1 - \frac{2b}{3c} + \frac{b^2}{3c^2}\right)\sqrt{1 - \frac{2}{c} + \frac{b}{c^2}}. \tag{2.3.13}$$

These curves are represented in Figure 2.3.2.

Here is the point where we impose the restriction $c > 1+\sqrt{3}$. It is necessary for the preservation of the topological equivalence of the domains of the (b, a)-plane where $\lambda_{1,2}$ have some well defined properties. These properties concern the sign of their real parts and their nature. For $c \in (0, 1+\sqrt{3}]$, one must consider several distinct situations, each one requiring an analysis similar to that developed in the present book. We have chosen the case $c > 1+\sqrt{3}$ owing to its relevance in physiological applications.

At $c = 1 + \sqrt{3}$, already the two points of intersection of C_1, C_2 and the Ob axis coincide with the point $(3+\sqrt{3}, 0)$. For still smaller c some loops determined by C_{1-4} disappear and some other points of intersection coincide. Mostly this occurs for $c = 1$.

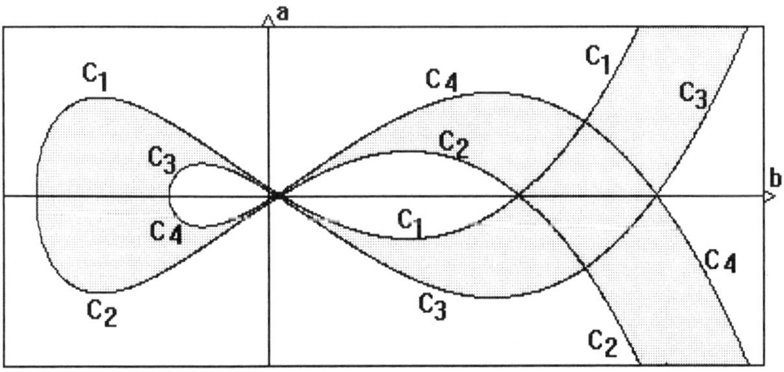

Fig. 2.3.2. The curves C_{1-4} delimiting, in the (b, a)-plane, the regions with complex eigenvalues (hatched).

The curves $C_{1,2}$ intersect the Ob-axis at the points of absicisae $-c^2-2c$ and $c^2-c\pm c\sqrt{c^2-2c-2}$. The curve C_1 intersects the Oa-axis at the point $\left(0,\sqrt{1+\frac{2}{c}}\right)$ and C_2 at $\left(0,-\sqrt{1+\frac{2}{c}}\right)$. The curves $C_{3,4}$ intersect the Ob-axis at the points of absicisae $-c^2+2c$ and $c^2+c\pm c\sqrt{c^2+2c-2}$. The curve C_3 intersects the $Oa-$axis at the point $\left(0,\sqrt{1-\frac{2}{c}}\right)$ and C_4 at the point $\left(0,-\sqrt{1-\frac{2}{c}}\right)$ (Figures 2.3.2 and 2.3.4).

The curves C_i, $i=\overline{1,4}$ delimit in the (b,a)-plane, the region where at least one of the equilibrium points has complex eigenvalues (Figure 2.3.2). Outside this region, all equilibria have real eigenvalues.

Consider only the case $a \geq 0$. Taking into account the expressions for x_i given by (2.2.3) and (2.2.4) it can be seen that, for $b<0$ and (b,a) in the region delimited by C_1 and C_3, the equilibrium (x_1,y_1) has complex eigenvalues (regions 1 and 2, in Figure 2.3.5).

If

$$b\in\left(0,c^2-c-c\sqrt{c^2-2c-2}\right)\cup\left(c^2-c+c\sqrt{c^2-2c-2},\infty\right)$$

and (b,a) is situated between C_1 and C_3, the equilibrium (x_0,y_0) has complex eigenvalues for the case when (b,a) is situated in the region delimited by S_1 with a single equilibrium (Figure 2.3.4, regions 1 and 2). The same conclusion is valid for (x_3,y_3) when (b,a) is situated in the region with three equilibria delimited by S_1 (Figure 2.3.7, regions 1 and 2) .

If

$$b\in\left(c^2+c-c\sqrt{c^2+2c-2},c^2+c+c\sqrt{c^2+2c-2}\right)$$

and (b,a) is situated between C_2 and C_4, then the equilibrium (x_2,y_2) has complex eigenvalues (Figure 2.3.6, regions 1 and 2).

Let us remark that for

$$b\in\left(c^2+c-c\sqrt{c^2+2c-2},c^2-c-c\sqrt{c^2-2c-2}\right)$$

and (b,a) situated in the curvilinear triangle delimited by C_4 , C_1 and the Ob-axis, both (x_2,y_2) and (x_3,y_3) have complex eigenvalues. The same conclusion is valid for

$$b\in\left(c^2-c+c\sqrt{c^2-2c-2},c^2+c+c\sqrt{c^2+2c-2}\right)$$

and (b,a) situated in the curvilinear triangle delimited by C_4, C_1 and the Ob-axis (Figures 2.3.6 and 2.3.7).

Let H' be the spatial curve obtained as the intersection of Σ with the cylinder B_3' and let $H_{1,2}$ be the curves obtained as projections of H' on the (b, a)-plane. They are defined by the following equations obtained by eliminating x between (2.2.1) and (2.3.9)

$$H_{1,2}: \ a = \pm\frac{b}{3}\left(-2 + \frac{3}{b} - \frac{b}{c^2}\right)\sqrt{1 - \frac{b}{c^2}}, \quad b \in (-\infty, c^2]. \tag{2.3.14}$$

The curves $H_{1,2}$ are of a crucial importance for our investigation of dynamic bifurcation. Parts of $H_{1,2}$ are Hopf bifurcation values and our local analyses starts with this kind of bifurcation. Since the Hopf bifurcation values correspond to purely imaginary eigenvalues, it follows that we are primarly interested in those parts of $H_{1,2}$ situated in the complex region delimited by the curves C_{1-4}. Whence the importance of having the same relative position of the points of intersection of Ob-axis, C_{1-4} and H_{1-2}.

This relative position keeps unchanged for $c > 1 + \sqrt{3}$ and is distroyed for $c \leq 1 + \sqrt{3}$.

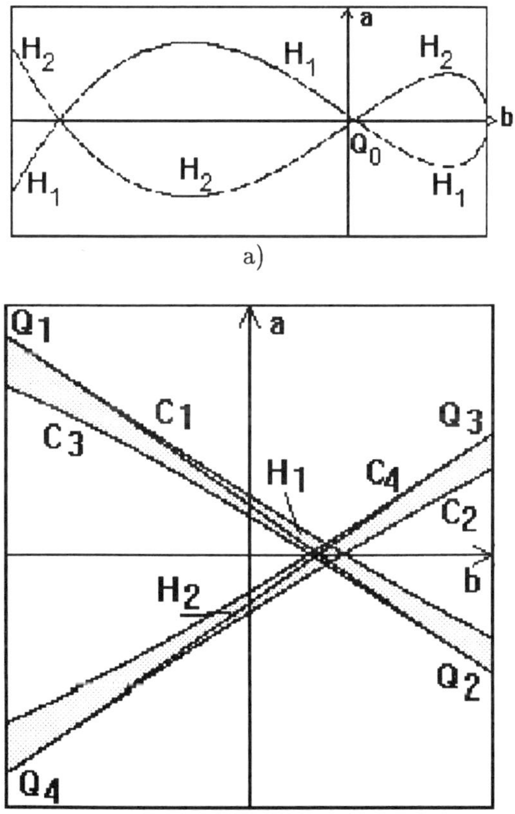

Fig. 2.3.3 a) The curves $H_{1,2}$; b) their portions corresponding to complex eigenvalues for $b \in (-c, c)$.

The curves $H_{1,2}$ intersect the Ob-axis at the points $\left(-c-\sqrt{c^4+3c^2},0\right)$, $(b_0,0)$ and $\left(c^2,0\right)$, where $b_0=-c^2+\sqrt{c^4+3c^2}$. The curve H_1 intersects the Oa-axis at $(0,1)$ and H_2 intersects the Oa-axis at $(0,-1)$ (Figure 2.3.3a). In the following, the point $(b_0,0)$ will be denoted by Q_0.

The curve H_1 is tangent to S_1 and C_1 at $Q_1\left(-c,\frac{2}{3}(c+1)\sqrt{1+\frac{1}{c}}\right)$ and to S_2 and C_3 at $Q_2\left(c,-\frac{2}{3}(c-1)\sqrt{1-\frac{1}{c}}\right)$, while H_2 is tangent to S_1 and C_4 at $Q_3\left(c,\frac{2}{3}(c-1)\sqrt{1-\frac{1}{c}}\right)$ and to S_2 and C_2 at $Q_4\left(-c,-\frac{2}{3}(c+1)\sqrt{1+\frac{1}{c}}\right)$ (Figure 2.3.3b).

For $(b,a)\in H_{1,2}$ at least one corresponding equilibrium point has purely imaginary or real eigenvalues with $\lambda_1=-\lambda_2$.

Let S' be the spatial curve obtained intersecting Σ with the surfaces $B'_{4,7}$. The projection of this curve, on the (b,a)-plane, obtained by eliminating x between (2.2.1) and (2.3.11), consists of the curves $S_{1,2}$ already defined by equations (2.2.5). For $(b,a)\in S_{1,2}$ there exists an equilibrium point having at least one zero eigenvalue.

Taking into account the conclusions from Section 2.2, for $(b,a)\in S_{1,2}$, $(b,a)\neq(1,0)$, the F–N system has a double equilibrium point with at least one zero eigenvalue, whilst for $(b,a)=(1,0)=S_1\cap S_2$, it has a triple equilibrium point with one zero eigenvalue.

Taking into account the expressions for x_i ,$i=0,1,2,3$, from (2.2.2), (2.2.3) and (2.2.4) we were able to find the regions of the (b,a)-plane corresponding to the equilibria (x_i,y_i) for which the eigenvalues have the same nature and Re λ_1, Re λ_2 keep a constant sign. These regions are represented in Figures 2.3.4-2.3.7. The numbers correspond to those of Figure 2.3.1. In the shadowed regions the investigated equilibrium does not exist.

Summing up the results from Figures 2.3.4-2.3.7, we can state that for (b,a) situated on H_1 or H_2, $b\in(-c,c)$, $b\neq b_0$, just one corresponding equilibrium point has purely imaginary eigenvalues. This equilibrium point is a candidate for a Hopf or degenerated Hopf bifurcation point.

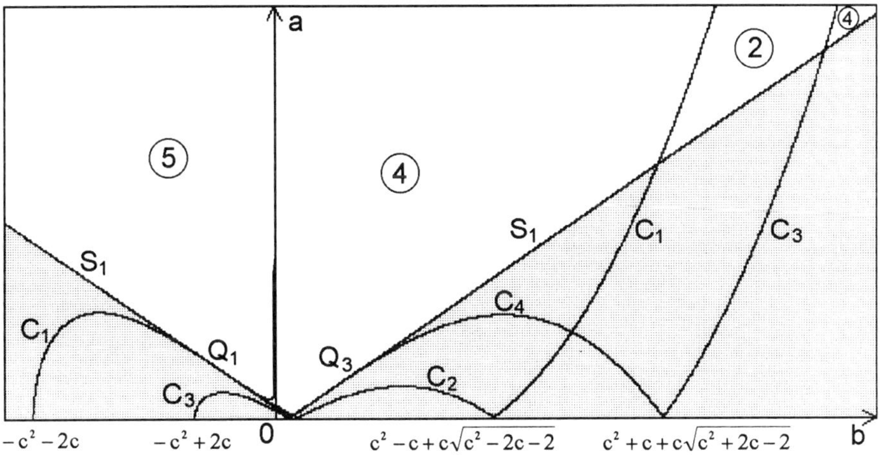

a)

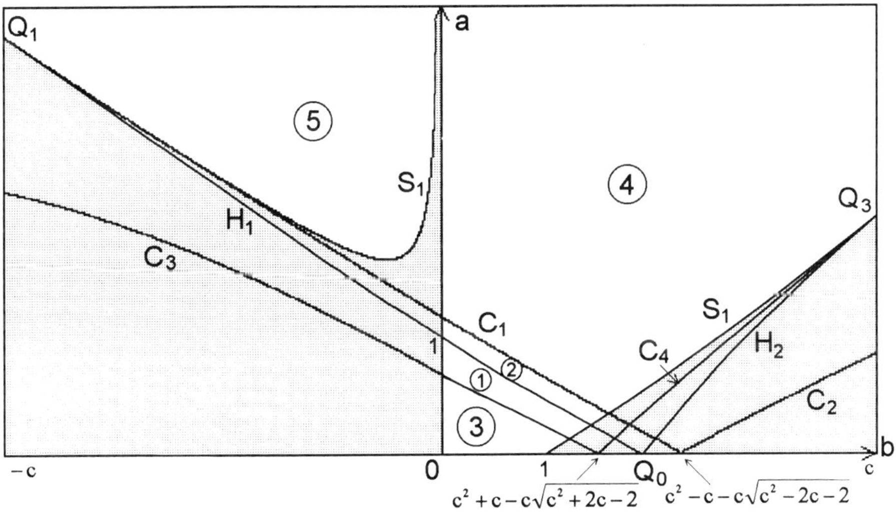

b)

Fig. 2.3.4. (a) Regions in the $(b,\, a)$-plane where the nature of the eigenvalues and the sign of Re $\lambda_{1,2}$ are preserved for the equilibrium point $(x_0,\, y_0)$;

(b) Zoom of Figure 2.3.4 (a) for $b \in [-c, c]$.

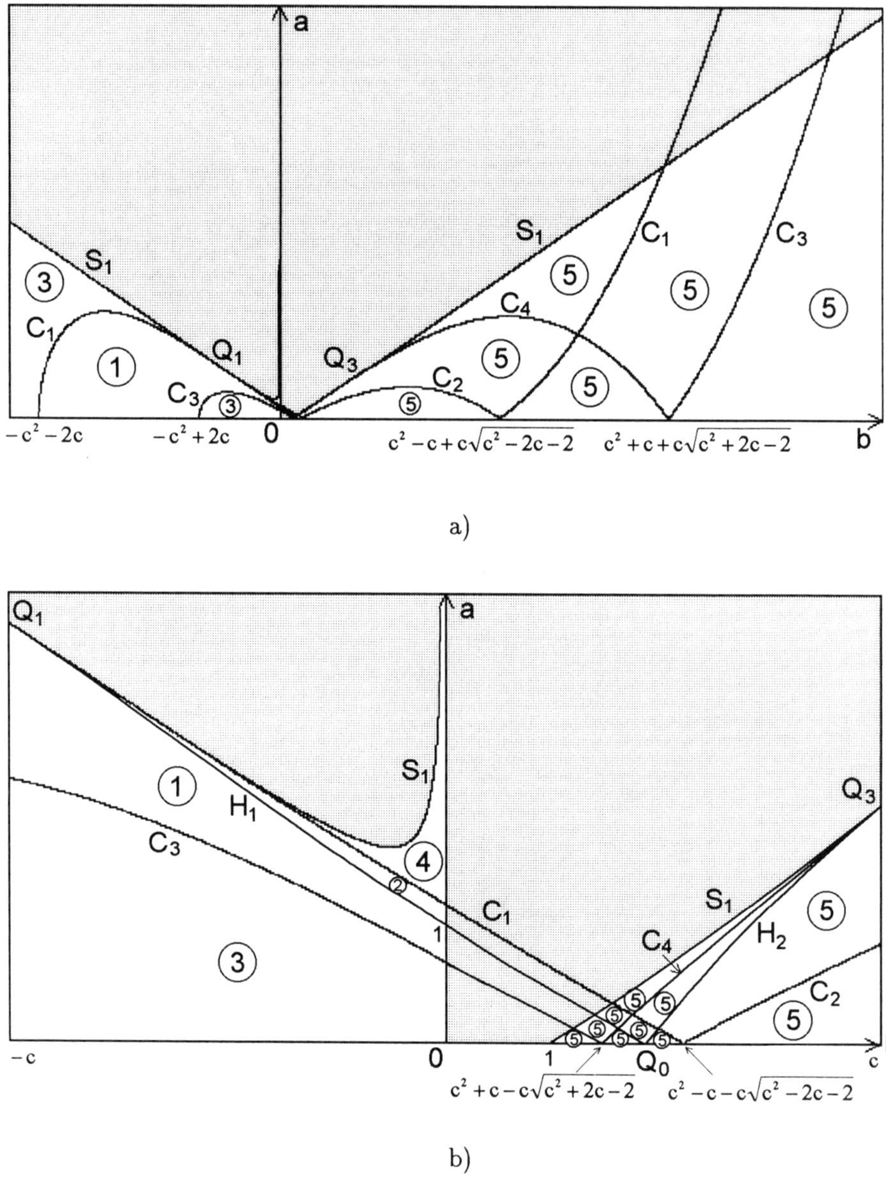

Fig. 2.3.5. (a) Regions in the $(b,\ a)$-plane where the nature of the eigenvalues and the sign of Re $\lambda_{1,2}$ are preserved for the equilibrium point $(x_1,\ y_1)$;

(b) Zoom of Figure 2.3.5 (a) for $b \in [-c, c]$.

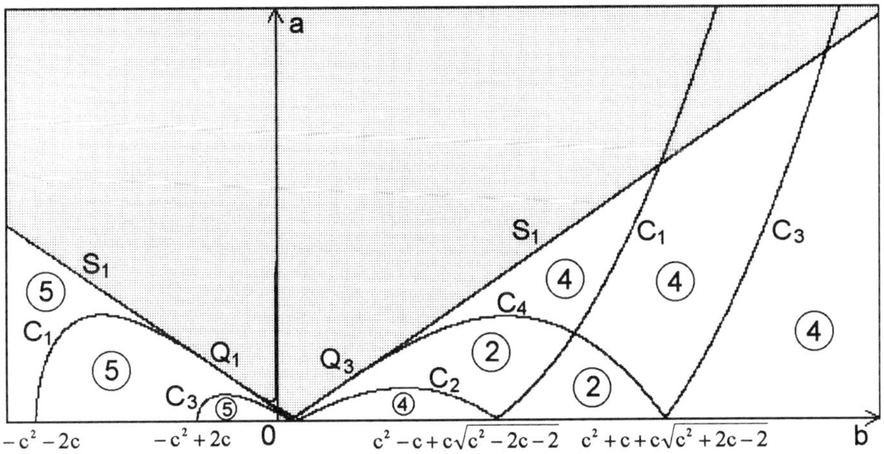

a)

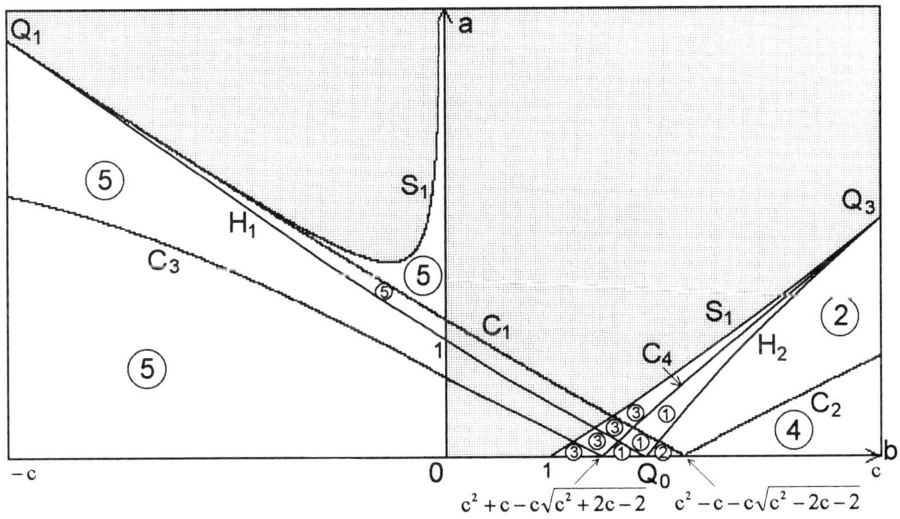

b)

Fig. 2.3.6. (a) Regions in the $(b,\, a)$-plane where the nature of the eigenvalues and the sign of Re $\lambda_{1,2}$ are preserved for the equilibrium point $(x_2,\, y_2)$;

(b) Zoom of Figure 2.3.6 (a) for $b \in [-c, c]$.

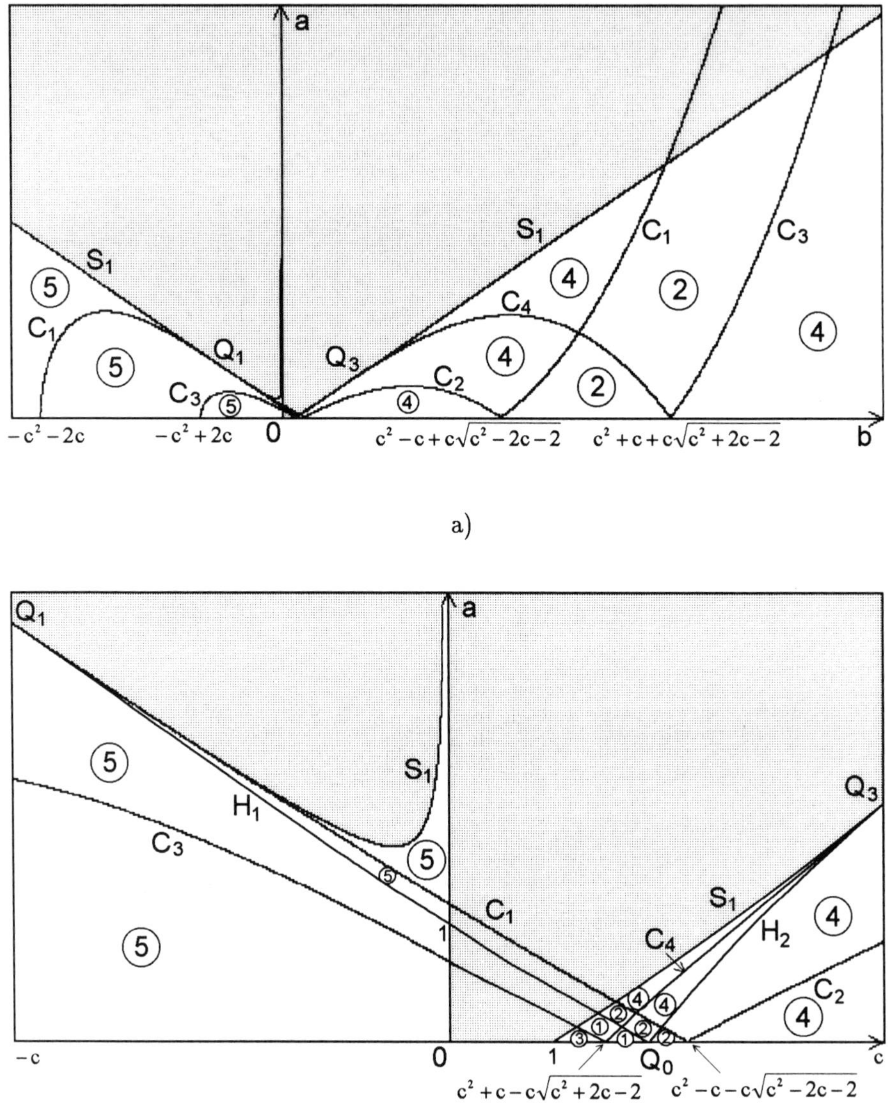

Fig. 2.3.7. (a) Regions in the $(b,\, a)$-plane where the nature of the eigenvalues and the sign of Re $\lambda_{1,2}$ are preserved for the equilibrium point $(x_3,\, y_3)$;
(b) Zoom of Figure 2.3.7 (a) for $b \in [-c, c]$.

2.3.4. THE VARIATION OF THE EIGENVALUES WITH RESPECT TO THE PARAMETERS

Taking into account the above discussion, it follows that the variation with respect to a, b of the eigenvalues is very different for different values of (b, a) in the parameter plane and branches of solutions. Taking into account the nature and sign of the eigenvalues, the (b, a)-plane will be decomposed into regions corresponding to different phase dynamics as it will be seen in Section 2.6.

As an example, for $a = 1$, $c = 5$ and the branch of solutions $x_1x_0x_3$ from Figure 2.2.2b, the variation of Im $\lambda_{1,2}$ with b is given in Figure 2.3.8a and the variation of Re $\lambda_{1,2}$ with b is given in Figure 2.3.8b. Figures 2.3.8a$'$, b$'$ are zooms of Figures 2.3.8a, b for $b \in (-3, 3)$.

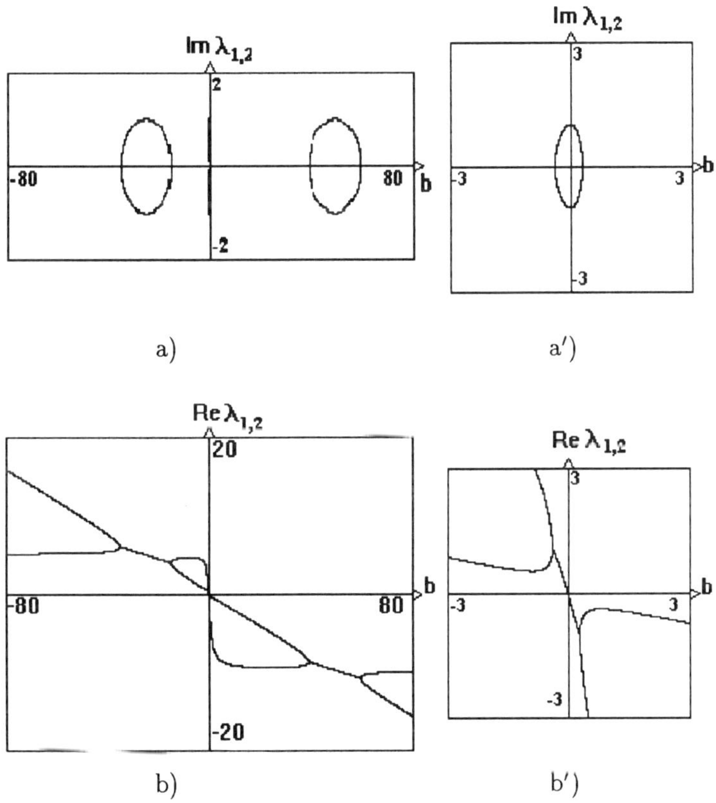

Fig. 2.3.8. The variation, with b, of the eigenvalues corresponding to the branch of equilibria $x_1x_0x_3$ for $a = 1$ and $c = 5$, as b increases from -80 to 80.

As $b \to -\infty$ we have Re $\lambda_2 \to 5.09$, whilst as $b \to \infty$ we have Re $\lambda_1 \to -10.09$. In Figures 2.3.8a and 2.3.8a$'$, Im λ_1 takes positive values. In Figures 2.3.8b and 2.3.8b$'$, for very small negative b we have Re $\lambda_1 =$

Re λ_2, then decreasing b we have Re $\lambda_1 >$ Re λ_2, then, for some range of negative b we have Re $\lambda_1 =$ Re λ_2, and for large negative b, Re $\lambda_1 >$ Re λ_2. Increasing b beyond $b = 0$, first we have Re $\lambda_1 =$ Re λ_2, then Re $\lambda_1 >$ Re λ_2, again Re $\lambda_1 =$ Re λ_2 and for very large b, we have Re $\lambda_1 >$ Re λ_2.

2.3.5. EIGENVECTORS AND EIGEN-DIRECTIONS

The results of this section are deduced taking into account the eigenvalues and eigenvectors of the matrix associated with the system linearized around an equilibrium [48].

Assume $\lambda_{1,2} \in \mathbf{R}$, $\lambda_1 \neq \lambda_2$. Then, from (2.3.1) and taking into account (2.3.3) we find the solution

$$\mathbf{u} = e^{\lambda_1 t} \left(-\left(b + c\lambda_1\right), 1\right)^{\mathrm{T}} \tag{2.3.15}$$

corresponding to λ_1 and the solution

$$\mathbf{v} = e^{\lambda_2 t} \left(-\left(b + c\lambda_2\right), 1\right)^{\mathrm{T}} \tag{2.3.16}$$

corresponding to λ_2, where $\left(-\left(b + c\lambda_1\right), 1\right)^{\mathrm{T}}$ and $\left(-\left(b + c\lambda_2\right), 1\right)^{\mathrm{T}}$ are the eigenvectors of the matrix associated with (2.3.1), corresponding to λ_1 and λ_2, respectively.

The solutions $\mathbf{u} = (u_1, u_2)$ and $\mathbf{v} = (v_1, v_2)$ form a basis for the solution set of (2.3.1). They coincide if (b, x_i) is situated on parabolas B_1 or B_2 or, equivalently, if (b, a) is situated on the curves $C_{1,2}$ or $C_{3,4}$ respectively.

In the (b, x)-plane and the inner regions delimited by B_1 and B_2, λ_1 and λ_2 are complex, outside those regions λ_1 and λ_2 are real and distinct, while on B_1 and B_2 they are real and equal. The corresponding regions, in the (b, a)-plane are: the regions between C_1 and C_3 or C_2 and C_4, the regions outside them and the regions on them. They are different for different x_i.

The eigen-directions, asymptotic to the trajectories as $t \to \infty$, are defined only for $\lambda_1, \lambda_2 \in \mathbf{R}$ therefore only on B_1 , B_2 and in the outer regions delimited by these curves and their correspondents in the (b, a)-plane. Because the eigen-directions of saddles are tangent to their stable and unstable manifolds, we shall determine the eigen-directions only for the hyperbolic nodes. (We recall that we are in the linear case.) They are

$$\frac{u_2}{u_1} = -\frac{1}{b + c\lambda_1} = -\frac{2}{c^2\left[1 - x_i^2 + \sqrt{(1 - x_i^2 + \frac{b}{c^2})^2 - \frac{4}{c^2}}\,\right]}, \tag{2.3.17}$$

$$\frac{v_2}{v_1} = -\frac{1}{b + c\lambda_2} = -\frac{2}{c^2\left[1 - x_i^2 - \sqrt{\left(1 - x_i^2 + \frac{b}{c^2}\right)^2 - \frac{4}{c^2}}\,\right]} \tag{2.3.18}$$

if

$$1 - x_i^2 + \sqrt{\left(1 - x_i^2 + \frac{b}{c^2}\right)^2 - \frac{4}{c^2}} \neq 0$$

and

$$1 - x_i^2 - \sqrt{\left(1 - x_i^2 + \frac{b}{c^2}\right)^2 - \frac{4}{c^2}} \neq 0.$$

The two square brackets in (2.3.17) and (2.3.18) vanish iff $x_i = 1$, $b = \pm 2c$, or $x_i = -1$, $b = \pm 2c$. Consequently, $a = \left(1 \mp \frac{4}{3}c\right)$ or $a = \left(-1 \pm \frac{4}{3}c\right)$ respectively. Since this vanishing implies that $\mathbf{u}$ and $\mathbf{v}$ are no longer linearly independent, the computation of eigen-directions is meaningless. The same conclusion holds if only one square bracket in (2.3.17) or (2.3.18) vanishes. In this case

$$x_i = \pm\sqrt{\frac{b}{2c^2} - \frac{2}{b} + 1}$$

and

$$a = \frac{b}{3}\left(\frac{b}{2c^2} + \frac{1}{b} - 2\right)\sqrt{\frac{b}{2c^2} - \frac{2}{b} + 1}.$$

If $\lambda_1 > \lambda_2 > 0$, the eigen-direction as $t \to \infty$ is (2.3.17) and as $t \to -\infty$ is (2.3.18).

If $\lambda_2 > \lambda_1 > 0$, the eigen-direction as $t \to \infty$ is (2.3.18) and as $t \to -\infty$ is (2.3.17).

If $\lambda_1 < \lambda_2 < 0$, the eigen-direction as $t \to \infty$ is (2.3.18) and as $t \to -\infty$ is (2.3.17).

If $\lambda_2 < \lambda_1 < 0$, the eigen-direction as $t \to \infty$ is (2.3.17) and as $t \to -\infty$ is (2.3.18).

The two eigen-directions u_2/u_1 and v_2/v_1 depend on a, b and c. For the same values of a, b and c they depend also on the equilibrium (x_i, y_i) at which they are computed.

When $\lambda_1 = \lambda_2 \neq 0$, (b, a) is on $C_{1,2}$, but not on $H_{1,2}$ and a single eigen-direction exists.

When the equilibrium is not hyperbolic, the information from the linear case can be used no longer.

2.4. STATIC BIFURCATION DIAGRAMS: PARTIAL DYNAMICAL CHARACTERIZATION

The general static bifurcation diagram is the surface Σ from the (b, a, x)-space, already introduced in Section 2.2 (Figure 2.2.1), whilst the one-parameter static bifurcation diagrams are those from Figure 2.2.2.

In order to complete the static bifurcation diagrams with dynamic information, obtained from linearization, about the type of the corresponding equilibrium points let us note that the intersections of the surfaces $B_i^{'}$, $i = \overline{1,7}$ with the planes a =constant are the same for every a. More precisely, they are the curves B_i, $i = \overline{1,7}$ from Figure 2.3.1. This is why the intersections of the surfaces Σ and B_i', $i = \overline{1,7}$ with planes a =constant are the one-parameter static bifurcation diagrams and the curves $B_i, i = \overline{1,7}$. These intersections are represented in Figure 2.4.1, where the static bifurcation diagrams are represented for the four situations $a = 0$, $a \in \left(0, \sqrt{3}\right)$, $a = \sqrt{3}$ and $a > \sqrt{3}$.

Only the segments $\overline{Q}_1\overline{Q}_2$ and $\overline{Q}_3\overline{Q}_4$ of B_3 were represented, because the part of B_3 which is not situated between B_1 and B_2 has no relevance for the nature and sign of the eigenvalues.

The representation in Figure 2.4.1 is qualitative.

For fixed a, the point of intersection of the one-parameter diagram and B_7 is denoted by P. As a varies, P describes a spatial curve the projection of which, on the (b, a)-plane, is S_1 with $b > 1$. This follows immediately by eliminating x between the equations (2.2.1) and (2.3.11). This shows that at P we have $x_1 = x_2$.

If a increases from 0 to infinity, the abscissa of P increases from 1 to infinity, while its ordinate decreases from 0 to -1. The point P is a fork bifurcation point for $a = 0$ and a lateral bifurcation point for $a > 0$.

For $a = \sqrt{3}$, the point $P_3\left(-\frac{1}{2}, \sqrt{3}\right)$ is the intersection of the one-parameter diagram with B_5 and for a fixed $a > \sqrt{3}$ let $P_3^{'}$, $P_3^{''}$ be the intersections of the one-parameter diagram with B_5. As a varies, $P_3^{'}$ and $P_3^{''}$ describe two spatial curves, whose projections on the (b, a)-plane are the branches of S_1 for $b \in \left(-\frac{1}{2}, 0\right)$ and $b \in \left(-\infty, -\frac{1}{2}\right)$, respectively. This shows that $P_3^{'}$ and $P_3^{''}$ are the points at which $x_1 = x_3$.

Define the limit points of the different branches of solutions $M\left(-\infty, 0\right)$, $V\left(-\infty, \sqrt{3}\right)$, $W\left(-\infty, -\sqrt{3}\right)$, $U\left(0, \infty\right)$, $Y\left(0, -\infty\right)$, $N\left(\infty, 0\right)$, $R\left(\infty, \sqrt{3}\right)$, $T\left(\infty, -\sqrt{3}\right)$.

As a increases from 0 to $\sqrt{3}$, the curves VU and MR approch the straight line $x = \sqrt{3}$ and intersect themselves for $a = \sqrt{3}$ at the point P_3, which is a point of transcritical bifurcation. If a continue increasing to infinity, the point P_3 splits into the points $P_3^{'}$ and $P_3^{''}$ situated on B_5. The abscissa of $P_3^{'}$ tends to 0 and its ordinate tends to infinity, while the abscissa of $P_3^{''}$ tends to $-\infty$ and its ordinate to 1.

For a fixed a, denote by P_1 and P_2 the intersections of the one-parameter diagram with B_3 for $x > 0$ and $x < 0$ respectively. Consequently, $P_1 \in \overline{Q}_1\,\overline{Q}_2$ and $P_2 \in \overline{Q}_4\,\overline{Q}_3$.

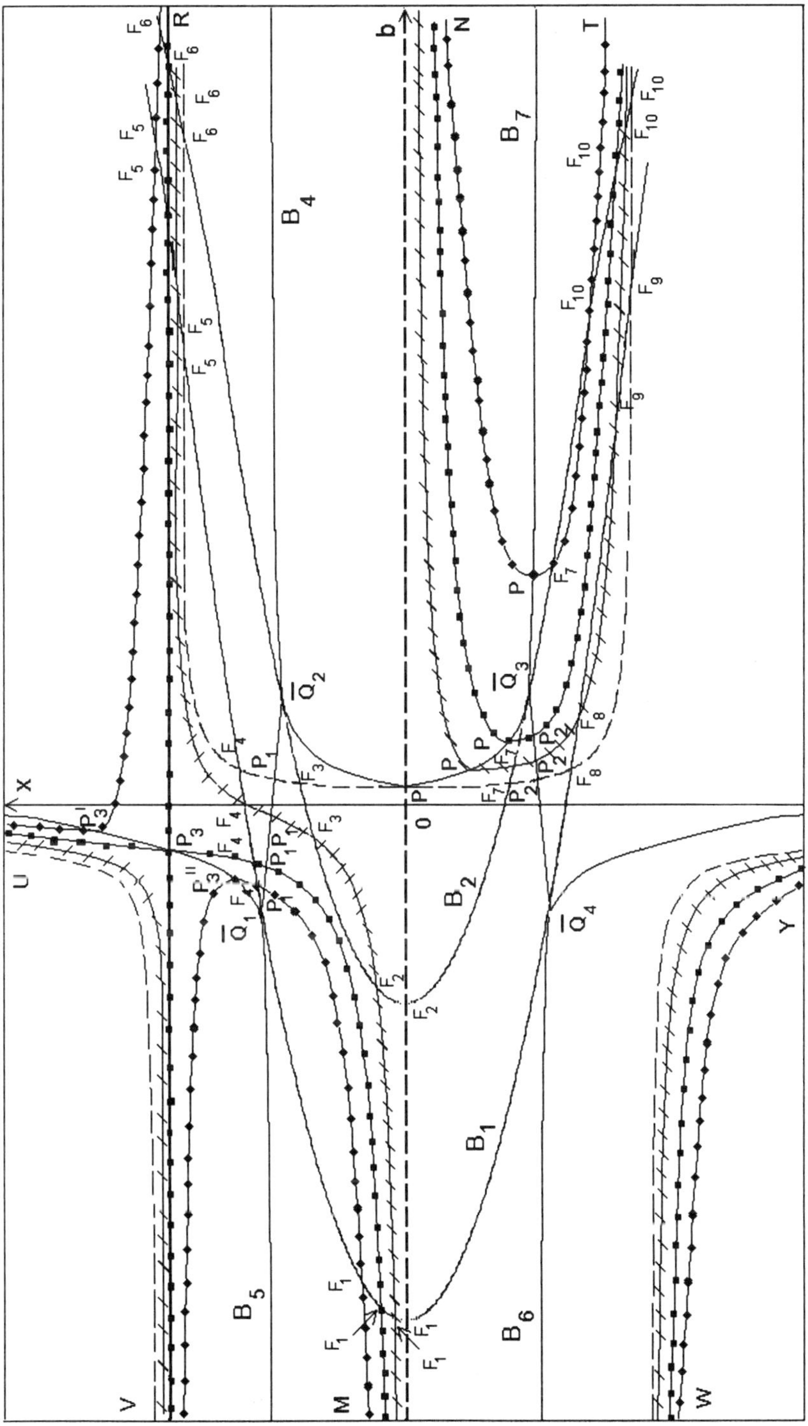

Fig. 2.4.1. Static bifurcation diagrams for a=0 (— — —), $0 < a < \sqrt{3}$ (⟋⟋⟋), $a=\sqrt{3}$ (—■—), $a > \sqrt{3}$ (—◆—) crossing regions 1-5 from Figure 2.3.1.

As a increases from 0 to $\frac{2}{3}(c+1)\sqrt{1+\frac{1}{c}}$, the point P_1 describes the region of B_3 situated between P_1 corresponding to $a = 0$ and $\overline{Q}_1$. As a increases from 0 to $\frac{2}{3}(c-1)\sqrt{1-\frac{1}{c}}$, the point P_2 describes the region of B_3 situated between P_2 corresponding to $a = 0$ and $\overline{Q}_3$.

The projections of these segments of B_3 on the (b, a)-plane are the curves $H_{1,2}$, for $b \in (-c, c)$ and $a \geq 0$.

Denote by F_1, F_4, F_5 the points of intersection of B_1 with the branch of solutions MR for $a \in \left[0, \sqrt{3}\right]$ or with the branches MP_3'' and $P_3'R$ for $a > \sqrt{3}$, and by F_2, F_3 and F_6 the points of intersection of B_2 with the mentioned branches of solutions.

Denote by F_8, F_9 the points of intersection of B_1 with the branch of solutions PT (that is x_2) and by F_7, F_{10} the points of intersection of B_2 with PT.

Remark that as a is increased from 0 to ∞ the abscissae of P, P_2, P_3', F_1, F_2, F_5, F_6, F_7, and F_8 are increasing, while the abscissae of P_1, P_3'', F_3, F_4, F_9, and F_{10} are decreassing, so that the segments $F_2\, F_3$ of B_2 and $F_8 F_9$ of B_1 become smaller and smaller, while the segment $F_4 F_5$ of B_1 becomes bigger and bigger.

The four one-parameter diagrams, compleated with the nature and signs of the eigenvalues for the different branches of equilibria, are presented in Figures 2.4.2-2.4.5, where we marked the branches of equilibria with $\lambda_{1,2} \in \mathbf{C}$, Re $\lambda_1 =$ Re $\lambda_2 > 0$ by $\circ$, those with $\lambda_{1,2} \in \mathbf{C}$, Re $\lambda_1 =$ Re $\lambda_2 < 0$ by $\bullet$, those with $\lambda_{1,2} \in \mathbf{R}$, $\lambda_{1,2} > 0$ by $\square$, those with $\lambda_{1,2} \in\mathbf{R}$, $\lambda_{1,2} < 0$ by $\blacksquare$ and those with $\lambda_{1,2} \in \mathbf{R}$, $\lambda_1\lambda_2 < 0$ by $\times$. These situations occur when the diagrams in Figure 2.4.1 cross the domains numbered by 1-5 in Figure 2.3.1.

Since around the nodes and foci the phase portraits are topologically equivalent, mathematically we do not distinguish between them. However, from the physiological viewpoint, the phenomena transient to foci are oscillating whilst the phenomena transient to nodes are nonoscillating. This is why, on the bifurcation diagram we mention the nature of the eigenvalues as well as the signs of their real parts.

Let us give dynamic details in the case (1), when $a = 0$ (Figure 2.4.2). Thus, equilibrium points (x_i, y_i) corresponding to b such that (b, x_i) is situated on VU, WY or PN have $\lambda_{1,2} \in\mathbf{R}$, $\lambda_1\ \lambda_2 < 0$, therefore they are saddles, those for which (b, x_i) is situated on the open segments MP, PP_1 or PP_2 have $\lambda_{1,2} \in\mathbf{C}$, Re $\lambda_{1,2} > 0$, or $\lambda_{1,2} \in\mathbf{R}_+$, hence they are repulsors, and those for which (b, x_i) is situated on P_1R or P_2T have $\lambda_{1,2} \in\mathbf{C}$, with Re $\lambda_{1,2} < 0$ or $\lambda_{1,2} \in\mathbf{R}$, $\lambda_{1,2} < 0$, therefore they are attractors.

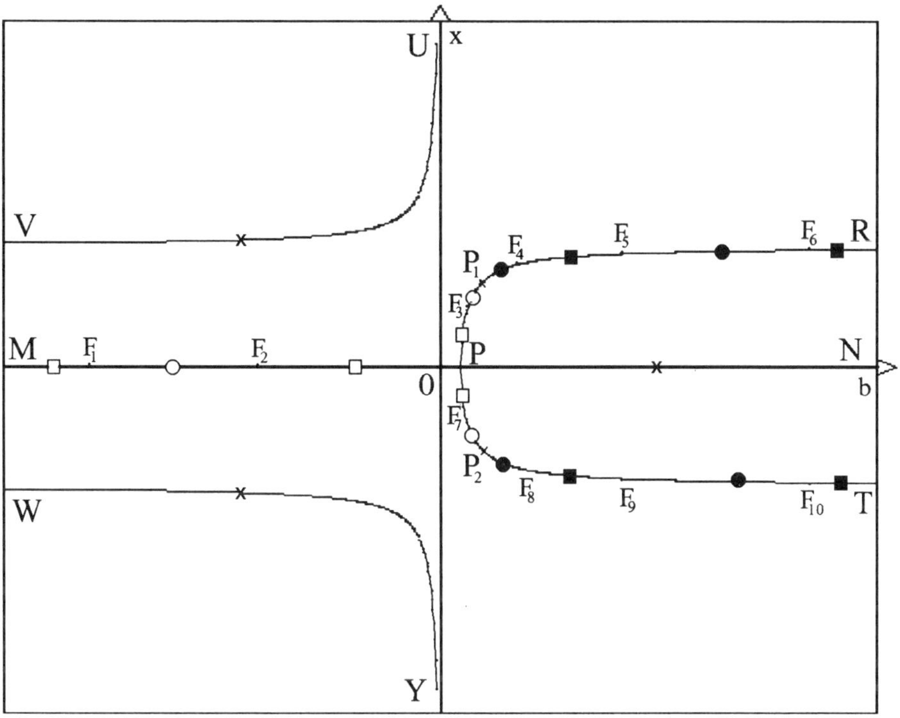

Fig. 2.4.2. Static bifurcation diagram for $a = 0$.

In the case (2), when $a \in (0, \sqrt{3})$ (Figure 2.4.3), the equilibria (x_i, y_i) corresponding to b such that (b, x_i) is situated on VU, WY or PN have $\lambda_{1,2} \in \mathbf{R}$, $\lambda_1\lambda_2 < 0$ (saddles), those for which (b, x_i) is situated on the open segments MP_1, or PP_2 have $\lambda_{1,2} \in \mathbf{C}$, Re $\lambda_{1,2} > 0$ or $\lambda_{1,2} \in \mathbf{R}_+$ (repulsors), and those for which (b, x_i) is situated on the open segments P_1R or P_2T have $\lambda_{1,2} \in \mathbf{C}$, Re $\lambda_{1,2} < 0$ or $\lambda_{1,2} \in \mathbf{R}_-$ (attractors).

It may be possible, for certain $a \in \left(0, \sqrt{3}\right)$, to have $F_2 \equiv F_3$ and for other value of a, to have $F_8 \equiv F_9$ so that the segments F_2F_3, F_8F_9 disappear and the segment F_1P_1 consists only of equilibria with $\lambda_{1,2} \in \mathbf{C}$, Re $\lambda_{1,2} > 0$, and P_2F_{10} of equilibria with $\lambda_{1,2} \in \mathbf{C}$, Re $\lambda_{1,2} < 0$.

In addition, if $c < 4$, at certain $a \in \left(0, \sqrt{3}\right)$ the points P, F_7, P_2 collide with $\overline{Q}_3$ (Figure 2.4.1) and P_2 disappears. If a continues increassing, the new segment PF_7 of x_2 consists of equilibria with $\lambda_{1,2} \in \mathbf{R}_-$, F_7F_{10} of equilibria with $\lambda_{1,2} \in \mathbf{C}$, Re $\lambda_{1,2} < 0$ and $F_{10}T$ of equilibria with $\lambda_{1,2} \in \mathbf{R}_-$.

If $c \geq 4$ the abscissa of P is always smaller than the abscissa of $\overline{Q}_3$, so the above situation does not take place.

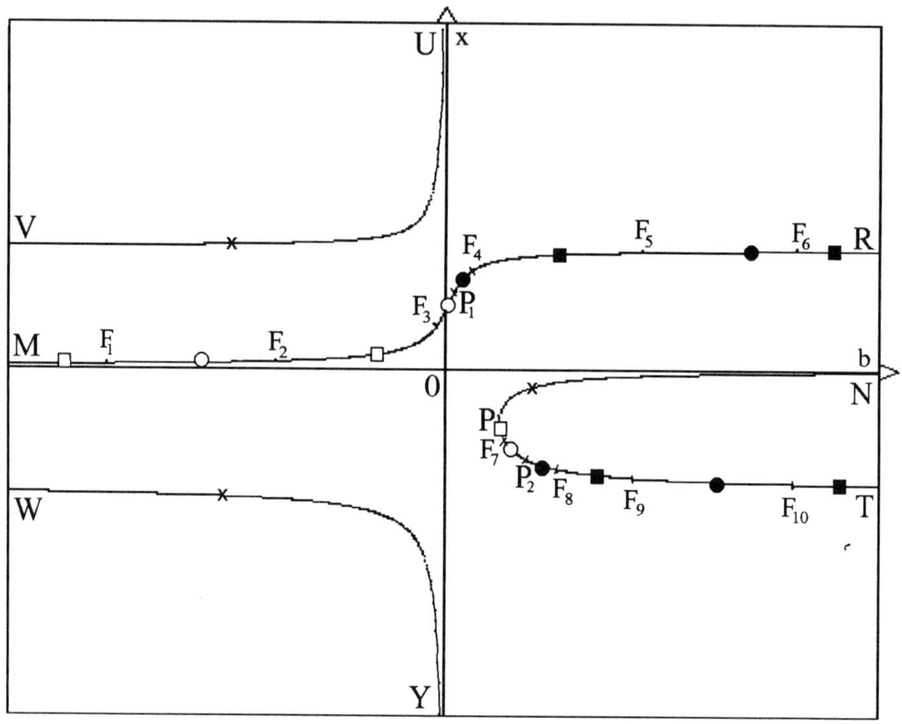

Fig. 2.4.3. Static bifurcation diagram for $a \in (0, \sqrt{3})$.

In the case (3), when $a = \sqrt{3}$ (Figure 2.4.4), the equilibria (x_i, y_i) corresponding to b such that (b, x_i) is situated on VP_3, P_3U, WY or PN have $\lambda_{1,2} \in \mathbf{R}$, $\lambda_1\lambda_2 < 0$ (saddles), those for which (b, x_i) is situated on the open segments MP_1 or PP_2 have $\lambda_{1,2} \in \mathbf{C}$, Re $\lambda_{1,2} > 0$ or $\lambda_{1,2} \in \mathbf{R}_+$ (repulsors), and those for which (b, x_i) is situated on the open segments P_1P_3, P_3R or P_2T have $\lambda_{1,2} \in \mathbf{C}$, Re $\lambda_{1,2} < 0$ or $\lambda_{1,2} \in \mathbf{R}_-$ (attractors).

Remark that the situation from Figure 2.4.4 holds only for $c > 4$. In the case $a = \sqrt{3}$, P has the coordinates $\left(4, -\frac{\sqrt{3}}{2}\right)$ so, if $c < 4$, the abscissa of P is greater than the abscissa c of $\overline{Q}_3$ (Figure 2.4.1), the point P_2 does not exist and the segments PF_7, $F_{10}T$ consist of equilibria with $\lambda_{1,2} \in \mathbf{R}_-$, while the segments F_7F_8 and F_9F_{10} consist of equilibria with $\lambda_{1,2} \in \mathbf{C}$, Re $\lambda_{1,2} < 0$.

If $c < 4$ but still smaller, the points F_8 and F_9 collide, so that the segment F_8F_9 disappears and F_7F_{10} consists of equilibria with $\lambda_{1,2} \in \mathbf{C}$, Re $\lambda_{1,2} < 0$. For a certain value $c < 4$, the points F_2 and F_3 collide also and the segment F_2F_3 disappears, so that for small c, the segment F_1P_1 consists only of equilibria with $\lambda_{1,2} \in \mathbf{C}$, Re $\lambda_{1,2} > 0$.

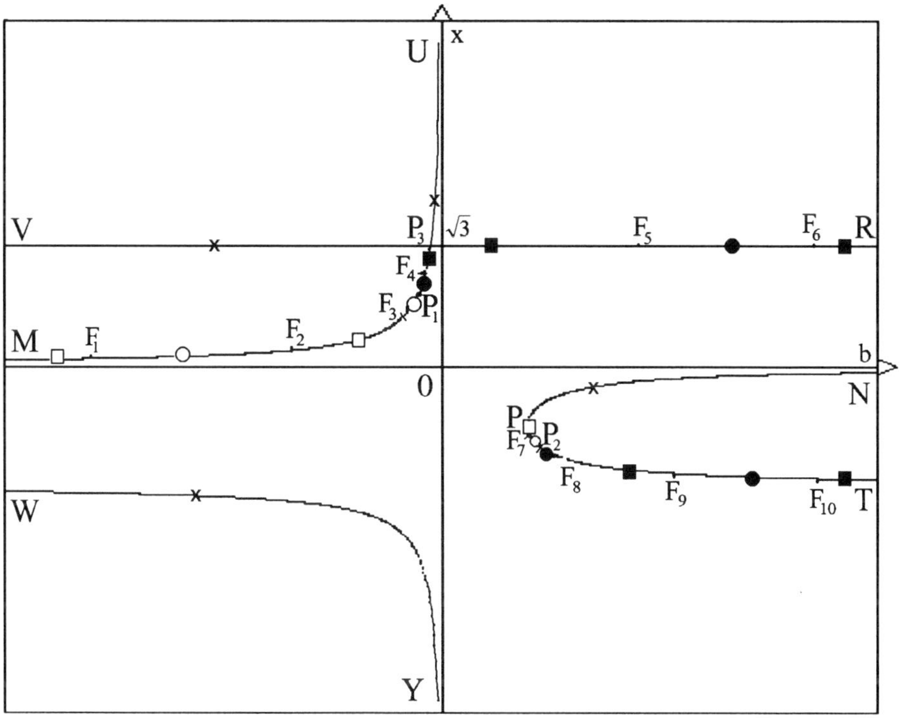

Fig. 2.4.4. Static bifurcation diagram for $a = \sqrt{3}$.

In the case (4), when $a > \sqrt{3}$ (Figure 2.4.5) the equilibria (x_i, y_i) corresponding to b such that (b, x_i) is situated on $P_3'U$, VP_3'', WY or PN have $\lambda_{1,2} \in \mathbf{R}$, $\lambda_1\lambda_2 < 0$ (saddles). Those equilibria for which (b, x_i) is situated on the open segments MP_1 or PP_2 have $\lambda_{1,2} \in \mathbf{C}$, Re $\lambda_{1,2} > 0$ or $\lambda_{1,2} \in \mathbf{R}_+$ (repulsors), and those for which (b, x_i) is situated on the open segments $P_3''P_1$, $P_3'R$ or P_2T have $\lambda_{1,2} \in \mathbf{C}$, Re $\lambda_{1,2} < 0$ or $\lambda_{1,2} \in \mathbf{R}_-$ (attractors).

Remark that if c is fixed and a is increased from $\sqrt{3}$ to $+\infty$ the segments $P_3''F_4$ and F_4P_1 become smaller and smaller until, for a certain value of a, the points P_3'', F_4 and P_1 collide with $\overline{Q}_1$ (Figure 2.4.1) and P_1 disappears. For a still larger a, the segment $P_3''F_4$ consists of equilibria with $\lambda_{1,2} \in \mathbf{R}_+$. If a continues increasing F_4 and F_1 collide, the segment F_1F_4 disappears and the segment $P_3''M$ consists only of equilibria with $\lambda_{1,2} \in \mathbf{R}_+$.

A similar situation takes place with the branch PT of solutions, when, at a certain value of a, the points P, F_7, P_2 collide with $\overline{Q}_3$ and for a still larger a, P_2 disappears and the new segment PF_7 consists of equilibria with $\lambda_{1,2} \in \mathbf{R}_-$. If a continues increasing, F_7 and F_{10} collide, the segment F_7F_{10} disappears and the segment PT consists only of equilibria with $\lambda_{1,2} \in \mathbf{R}_-$.

In addition, if c is fixed, but $c < 4$, the situation from Figure 2.4.5 does not hold any longer because the abscissa of P is greater than the abscissa

c of $\overline{Q}_3$. In this situation the point P_2 does not exist and the segments PF_7 and F_8F_9 consist of equilibria with $\lambda_{1,2} \in \mathbf{R}_-$, whilst the segments F_7F_8, F_9F_{10} consist of equilibria with $\lambda_{1,2} \in \mathbf{C}$, Re $\lambda_{1,2} < 0$.

If a is increased, F_8 collides with F_9 and the segment F_8F_9 disappears, so that F_7F_{10} consists of equilibria with $\lambda_{1,2} \in \mathbf{C}$, Re $\lambda_{1,2} < 0$. If a increases further, F_7 collides with F_{10} and the segment F_7F_{10} disappears, so that the segment PT consists only of equilibria with $\lambda_{1,2} \in \mathbf{R}_-$.

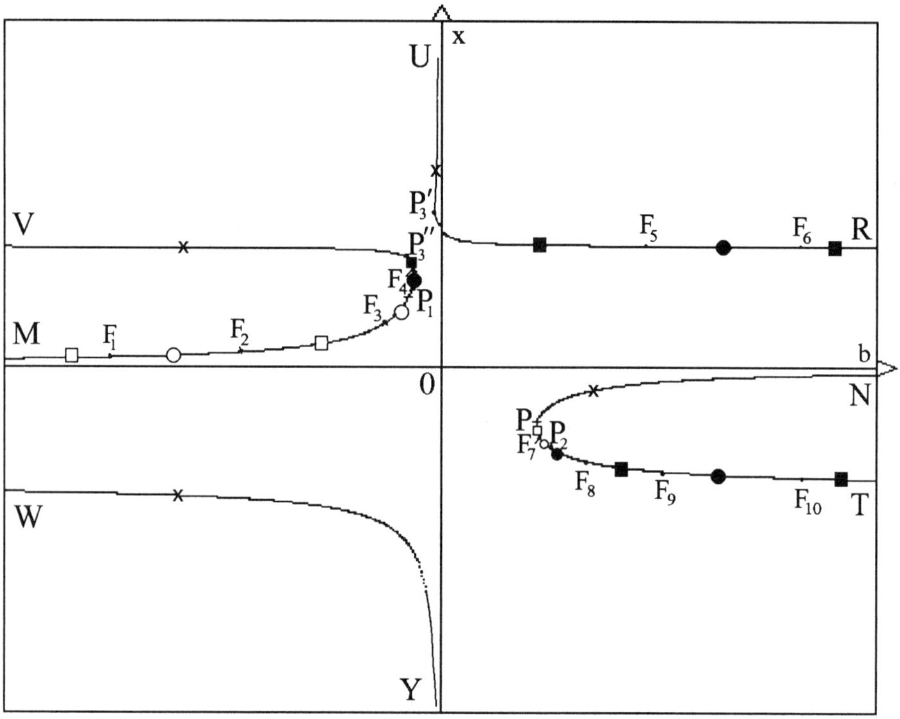

Fig. 2.4.5. Static bifurcation diagram for $a > \sqrt{3}$.

For all diagrams in Figures 2.4.2-2.4.5, the equilibria corresponding to the points F_1, F_2, F_3 and F_7 have $\lambda_1 = \lambda_2 \in \mathbf{R}_+$ (repulsors), while those corresponding to F_4, F_5, F_6, F_8, F_9, F_{10} have $\lambda_1 = \lambda_2 \in \mathbf{R}_-$ (attractors). In addition, the double equilibria corresponding to P, P_3, P_3' and P_3'' have one eigenvalue equal to zero, while the other one is negative for P_3, P_3' and P_3'' and pozitive for P. Exceptional cases are those when P_3'' is at $\overline{Q_1}$ or P is at $\overline{Q}_3$. In these cases both eigenvalues are zero.

Finally, the equilibria corresponding to P_1 or P_2 have $\lambda_{1,2} \in \mathbf{C}$, Re $\lambda_1 =$ Re $\lambda_2 = 0$, so these are possible Hopf bifurcation equilibria.

A similar dynamic characterization of equilibria, but discussed in the (b, a)-plane, will be presented in Section 2.6.

2.5. ASYMPTOTIC BEHAVIOUR OF THE STATIC BIFURCATION DIAGRAMS AS $c \to \infty$.

The results of this section are deduced by assuming that a and b are fixed, while c is variable. That is, they are complementary to those of the above sections. We shall analyze only those functions from Section 2.4 and, consequently from Section 2. 3, which depend on c. For the sake of simplicity, we shall understand that $c \to \infty$.

A first result, deduced from (2.3.3), is that $\lambda_1 \to \pm\infty, \lambda_2 \to 0$ for $1 - x_i^2 \gg \frac{1}{c}$. The sign of λ_1 is given by the sign of $1 - x_i^2$, while that of λ_2 by the sign of $\dfrac{1 - b\left(1 - x_i^2\right)}{\lambda_1}$. For $1 - x_i^2 \ll \dfrac{1}{c^2}$ we have $\lambda_{1,2} \sim \pm i$. For $\dfrac{1}{c^2} \ll 1 - x_i^2 \ll \frac{1}{c}$, we have $\lambda_{1,2} \sim \pm i$ and for $1 - x_i^2 \sim \frac{1}{c}$ we have $\lambda_{1,2} \sim \overline{\lambda}_{1,2} \pm i$, where $\overline{\lambda}_{1,2} = O(1)$.

In the (b, x)-plane, the vertex of the parabolas B_1 and B_2 tend to $-\infty$ and that of B_3, to $+\infty$. The intersection points of B_1 and B_2 with the Oa-axis tend to ± 1, $\overline{Q}_1$ tends to $(-\infty, 1)$, $\overline{Q}_2$ tends to $(\infty, 1)$, $\overline{Q}_3$ tends to $(\infty, -1)$ and $\overline{Q}_4$ tends to $(-\infty, -1)$.

For a fixed at some value, $P_1 \to (b_{P_1}, 1)$. If, in addition, $a \to \infty$, then $P_1 \to (-\infty, 1)$. Similarly, for fixed a, $P_2 \to (b_{P_2}, -1)$ and if, in addition, $a \to \infty$, it follows $P_2 \to (\infty, -1)$.

As P, P_3, P_3' , P_3'' do not depend on c, they keep unchanged as c is increased. The ordinate of P_2 tends to -1, such that the distance between P and P_2 grows. Similarly, the distance between P_1 and P_3'' grows.

In the (b, a)-plane, the curves C_i, $i = \overline{1,4}$ delimiting the region corresponding to complex eigenvalues, have the following asymptotic behaviour for large c: the intersection points of $C_{1,3}$ and $C_{2,4}$ with the Oa-axis become $(0,1)$ and $(0,-1)$ respectively. The curves C_1 and C_2 and the curves C_3 and C_4 are intersecting themselves on the Ob-axis for $-\infty$, $\frac{3}{2}$, ∞.

The curves C_1 and C_2 tend asymptolically to the asymptote A_1 of $S_{1,2}$, while C_3 and C_4 tend to the asymptote A_2 of $S_{1,2}$.

The curve H_1 tends to A_1, H_2 tends to A_2 and Q_0 tends to $\left(\frac{3}{2}, 0\right)$.

As a consequence, the region with complex eigenvalues represented in Figure 2.3.2 is narrowing around $H_{1,2}$ (Figure 2.5.1), for $c = \infty$ reducing to $H_{1,2}$. The important conclusion is that always the Hopf bifurcation holds.

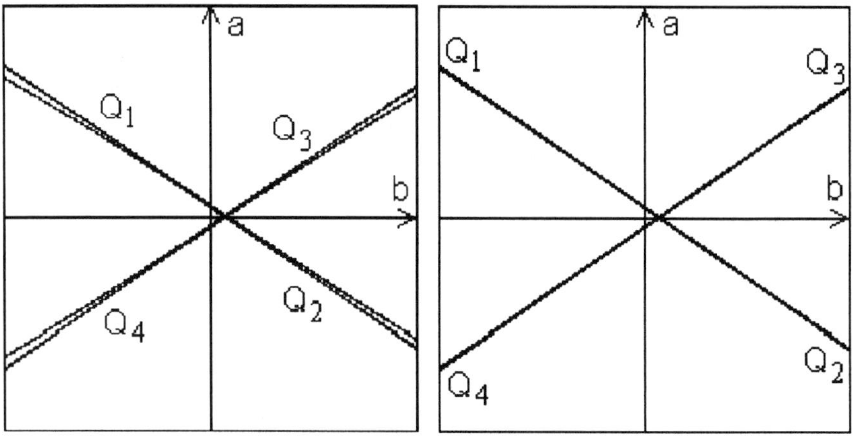

a) b)

Fig. 2.5.1. The narrow region corresponding to complex eigenvalues, for $b \in (-20, 20)$ and (a) $c = 10$, (b) $c = 20$.

2.6. TYPES OF HYPERBOLIC EQUILIBRIA

Taking into account the discussion in the (b, a)-plane concerning the nature of the eigenvalues and the sign of their real parts (Section 2.3.3), the nature of the hyperbolic equilibria can be deduced.

Let $\tilde{b} \in \left(1, \frac{3}{2}\right)$ be the abscissa of the intersection point of H_1 and S_1. More precisely, we have $\tilde{b} = \dfrac{-3c^2 + \sqrt{9c^4 + 16c^2}}{2}$.

In Figure 2.6.1 the domain 1 is delimited by Ob, with $b < 0$, Oa, $a \in (0, 1)$, H_1 for $b \in (-c, 0)$ and S_1 for $b \in (-\infty, -c]$; domain 2 is delimited by S_1 and H_1, with $b \in (-c, 0)$ and Oa, with $a > 1$; domain 3 is situated above S_1, for $b < 0$; domain 4 is delimited by H_1 with $b \in \left(0, \tilde{b}\right)$, S_1 with $b > \tilde{b}$ and Oa with $a > 1$; domain 5 is delimited by Oa with $a \in (0, 1)$, H_1 with $b \in \left(0, \tilde{b}\right)$, S_1 for $b \in \left(1, \tilde{b}\right)$ and Ob with $b \in (0, 1)$; domain 6 is delimited by S_1, with $b \in \left(1, \tilde{b}\right)$, H_1 with $b \in \left(\tilde{b}, b_0\right)$ and Ob, $b \in (1, b_0)$; domain 7 is delimited by S_1, with $b \in \left(\tilde{b}, c\right)$, H_2 with $b \in (b_0, c)$ and H_1 with $b \in \left(\tilde{b}, b_0\right)$; domain 8 is delimited by Ob with $b > b_0$, H_2 with $b \in (b_0, c)$ and S_1 with $b \geq c$.

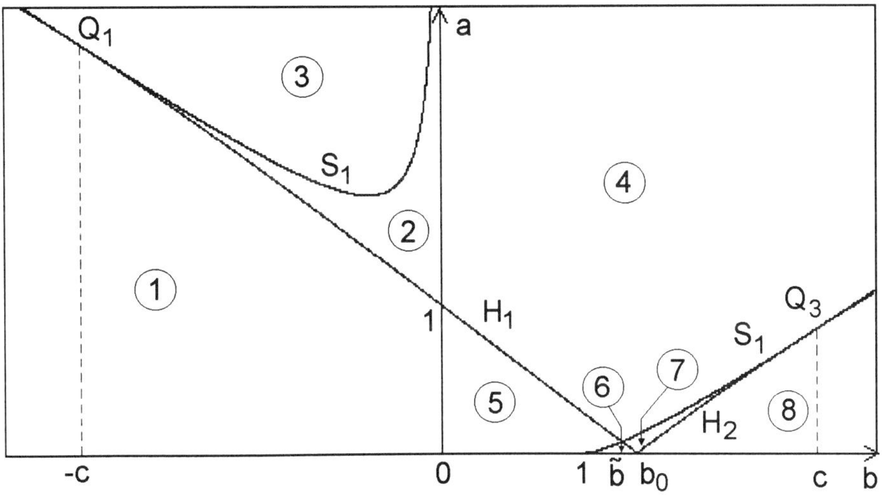

Fig. 2.6.1. The domains of the (b, a)-plane where the signs of Re $\lambda_{1,2}$, corresponding to an equilibrium point, are preserved.

For (b, a) in domain 1 or on the Ob-axis with $b < 0$, (x_1, y_1) is a repulsor, while (x_2, y_2) and (x_3, y_3) are saddle points.

For (b, a) in domain 2, (x_1, y_1) is an attractor, while (x_2, y_2) and (x_3, y_3) are saddle points.

For (b, a) in domain 3, a single saddle equilibrium point (x_0, y_0) exists.

For (b, a) in domain 4 or on the Oa-axis with $a > 1$, a single equilibrium point (x_0, y_0) exists and it is an attractor.

For (b, a) in domain 5 or on the Oa-axis with $a \in [0, 1)$ or on the Ob-axis with $b \in [0, 1)$, a single equilibrium point (x_0, y_0) exists and it is a repulsor.

For (b, a) in domain 6 or on the Ob-axis, with $b \in (1, b_0)$, (x_1, y_1) is a saddle point, while (x_2, y_2) and (x_3, y_3) are repulsors.

For (b, a) in domain 7, (x_1, y_1) is a saddle point, (x_2, y_2) is a repulsor, while (x_3, y_3) is an attractor.

For (b, a) in domain 8 or on the Ob-axis, with $b > b_0$, (x_1, y_1) is a saddle point, while (x_2, y_2) and (x_3, y_3) are attractors.

For (b, a) on $H_{1,2}$ with $b \in [-c, c]$ or on $S_{1,2}$, at least one equilibrium point is non-hyperbolic.

More precisely, for $(b, a) \in S_1$ and $b < 0$, $b \neq -c$ the equilibria (x_1, y_1) and (x_3, y_3) coincide, resulting an equilibrium with one zero eigenvalue, while the other eigenvalue is positive for $b < -c$ and negative for $b \in (-c, 0)$. The other equilibrium, (x_2, y_2), is a saddle point.

For $(b, a) \in S_1$ and $b > 1$, $b \neq c$ the equilibria (x_1, y_1) and (x_2, y_2) coincide, resulting an equilibrium with one zero eigenvalue, while the other eigenvalue is positive for $b \in (1, c)$ and negative for $b > c$. The other equilibrium, (x_3, y_3), is a repulsor if $b \in \left(1, \tilde{b}\right)$ and an attractor if $b > \tilde{b}$.

The curves S_1 and S_2 intersect themselves at Q, corresponding to $a = 0$ and $b = 1$, where $x_1 = x_2 = x_3 = 0$, $\lambda_1 = c - \frac{1}{c}$ and $\lambda_2 = 0$.

The inspection of Figures 2.3.4b-2.3.7b shows us that to any point of $H_{1,2}$, with $b \in (-c, c)$ except for Q_0, only one equilibrium corresponds for which $\lambda_{1,2} \in \mathbf{C}$, Re $\lambda_1 =$ Re $\lambda_2 = 0$.

If (b, a) is on H_1 and $b \in (-c, 0)$, then (x_1, y_1) has purely imaginary eigenvalues, whilst (x_2, y_2) and (x_3, y_3) are saddle points. If (b, a) is on H_1 and $b \in \left[0, \tilde{b}\right)$, then (x_0, y_0) has purely imaginary eigenvalues.

If (b, a) is at $H_1 \cap S_1$, then $b = \tilde{b}$, $x_1 = x_2$, one eigenvalue is zero, and the other is positive, whilst the eigenvalues corresponding to (x_3, y_3) are purely imaginary.

If (b, a) is on H_1 and $b \in \left(\tilde{b}, b_0\right)$, then (x_3, y_3) is a non-hyperbolic point, having purely imaginary eigenvalues, whilst (x_1, y_1) is a saddle and (x_2, y_2) is a repulsor. If (b, a) is on H_2 and $b \in (b_0, c)$, then (x_2, y_2) has purely imaginary eigenvalues, whilst (x_1, y_1) is a saddle point and (x_3, y_3) is an attractor.

In this way, in spite of the fact that for (b, a) situated on some parts of H_1 and H_2, two or three equilibria may correspond, except for Q_0 only one of them may be a Hopf bifurcation point.

For $b < -c$ and $b > c$, the saddles corresponding to (b, a) on $H_{1,2}$ are neutral saddles.

At Q_0 we have $b = -c^2 + \sqrt{c^4 + 3c^2}$, $a = 0$, $x_1 = 0$, $x_2 = -\sqrt{\frac{2c - \sqrt{c^2+3}}{c}}$, $x_3 = \sqrt{\frac{2c - \sqrt{c^2+3}}{c}}$, $\lambda_{1,2}(x_1) = c - \frac{1}{2}\sqrt{c^2+3} \pm \frac{1}{2}\sqrt{c^2-1}$ and $\lambda_{1,2}(x_2) = \lambda_{1,2}(x_3) = \pm i\sqrt{2}\sqrt{\sqrt{c^4+3c^2} - (c^2+1)}$, hence (x_1, y_1) is a saddle point and there are two candidates for the Hopf bifurcation equilibria, namely (x_2, y_2) and (x_3, y_3).

At Q_1 we have $b = -c$, $a = \frac{2}{3}(c+1)\sqrt{1+\frac{1}{c}}$, $x_1 = x_3 = \sqrt{1+\frac{1}{c}}$, $x_2 = -2\sqrt{1+\frac{1}{c}}$, $\lambda_{1,2}(x_1) = 0$ and $\lambda_{1,2}(x_2) = -\frac{3}{2}(c+1) \pm \frac{1}{2}\sqrt{9c^2+30c+21}$, hence (x_2, y_2) is a saddle point and the candidate for the Bogdanov–Takens bifurcation point is the double equilibrium.

At Q_3 we have $b = c$, $a = \frac{2}{3}(c-1)\sqrt{1-\frac{1}{c}}$, $x_1 = x_2 = -\sqrt{1-\frac{1}{c}}$, $x_3 = 2\sqrt{1-\frac{1}{c}}$, $\lambda_{1,2}(x_1) = 0$ and $\lambda_{1,2}(x_3) = -\frac{3}{2}(c-1) \pm \frac{1}{2}\sqrt{9c^2-30c+21}$, hence (x_3, y_3) is an attractor and the candidate for the Bogdanov–Takens bifurcation point is the double equilibrium point, too.

Conversely, if $\lambda_{1,2} \in C$, Re $\lambda_{1,2}(b,a) = 0$ it follows that $(b,a) \in H_{1,2}$; λ_1 and/or $\lambda_2 = 0$ implies $(b,a) \in S_{1,2}$ and $\lambda_1 = \lambda_2 = 0$ implies $(b,a) \in \{Q_i, i = \overline{1,4}\}$.

Thus Q_1, Q_2, Q_3, and Q_4 are the only points of the (b,a)-plane at which $\lambda_1 = \lambda_2 = 0$.

We mention that in the case of non-hyperbolic equilibria (such as saddle–nodes, Hopf, Bogdanov–Takens or Bautin bifurcation points) the nature and signs of $\lambda_{1,2}$ are not sufficient to characterize the type of the equilibrium, additional information of nonlinear nature beeing necessary. It will be provided in Section 2.7 for the non-hyperbolic equilibria corresponding to $S_{1,2}$, $b \neq \pm c$ and in Chapter 3 for the rest of non-hyperbolic equilibria.

2.7. THE CENTER MANIFOLD AND THE SADDLE–NODE BIFURCATION

2.7.1. CASE OF POSITIVE b

Although the determination of the nature of the non-hyperbolic equilibria cannot be done by using information from the linearized equation, we determine in this chapter the nature of the points of the curves $S_{1,2}$, taking into account the fact that $S_{1,2}$ are curves of static bifurcation.

If $(b,a) \in S_1$ and $b \geq 1$, then $a = \frac{2}{3}b\sqrt{\left(1-\frac{1}{b}\right)^3}$ and the double equilibrium point has the coordinates $x_1 = x_2 = -\sqrt{1-\frac{1}{b}}$, $y_1 = y_2 = \frac{1}{b}\left(a + \sqrt{1-\frac{1}{b}}\right)$. The eigenvalues of the system linearized around it are $\lambda_1 = 0$, $\lambda_2 = \frac{c}{b} - \frac{b}{c}$.

The translation

$$\begin{aligned} \psi_1 &= x + \sqrt{1-\tfrac{1}{b}}, \\ \psi_2 &= y - \tfrac{1}{b}\left(a + \sqrt{1-\tfrac{1}{b}}\right), \end{aligned} \tag{2.7.1}$$

transforms (1.1.17) into the system

$$\begin{aligned} \dot{\psi}_1 &= \tfrac{c}{b}\psi_1 + c\psi_2 + c\sqrt{1-\tfrac{1}{b}}\psi_1^2 - \tfrac{c}{3}\psi_1^3, \\ \dot{\psi}_2 &= -\tfrac{1}{c}\psi_1 - \tfrac{b}{c}\psi_2, \end{aligned} \tag{2.7.2}$$

which has $(0,0)$ as an equilibrium point. Two eigenvectors corresponding to the eigenvalues λ_1 and λ_2 may be chosen $\begin{pmatrix} b \\ -1 \end{pmatrix}$ and $\begin{pmatrix} c^2 \\ -b \end{pmatrix}$, respectively. If $b \neq c$, these eigenvectors form a basis in $\mathbf{R}^2$ and, using the transformation

$$\begin{aligned} \psi_1 &= bu + c^2 v, \\ \psi_2 &= -u - bv, \end{aligned} \tag{2.7.3}$$

the system (2.7.2) is written in the basis of eigenvectors as

$$\begin{aligned} \dot{u} &= g\left(u, v\right), \\ \dot{v} &= \alpha v + h\left(u, v\right), \end{aligned} \tag{2.7.4}$$

where

$$\begin{aligned} g\left(u, v\right) &= -\frac{b}{c^2 - b^2}\left(bu + c^2 v\right)^2 \left[c\sqrt{1 - \frac{1}{b}} - \frac{c}{3}\left(bu + c^2 v\right)\right], \\ \alpha &= \frac{c^2 - b^2}{bc}, \\ h\left(u, v\right) &= \frac{1}{c^2 - b^2}\left(bu + c^2 v\right)^2 \left[c\sqrt{1 - \frac{1}{b}} - \frac{c}{3}\left(bu + c^2 v\right)\right]. \end{aligned} \tag{2.7.5}$$

Thus the linearized system of (2.7.4) around the equilibrium $(0, 0)$ is defined by the matrix

$$A = \begin{pmatrix} 0 & 0 \\ 0 & \alpha \end{pmatrix}. \tag{2.7.6}$$

The center manifold of system (2.7.4) can be locally represented as the graph of a smooth function $v = \mathcal{V}(u)$, tangent at the origin to the critical eigenspace corresponding to $\lambda_1 = 0$.

Substituting $v = \mathcal{V}(u)$ in the second equation (2.7.4) it follows

$$\frac{\partial \mathcal{V}}{\partial u} \cdot \dot{u} = \alpha \mathcal{V}\left(u\right) + h\left(u, \mathcal{V}\left(u\right)\right). \tag{2.7.7}$$

Moreover, on the center manifold we have

$$\dot{u} = g\left(u, \mathcal{V}\left(u\right)\right). \tag{2.7.8}$$

Thus we obtain a first order partial differential equation in $\mathcal{V}$

$$\frac{\partial \mathcal{V}}{\partial u} g\left(u, \mathcal{V}\left(u\right)\right) = \alpha \mathcal{V}\left(u\right) + h\left(u, \mathcal{V}\left(u\right)\right) \tag{2.7.9}$$

with $\mathcal{V}\left(0\right) = 0$ and the tangency condition $\frac{\partial \mathcal{V}}{\partial u}\left(0\right) = 0$.

We approximate $\mathcal{V}\left(u\right)$ by its Taylor series expansion and look for its second-order term representing the center manifold.

Thus, replacing $\mathcal{V}(u) = \beta u^2 + \gamma u^3 + \ldots$ into (2.7.9) and matching the resulted asymptotic series from the left and the right-hand side, we obtain

$$\beta = -\frac{b^3 c^2}{(c^2 - b^2)^2}\sqrt{1 - \frac{1}{b}} \tag{2.7.10}$$

so

$$v = \mathcal{V}(u) = \beta u^2 + O\left(u^3\right) \tag{2.7.11}$$

is the center manifold.

The flow on the center manifold is determined by equation (2.7.8), which is written in the form

$$\begin{aligned} \dot{u} &= -\frac{b^3 c}{c^2 - b^2}\sqrt{1 - \frac{1}{b}}u^2 \\ &+\frac{1}{3}\frac{b^4 c}{c^2 - b^2}\left[1 + \frac{6bc^4}{(c^2 - b^2)^2}\left(1 - \frac{1}{b}\right)\right]u^3 + O\left(u^4\right). \end{aligned} \tag{2.7.12}$$

If the coefficient of u^2 is nonzero, the double equilibrium is a saddle–node, partially attractive if the nonzero eigenvalue is negative or partially repulsive, if it is positive.

Thus, for a fixed $b \in (1, c)$, the non-hyperbolic equilibrium point of S_1 is a partially repulsive saddle–node, whilst if $b > c$, it is a partially attractive saddle–node.

If the coefficient of u^2 is equal to zero, we get $b = 1$ and (2.7.12) becomes

$$\dot{u} = \frac{c}{3(c^2 - 1)}u^3 + O\left(u^4\right). \tag{2.7.13}$$

In this case, the stability of the equilibrium point can be determined considering the third order term. The coefficient of u^3 is positive, therefore the solution $u = 0$ of equation (2.7.13) is repulsive. Then by Theorem 1.3.8 the solution $(0, 0)$ of the system (2.7.2) is also repulsive. Thus, the triple equilibrium point of system (1.1.17) corresponding to $b = 1, a = 0$ is a non-hyperbolic repulsor.

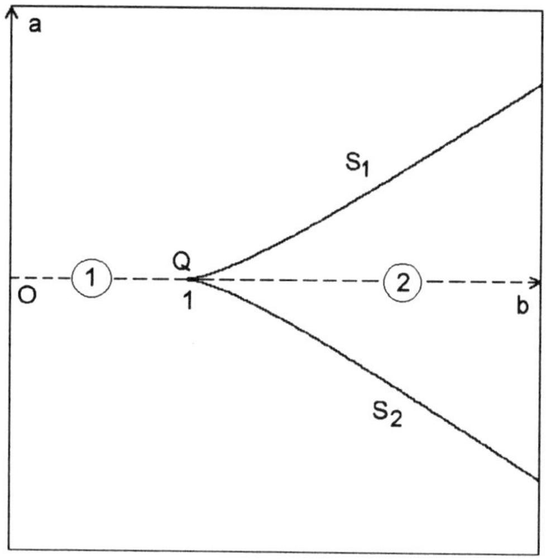

Fig. 2.7.1. The point Q.

In Figure 2.7.1, we represented the point $Q(1, 0)$ of the parameter plane and the curves $S_{1,2}$, intersecting at Q. In Figure 2.7.2, some phase portraits corresponding to regions 1 and 2 from Figure 2.7.1 and to the point Q are represented in the neighborhood of the origin of the phase plane.

Thus, for (b, a) in domain 1 of Figure 2.7.1, the system possesses a single equilibrium point, which is a hyperbolic repulsor. For the parameters in Figure 2.7.2a, this repulsor is the origin of the phase plane.

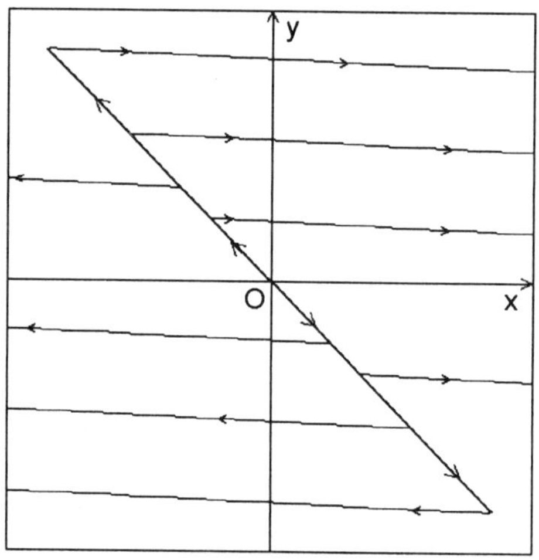

a) $a = 0, b = 0.98$

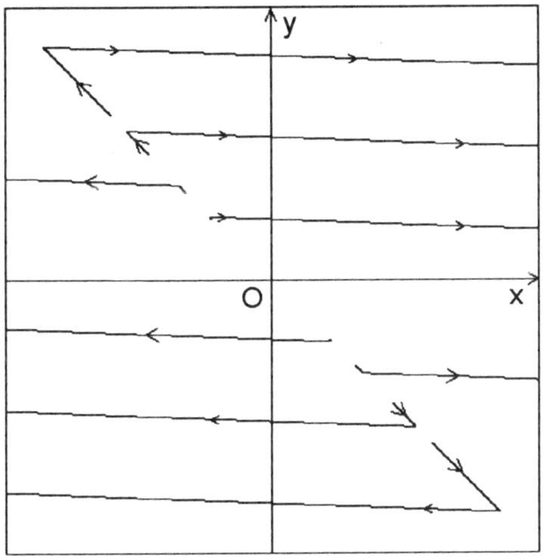

b) $a = 0, b = 1$

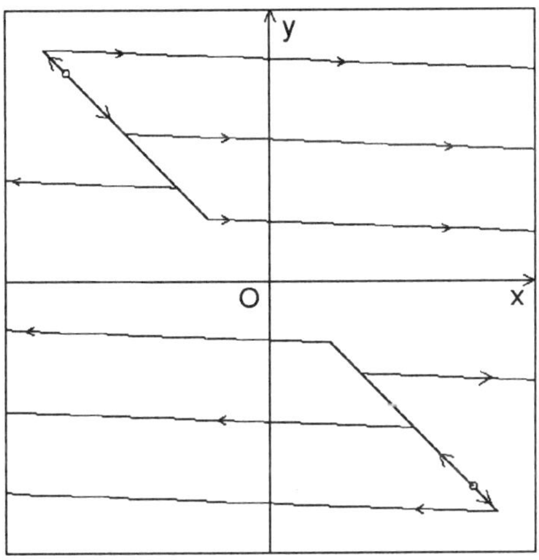

c) a=0, b=1.002

Fig. 2.7.2. Phase portraits corresponding to: (a) domain 1; (b) the point Q; and (c) domain 2 from Figure 2.7.1 and $x, y \in (-0.1, 0.1)$.

For (b, a) in domain 2 of Figure 2.7.1 the system possesses three equilibria, namely two hyperbolic repulsors and a saddle. For the parameters in Figure 2.7.2c the saddle point is the origin of the phase plane.

At the point Q, the single equilibrium point is the origin, which is a non-hyperbolic repulsor (Figure 2.7.2b). Unlike the hyperbolic case, where approach to the repulsive equilibrium is exponentially fast as $t \to -\infty$, it takes a long time to approach the non-hyperbolic equilibrium point.

The phase trajectories in Figures 2.7.2a,b,c are represented for the same number of integration steps and for the same initial conditions.

2.7.2. CASE OF NEGATIVE b

If $b < 0$, $b \neq -c$ and $(b,a) \in S_1$, then $a = -\frac{2}{3}b\sqrt{(1-\frac{1}{b})^3}$, $x_1 = x_3 = \sqrt{1-\frac{1}{b}}$ and $y_1 = y_3 = \frac{1}{b}\left(a - \sqrt{1-\frac{1}{b}}\right)$. Similarly to Section 2.7.1, we have

$$\begin{aligned} \psi_1 &= x - \sqrt{1-\tfrac{1}{b}}, \\ \psi_2 &= y - \tfrac{1}{b}\left(a - \sqrt{1-\tfrac{1}{b}}\right), \end{aligned} \tag{2.7.14}$$

$$\begin{aligned} \dot{\psi}_1 &= \tfrac{c}{b}\psi_1 + c\psi_2 - c\sqrt{1-\tfrac{1}{b}}\psi_1^2 - \tfrac{c}{3}\psi_1^3, \\ \dot{\psi}_2 &= -\tfrac{1}{c}\psi_1 - \tfrac{b}{c}\psi_2, \end{aligned} \tag{2.7.15}$$

$$\begin{aligned} \psi_1 &= bu + c^2 v, \\ \psi_2 &= -u - bv, \end{aligned} \tag{2.7.16}$$

$$\begin{aligned} \dot{u} &= g(u,v), \\ \dot{v} &= \alpha v + h(u,v), \end{aligned} \tag{2.7.17}$$

where

$$\begin{aligned} g(u,v) &= \frac{b}{c^2-b^2}\left(bu+c^2v\right)^2\left[c\sqrt{1-\frac{1}{b}} + \frac{c}{3}\left(bu+c^2v\right)\right], \\ \alpha &= \frac{c^2-b^2}{bc} \\ h(u,v) &= -\frac{(bu+c^2v)^2}{c^2-b^2}\left[c\sqrt{1-\frac{1}{b}} + \frac{c}{3}\left(bu+c^2v\right)\right]. \end{aligned} \tag{2.7.18}$$

Thus the center manifold has the following quadratic approximation

$$v = \mathcal{V}(u) = \frac{b^3c^2}{(c^2-b^2)^2}\sqrt{1-\frac{1}{b}}u^2 + O(u^3) \tag{2.7.19}$$

and the restriction of the system to the center manifold is given by

$$\begin{aligned}\dot{u} &= \frac{b^3c}{c^2-b^2}\sqrt{1-\frac{1}{b}}u^2 + \\ &\frac{1}{3}\frac{b^4c}{c^2-b^2}\left[1+\frac{6bc^4}{(c^2-b^2)^2}\left(1-\frac{1}{b}\right)\right]u^3 + O\left(u^4\right).\end{aligned} \tag{2.7.20}$$

As the coefficient of u^2 is nonzero for every $b < 0$, $b \neq -c$, the double equilibria corresponding to S_1 are all saddle–nodes, partially repulsive for $b \in (-\infty, -c)$ and partially attractive for $b \in (-c, 0)$.

CHAPTER 3

DYNAMIC BIFURCATION FOR THE FITZHUGH–NAGUMO MODEL

In this chapter the dynamic bifurcation for the F–N system (1.1.17) is studied. In the (b, a)-parameter plane the oscillatory regimes are related to the Hopf bifurcation.

In Section 3.1 the Hopf bifurcation values (b_H, a_H) situated on the curves $H_{1,2}$ are deduced.

In Section 3.2 the existence of the Bogdanov–Takens bifurcation is proved. Using the normal form method, in the parameter plane the approximate curves BT_i, $i = \overline{1,4}$ of the homoclinic bifurcation values are determined.

In Section 3.3 these last theoretical results are regained by numerical computations. The last ones are also used in the regions where the local theory cannot yield any longer the curves of homoclinic bifurcation. The curves $BTn_{1,2}$, existing for $b < 0$, emerge from the points $Q_{1,4}$ and run up to the point Q_5. The curves $BTn_{3,4}$, existing for $b > 0$, emerge at the points $Q_{3,2}$, cross the Ob-axis at Q_6 and continue up to the points $Q_{8,7}$. The points Q_5 and Q_6 belong to the Ob-axis. At the point Q_6 the bifurcation occurs owing to the transformation of one limit cycle into a double homoclinic orbit. From the double homoclinic point Q_6 two new curves $D_{1,2}$ of homoclinic bifurcation emerge. The related saddle–node homoclinic bifurcation and saddle–node separatrix loop bifurcation are shown.

In Section 3.4, using the stable and unstable manifolds of saddles, the curves K_i, $i = \overline{1,4}$ of breaking saddle connection bifurcation are obtained. They emerge from Q_5 and continue up to the points $Q_{11,12}$, $Q_{15,16}$. A local numerical investigation around these points allowed us to emphasize new types of bifurcations, named by us saddle–node–saddle connection bifurcation and saddle–node–saddle with separatrix connection bifurcation. An interesting analogy between the bifurcations described in Sections 3.3 and 3.4 is made.

In Section 3.5 the degenaracy of the Hopf bifurcation into the Bautin bifurcation is analyzed and the set of the Bautin bifurcation values is determined theoretically and numerically. For each fixed c, two Bautin bifurcation points $Q_{17,18}$ are found. The curves $Ba_{1,2}$ emerging from these points are determined. For the parameters situated on the curves $Ba_{1,2}$, the F–N model admits a non-hyperbolic limit cycle.

3.1. HOPF BIFURCATION

3.1.1. LOCUS OF HOPF BIFURCATION POINTS AND VALUES

Let (x_H, y_H) be an equilibrium point corresponding to the parameters (b_H, a_H, c_H). Taking into account Theorem 1.3.2 and the expression (2.3.3) of the eigenvalues of the F–N system linearized around an equilibrium point, a necessary condition for (x_H, y_H) to be a Hopf bifurcation point reads

$$1 - x^2 - \frac{b}{c^2} = 0, \tag{3.1.1}$$

$$\left(1 - x^2 + \frac{b}{c^2}\right)^2 - \frac{4}{c^2} < 0, \tag{3.1.2}$$

where, for the sake of simplicity, the index H has been dropped.

From (3.1.1) and (3.1.2) it follows immediately that $b \in (-c, c)$. Indeed, (3.1.1) implies $1 - x^2 = \frac{b}{c^2}$. Replacing this expression in (3.1.2) we get $b^2 < c^2$, whence $b \in (-c, c)$.

In the following we shall find all the Hopf bifurcation points and values for fixed c, with $c > 1 + \sqrt{3}$ [46]. Let us remark that it is sufficient to determine (b_H, a_H, x_H). Indeed, once determined x_H, then y_H follows from (2.1.1).

The conditions (3.1.1) and (3.1.2) show that in (b, a, x)-space it is necessary that the Hopf bifurcation points be situated on the curve H'. In Section 2.3 this curve was defined as the intersection of the cylinder B_3' with the surface Σ.

In addition, the Hopf bifurcation points must belong to the region of complex eigenvalues delimited by the cylinders B_1' and B_2'.

For the equilibria (x_i, y_i) such as $(b, a, x_i) \in H'$, $b < -c$ or $b > c$, the eigenvalues $\lambda_{1,2}$ are real and $\lambda_1 + \lambda_2 = 0$.

The Hopf bifurcation values are the projections on the (b, a)-plane of the Hopf bifurcation points (b_H, a_H, x_H) and are situated on the curves $H_{1,2}$ of equation (2.3.14) (Figure 2.3.3). The curves $H_{1,2}$ do not consist entirely (for all $b \leq c^2$) of Hopf bifurcation values. As shown in the above it is necessary that $b \in (-c, c)$. Therefore, the Hopf bifurcation values must be situated on H_1 between Q_1 and Q_2 and on H_2 between Q_3 and Q_4.

To one point of $H_{1,2}$ one, two or three equilibria of the F–N system may correspond. This is why, the points (b, a) of $H_{1,2}$ such that $b < -c$ or $b > c$ are not Hopf bifurcation values even if they are situated in the complex region delimited by the curves $C_{\overline{1,4}}$. The dynamical system corresponding to such parameters (b, a) has one equilibrium point (x_i, y_i) such that $(b, a, x_i) \in H'$ (so $\lambda_{1,2} \in \mathbf{R}$, $\lambda_1 + \lambda_2 = 0$) and another equilibrium point (x_k, y_k) such that (b, a, x_k) is situated between the cylinders B_1', B_2', but not on H' (so $\lambda_{1,2} \in \mathbf{C}$, but Re $\lambda_1 =$ Re $\lambda_2 \neq 0$).

If $b \in (-c, c)$, $b \neq b_0$, a single equilibrium point corresponding to a point of $H_{1,2}$, has complex eigenvalues, with zero real parts.

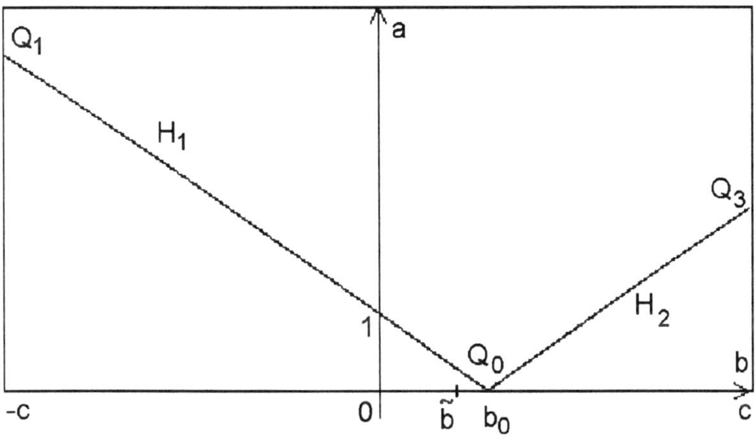

Fig. 3.1.1. Regions of the curves $H_{1,2}$ corresponding to different equilibria which are Hopf bifurcation points.

Investigation (Section 2.4) of the static bifurcation diagrams revealed that their points P_1 and P_2 are the only ones satisfaying (3.1.1) and (3.1.2).

If a is increased from 0 to $\frac{2}{3}(c+1)\sqrt{1+\frac{1}{c}}$, the projection of the point P_1 on the (b, a)-plane runs over H_1 between Q_0 and Q_1 (Figure 3.1.1).

If a is increased from 0 to $\frac{2}{3}(c-1)\sqrt{1-\frac{1}{c}}$, the projection of the point P_2 on the (b, a)-plane runs over H_2 between Q_0 and Q_3.

If $a = 0$, the common projection on the (b, a)-plane of P_1 and P_2 is the point Q_0.

Consider $\tilde{b} \in \left(1, \frac{3}{2}\right)$ the abscissa of the point of intersection of H_1 and S_1 already introduced in Section 2.6. The F–N system, with $a \geq 0$, possesses three equilibria for the points of H_1, with $b \in (-c, 0) \cup (\tilde{b}, b_0)$ and of H_2 with $b \in (b_0, c)$, one equilibrium point for H_1, with $b \in \left[0, \tilde{b}\right)$ and two equilibria for H_1, with $b = \tilde{b}$ (where $x_1 = x_2$) (Figure 3.1.1). Among them, according to Section 2.6, the Hopf bifurcation points are expected to be the equilibria (x_1, y_1) for parameters corresponding to H_1, with $b \in (-c, 0)$, the equilibria

(x_0, y_0) for H_1, $b \in \left[0, \tilde{b}\right)$, the equilibria (x_3, y_3) for H_1, with $b \in \left[\tilde{b}, b_0\right)$ and (x_2, y_2) for H_2, with $b \in (b_0, c)$. The expressions of x_i, $i = 0, 1, 2, 3$ were given in Section 2.2.

The point $Q_0\,(b_0, 0)$ could correspond simultaneously to two Hopf bifurcation points: (x_2, y_2) and (x_3, y_3).

In Section 3.5.1 we show that, indeed, all points of the curves $H_{1,2}$ with $b \in (-c, c)$ are Hopf bifurcation values. Two of them are degenerate. The proof will be based on the normal form method. In this way we avoid the Hopf theorem for the case of a two-dimensional parameter. However, even in the absence of this result, assuming that only one of the parameters a or b varies, the graphs of the eigenvalues Im $\lambda_{1,2}$ versus Re $\lambda_{1,2}$ indicates (by the Hopf Theorem 1.3.2) that the Hopf bifurcation takes place for those values of the parameters at which the graphs intersect the Im λ-axis. For instance, in Figure 3.1.2 we plotted the points (Re λ_i, Im λ_i), $i = 1, 2$ for $a = 1$, $c = 5$ and b variable. In Figure 2.3.8 we plotted Re λ_i and Im λ_i in terms of b. Obviously, the Hopf bifurcation occurs for $b = 0$.

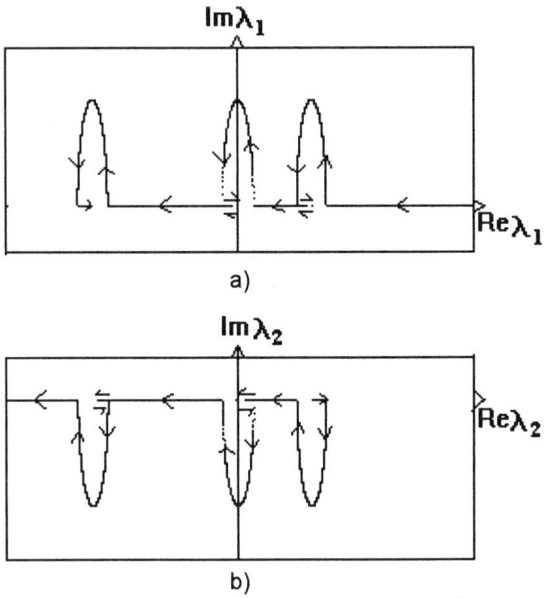

Fig. 3.1.2. The case $a = 1$, $c = 5$ and the branch of solutions $x_1 x_0 x_3$.
(a) Im λ_1 vs. Re λ_1 on the domain $(-15, 15) \times (-0.3, 1.5)$.
As $b \to \infty$, (Re λ_1,Im λ_1) $\to (-10.09, 0)$;
(b) Im λ_2 vs. Re λ_2 on the domain $(-15, 15) \times (-1.5, 0.3)$. As $b \to -\infty$, (Re λ_2, Im λ_2) $\to (5.09, 0)$. The arrows indicate the increase of b.

The analytical support of this assertion is given by Theorem 1.3.2. Indeed, for b =constant, $b \neq 0$, we have the relation

$$\frac{\partial \mathrm{Re}\lambda}{\partial a} = -\frac{c x_i}{b\left(x_i^2 - 1 + \frac{1}{b}\right)}, \quad b \neq 0 \tag{3.1.3}$$

obtained from (2.3.3). We have taken into account the fact that x_i satisfies (2.2.1) and, hence, we have

$$\left[3x_i^2 - 3\left(1 - \frac{1}{b}\right)\right]\frac{\partial x_i}{\partial a} - \frac{3}{b} = 0. \tag{3.1.4}$$

Since for the points (b, a) of $H_{1,2}$ we have

$$x_i = \pm\sqrt{1 - \frac{b}{c^2}},$$

then (3.1.3) becomes

$$\frac{\partial \mathrm{Re}\lambda}{\partial a} = \pm\frac{c^2\sqrt{c^2 - b}}{c^2 - b^2} \text{ for } b \in (-c, c)\,, \quad b \neq 0. \tag{3.1.5}$$

The relation (3.1.5) shows that for points of $H_{1,2}$ we have $\frac{\partial \mathrm{Re}\lambda}{\partial a} \neq 0$. Since, in addition, we have Re $\lambda = 0$ and Im $\lambda \neq 0$, it means that the hypotheses of the Hopf Theorem 1.3.2 are fullfiled.

For $b = 0$ the system (1.1.17) has a single equilibrium point $(x_0, y_0) = \left(a, \frac{a^3}{3} - a\right)$. The corresponding linearized system has the eigenvalues

$$\lambda_{1,2} = \frac{c\left(1 - a^2\right) \pm \sqrt{c^2\left(1 - a^2\right)^2 - 4}}{2}.$$

Therefore $\frac{\partial \mathrm{Re}\lambda}{\partial a} = -a\ c$, if $c^2\left(1 - a^2\right)^2 - 4 < 0$. The intersection of $H_{1,2}$ with the Oa-axis occurs for $a = \pm 1$, hence $\frac{\partial \mathrm{Re}\lambda}{\partial a} = \pm c \neq 0$ and Im $\lambda = \pm 1 \neq 0$. Therefore the Hopf Theorem holds for $b = 0$, too.

Similarly, for a =constant and variable $b \neq 0$, from (2.2.1) we have

$$\left[3x_i^2 - 3\left(1 - \frac{1}{b}\right)\right]\frac{\partial x_i}{\partial b} - \frac{3x_i}{b^2} + \frac{3a}{b^2} = 0, \tag{3.1.6}$$

therefore

$$\begin{aligned} \frac{\partial \mathrm{Re}\lambda}{\partial b} &= \frac{-c\ x_i\left(x_i - a\right)}{b^2\left(x_i^2 - \left(1 - \frac{1}{b}\right)\right)} - \frac{1}{2c} \\ &= \frac{-c^3 + bc \pm ac^2\sqrt{c^2 - b}}{b\left(c^2 - b^2\right)} - \frac{1}{2c} \neq 0. \end{aligned} \tag{3.1.7}$$

As a consequence, again the Hopf Theorem 1.3.2 holds.

The point Q_0, defined as the middle intersection point of $H_{1,2}$ and the Ob-axis, is a special Hopf bifurcation value. The parameters b and a corresponding to Q_0 are Hopf bifurcation values both for (x_2, y_2) and (x_3, y_3).

For example, for $a = 0$ and increasing b, as Q_0 is crossed, the equilibria (x_2, y_2) and (x_3, y_3) which were repulsors, become attractors and two repulsive limit cycles occur simultaneously around them and inside the attactive limit cycle already existing for $b < b_0$ (Figure 3.1.3).

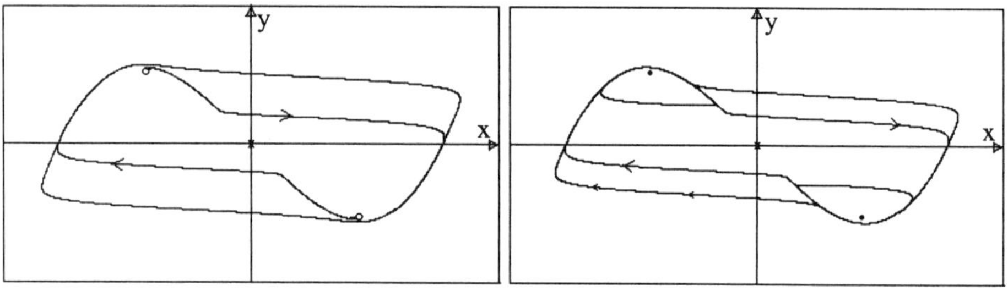

a) $a = 0, \quad b = 1.4$ b) $a = 0, \quad b = 1.47114$

Fig. 3.1.3. Hopf bifurcation at Q_0 for $a = 0$.

If b continues increasing further, these limit cycles develop until they touch the origin, which is a saddle point. Thus, a double homoclinic bifurcation occurs. It will be presented in the Section 3.3.

The Hopf bifurcation is a local phenomenon around an equilibrium point. Therefore, for the values of the parameters b and a corresponding to Q_0, we have, in fact, two such local phenomena.

Later we shall see that the points $Q_{\overline{1,4}}, Q_{17}$ and Q_{18} situated on $H_{1,2}$ are not genuine Hopf bifurcation values.

Remark that, although the Hopf bifurcation has the codimension equal to 1, the case of two variable parameters is not equivalent to the case of a single variable parameter. Indeed, as we already mentioned, on each curve H_1 and H_2 with $b \in (-c, c)$ there exists a point corresponding to a codimension-two bifurcation, namely the Bautin bifurcation (Section 3.5).

3.1.2. AN EQUIVALENT FORM OF THE F–N SYSTEM

For an easier derivation by normal form method of the degenerated Hopf bifurcations of Bogdanov–Takens and Bautin type, we shall perform some transformations of the F–N system (1.1.17). This method involves asymptotic expansions around the origin of the phase space, where the bifurcation value is the origin of the parameter space. Thus, our transformations will carry the bifurcation equilibria and parameters to the corresponding origin. Therefore, consider an equilibrium point (x^*, y^*) of (1.1.17).

The translation

$$\begin{aligned}\psi_1 &= x - x^*(b,a),\\ \psi_2 &= y - y^*(b,a).\end{aligned} \tag{3.1.8}$$

transforms (1.1.17) into the system

$$\begin{aligned}\dot{\psi}_1 &= c\left(1-(x^*(b,a))^2\right)\psi_1 + c\,\psi_2 - c\,x^*(b,a)\,\psi_1^2 - c\frac{\psi_1^3}{3},\\ \dot{\psi}_2 &= -\frac{\psi_1}{c} - b\frac{\psi_2}{c}.\end{aligned} \tag{3.1.9}$$

and carries the equilibrium point (x^*, y^*) of F–N system into the equilibrium $(0,0)$ of (3.1.9).

Let $P^* = (b^*,\, a^*)$ be a point in the (b,a) parameter plane. Using the translation

$$\begin{aligned}\alpha_1 &= b - b^*,\\ \alpha_2 &= a - a^*,\end{aligned} \tag{3.1.10}$$

the new parameters α_1 and α_2 are introduced. The point $(b^*,\, a^*)$ of the (b,a)-plane corresponds to the point $(0,0)$ in the (α_1,α_2)-plane.

In this way the abscissa $x^*(b,a)$ of an arbitrary equilibrium point becomes a function of α_1, α_2, b^* and a^*. It is denoted by $x^*(\alpha_1,\alpha_2,b^*,a^*)$ or, equivalently, by $x^*(\boldsymbol{\alpha}, P^*)$, where $\boldsymbol{\alpha} = (\alpha_1,\alpha_2)$. For an easier writing, we introduce the notation

$$\begin{aligned}E_1 &= E_1(\boldsymbol{\alpha}) = c\left(1-(x^*)^2\right),\\ E_2 &= E_2(\boldsymbol{\alpha}) = c\,x^*,\\ E_3 &= E_3(\boldsymbol{\alpha}) = -\frac{1}{c}(\alpha_1 + b^*).\end{aligned} \tag{3.1.11}$$

Then (3.1.9) becomes

$$\begin{aligned}\dot{\psi}_1 &= E_1\psi_1 + c\psi_2 - E_2\psi_1^2 - c\frac{\psi_1^3}{3},\\ \dot{\psi}_2 &= -\frac{\psi_1}{c} + E_3\psi_2.\end{aligned} \tag{3.1.12}$$

This is the system dealt with in the following.

Linearizing (3.1.12) in the neighborhood of the equilibrium point $(\psi_1,\psi_2) = (0,0)$, we obtain the system

$$\begin{aligned}\dot{X}_1 &= E_1X_1 + cX_2,\\ \dot{X}_2 &= -\frac{X_1}{c} + E_3X_2,\end{aligned} \tag{3.1.13}$$

whose eigenvalues are

$$\lambda_{1,2}(\boldsymbol{\alpha}) = \frac{E_1 + E_3 \pm \sqrt{(E_1 - E_3)^2 - 4}}{2}. \tag{3.1.14}$$

In the region of the (b, a)- plane where $(E_1 - E_3)^2 - 4 < 0$, we have

$$\lambda_{1,2}(\boldsymbol{\alpha}) = \mu(\boldsymbol{\alpha}) \pm i\,\omega(\boldsymbol{\alpha}), \tag{3.1.15}$$

where $\mu(\boldsymbol{\alpha})$, $\omega(\boldsymbol{\alpha}) \in \mathbf{R}$ and

$$\begin{aligned} \mu(\boldsymbol{\alpha}) &= \frac{c^2\left(1 - (x^*)^2\right) - b^* - \alpha_1}{2c}, \\ \omega(\boldsymbol{\alpha}) &= \frac{1}{2c}\sqrt{4c^2 - [c^2(1 - (x^*)^2) + \alpha_1 + b^*]^2}. \end{aligned} \tag{3.1.16}$$

3.2. BOGDANOV–TAKENS BIFURCATION

3.2.1. BOGDANOV–TAKENS BIFURCATION AT Q_1

In this section we obtain the following result.

Theorem 3.2.1 [124] *For a fixed c, the system (1.1.17) has for Q_1 a codimension-two Bogdanov–Takens bifurcation. From Q_1, a curve of homoclinic bifurcation values emerges. In the neighborhood of Q_1 it is approximated by the curve BT_1 of equation*

$$a = \frac{7b^2 - 10bc^2 + 3c^2}{15c^3}\sqrt{\frac{7b^2 + 5c^2b - 12c^2}{5b}}, \quad b \in (-c, 0). \tag{3.2.1}$$

To the point Q_1 an equilibrium point with both eigenvalues equal to zero and a saddle equilibrium correspond. In order to obtain the normal form of the F–N system, corresponding to the Bogdanov–Takens bifurcation at Q_1, let us remark that if, in particular, the point P^* (from Section 3.1.2) coincides with Q_1, we have

$$\begin{aligned} b^* &= -c, \\ a^* &= \tfrac{2}{3}(c+1)\sqrt{1 + \tfrac{1}{c}}. \end{aligned} \tag{3.2.2}$$

In the notation of Section 3.1.2 and according to Section 2.6, the equilibrium point (x^*, y^*), with both eigenvalues equal to zero, corresponding to Q_1, has the abscissa

$$x^*(\mathbf{0}, Q_1) = \sqrt{1 + \frac{1}{c}}, \tag{3.2.3}$$

whilst the abscissae of the equilibrium points which are close to Q_1 are $x^*(\boldsymbol{\alpha}, Q_1)$, where $\boldsymbol{\alpha}$ is small.

For such points $P\left(\alpha_1 - c, \alpha_2 + \frac{2(c+1)}{3}\sqrt{1+\frac{1}{c}}\right)$ the system (1.1.17) is written in the form (3.1.12) where

$$\begin{aligned} E_1 &= c\left(1 - (x^*)^2(\boldsymbol{\alpha}, Q_1)\right), \\ E_2 &= c\, x^*(\boldsymbol{\alpha}, Q_1), \\ E_3 &= 1 - \frac{\alpha_1}{c}. \end{aligned} \tag{3.2.4}$$

For $\boldsymbol{\alpha} = 0$, the matrix associated with (3.1.13) has the form

$$\mathbf{A}_0 = \begin{pmatrix} -1 & c \\ -\frac{1}{c} & 1 \end{pmatrix}. \tag{3.2.5}$$

As $\lambda_1(\mathbf{0}) = \lambda_2(\mathbf{0}) = 0$, then $\lambda_1(\mathbf{0})$ is a double eigenvalue. Its eigenspace does not generate $\mathbf{R}^2$. However, we can construct a base of $\mathbf{R}^2$ with the eigenvector $\mathbf{v}_0$ and the generalized eigenvector $\mathbf{v}_1$ corresponding to the eigenvalue $\lambda_1(\mathbf{0})$ for the matrix $\mathbf{A}_0$. These vectors are

$$\begin{aligned} \mathbf{v}_0 &= \left(\frac{c}{r_1}, \frac{1}{r_1}\right)^{\mathrm{T}}, \\ \mathbf{v}_1 &= \left(r_2, \frac{1}{r_1} + \frac{r_2}{c}\right)^{\mathrm{T}}, \end{aligned} \tag{3.2.6}$$

where r_1 and r_2 are arbitrary real numbers. So we have $\mathbf{A}_0\mathbf{v}_0 = \mathbf{0}$, $\mathbf{A}_0\mathbf{v}_1 = \mathbf{v}_0$.

Let $\mathbf{w}_1$ and $\mathbf{w}_0$ be the eigenvector and the generalized eigenvector for $\mathbf{A}_0^{\mathrm{T}}$, corresponding to the eigenvalue $\lambda_1(\mathbf{0})$. They are

$$\begin{aligned} \mathbf{w}_1 &= (s_1, -c\, s_1)^{\mathrm{T}}, \\ \mathbf{w}_0 &= (s_2, -c(s_1 + s_2))^{\mathrm{T}}, \end{aligned} \tag{3.2.7}$$

where s_1 and s_2 are arbitrary real numbers and $\mathbf{A}_0^{\mathrm{T}}\mathbf{w}_1 = \mathbf{0}$, $\mathbf{A}_0^{\mathrm{T}}\mathbf{w}_0 = \mathbf{w}_1$.

From the normalisation condition $\langle \mathbf{v}_0, \mathbf{w}_0 \rangle = 1$, it follows $s_1 = -r_1/c$ and $\langle \mathbf{v}_1, \mathbf{w}_1 \rangle = 1$ is identically satisfied.

From the orthogonality condition $\langle \mathbf{v}_1, \mathbf{w}_0 \rangle = 0$, it follows that $s_2 = (c + r_1 r_2) r_1 / c^2$ and $\langle \mathbf{v}_0, \mathbf{w}_1 \rangle = 0$ is identically satisfied.

Here $\langle , \rangle$ is the usual scalar product in $\mathbf{R}^2$.

In the following some transformations (diffeomorphisms) are carried out, in order to put (3.1.12) into its normal form.

The first transformation uses the eigenvectors. In order to simplify the computations, we shall choose $r_1 = 1$, $r_2 = 0$, such that $s_1 = -\frac{1}{c}$, $s_2 = \frac{1}{c}$. It follows

$$\begin{aligned} \mathbf{v}_0 &= (c,1)^{\mathrm{T}}, \qquad \mathbf{v}_1 = (0,1)^{\mathrm{T}}, \\ \mathbf{w}_0 &= \left(\frac{1}{c},0\right)^{\mathrm{T}}, \quad \mathbf{w}_1 = \left(-\frac{1}{c},1\right)^{\mathrm{T}}. \end{aligned} \tag{3.2.8}$$

Define the first linear transformation by $\gamma_1 = \langle \boldsymbol{\psi}, \mathbf{w}_0 \rangle$, $\gamma_2 = \langle \boldsymbol{\psi}, \mathbf{w}_1 \rangle$, i.e.,

$$\gamma_1 = \frac{\psi_1}{c}, \quad \gamma_2 = -\frac{\psi_1}{c} + \psi_2. \tag{3.2.9}$$

The inverse transformation reads $\psi = \gamma_1 \mathbf{v}_0 + \gamma_2 \mathbf{v}_1$, i.e.,

$$\psi_1 = c\gamma_1, \quad \psi_2 = \gamma_1 + \gamma_2. \tag{3.2.10}$$

In this way, up the terms of $O\left(\|\gamma\|^3\right)$, (3.1.12)becomes

$$\begin{aligned} \dot{\gamma}_1 &= (1+E_1)\,\gamma_1 + \gamma_2 - cE_2\gamma_1^2, \\ \dot{\gamma}_2 &= (E_3 - E_1 - 2)\,\gamma_1 + (E_3 - 1)\,\gamma_2 + cE_2\gamma_1^2 \end{aligned} \tag{3.2.11}$$

and has the matrix $\begin{pmatrix} 0 & 1 \\ 0 & 0 \end{pmatrix}$ associated with the linearized system around $(\gamma_1, \gamma_2) = (0,0)$. This matrix characterizes the Bogdanov–Takens bifurcations.

Using the nonlinear transformation

$$\begin{aligned} u_1 &= \gamma_1, \\ u_2 &= (1+E_1)\,\gamma_1 + \gamma_2 - cE_2\gamma_1^2 \end{aligned} \tag{3.2.12}$$

and its inverse

$$\begin{aligned} \gamma_1 &= u_1, \\ \gamma_2 &= (-1-E_1)\,u_1 + u_2 + cE_2u_1^2, \end{aligned} \tag{3.2.13}$$

and neglecting the terms of order greater than two (as for the next transformations and equations to follow), the system (3.2.11) becomes

$$\begin{aligned} \dot{u}_1 &= u_2, \\ \dot{u}_2 &= (-E_1E_3 - 1)\,u_1 + (E_1 + E_3)\,u_2 + cE_2E_3u_1^2 - 2cE_2u_1u_2. \end{aligned} \tag{3.2.14}$$

Finally, with the transformation

$$\begin{aligned} u_1 &= v_1 + \delta, \\ u_2 &= v_2, \end{aligned} \tag{3.2.15}$$

and, correspondingly,

$$\begin{aligned} v_1 &= u_1 - \delta, \\ v_2 &= u_2, \end{aligned} \tag{3.2.16}$$

the system (3.2.14) becomes

$$\begin{aligned} \dot{v}_1 &= v_2, \\ \dot{v}_2 &= (-E_1E_3 - 1)\,\delta + cE_2E_3\delta^2 \\ &\quad + (-E_1E_3 - 1 + 2cE_2E_3\delta)\,v_1 \\ &\quad + (E_1 + E_3 - 2cE_2\delta)\,v_2 + cE_2E_3v_1^2 - 2cE_2v_1v_2. \end{aligned} \tag{3.2.17}$$

The constant δ is chosen such that the coefficient of v_2 in $(3.2.17)_2$ be equal to 0, i.e., $\delta = (E_1 + E_3)/(2cE_2)$. Then (3.2.17) reads

$$\begin{aligned} \dot{v}_1 &= v_2, \\ \dot{v}_2 &= h_{00} + h_{10}v_1 + \frac{h_{20}}{2}v_1^2 + h_{11}v_1v_2, \end{aligned} \tag{3.2.18}$$

where h_{00}, h_{10}, h_{11} and h_{20} are the following functions of $\boldsymbol{\alpha}$

$$\begin{aligned} h_{00} &= \frac{E_3\left(E_3^2 - E_1^2\right) - 2\left(E_1 + E_3\right)}{4cE_2}, \\ h_{10} &= E_3^2 - 1, \\ h_{20} &= 2cE_2E_3, \\ h_{11} &= -2cE_2. \end{aligned} \tag{3.2.19}$$

For a new time scale $t = \left|\frac{B(\boldsymbol{\alpha})}{A(\boldsymbol{\alpha})}\right|\tau$, where

$$\begin{aligned} A(\boldsymbol{\alpha}) &= \frac{h_{20}(\boldsymbol{\alpha})}{2}, \\ B(\boldsymbol{\alpha}) &= h_{11}(\boldsymbol{\alpha}), \end{aligned} \tag{3.2.20}$$

and with the aid of another transformation

$$\begin{aligned} \eta_1 &= \frac{A(\boldsymbol{\alpha})}{B^2(\boldsymbol{\alpha})}v_1, \\ \eta_2 &= \operatorname{sign}\left(\frac{B(\boldsymbol{\alpha})}{A(\boldsymbol{\alpha})}\right)\frac{A^2(\boldsymbol{\alpha})}{B^3(\boldsymbol{\alpha})}v_2, \end{aligned} \tag{3.2.21}$$

the system (3.2.18) has the normal form

$$\begin{aligned} \dot{\eta}_1 &= \eta_2, \\ \dot{\eta}_2 &= \beta_1 + \beta_2\eta_1 + \eta_1^2 + s\eta_1\eta_2, \end{aligned} \tag{3.2.22}$$

where

$$\begin{aligned} s &= \operatorname{sign}\left(\frac{B(\mathbf{0})}{A(\mathbf{0})}\right), \\ \beta_1 &\equiv \beta_1(\boldsymbol{\alpha}) = \frac{B^4(\boldsymbol{\alpha})}{A^3(\boldsymbol{\alpha})} h_{00}(\boldsymbol{\alpha}), \\ \beta_2 &\equiv \beta_2(\boldsymbol{\alpha}) = \frac{B^2(\boldsymbol{\alpha})}{A^2(\boldsymbol{\alpha})} h_{10}(\boldsymbol{\alpha}). \end{aligned} \tag{3.2.23}$$

The transformation (3.2.21) and the equations (3.2.22) make a sense if the following restrictions are fullfield

$$A(\boldsymbol{\alpha}) \neq 0, \quad A(\mathbf{0}) \neq 0, \quad B(\boldsymbol{\alpha}) \neq 0.$$

In our case all these relations hold. Indeed, $A(\alpha) \neq 0$ for $\alpha_1 \neq c$ and $a \neq 0$, whilst $A(\mathbf{0}) = c^2\sqrt{1+\frac{1}{c}} > 0$, $B(\alpha) = -2cE_2 \neq 0$ for $a \neq 0$ and $B(\mathbf{0}) = -2c\sqrt{1+\frac{1}{c}} < 0$.

It also follows that $s = -1$.

According to Section 1.3.3, in the new (β_1, β_2) parameter plane, a curve of homoclinic bifurcation emerges from the origin and, up to higher order terms, it has the equation (1.3.17). In our case it reads

$$25A(\boldsymbol{\alpha}) h_{00}(\boldsymbol{\alpha}) + 6h_{10}^2(\boldsymbol{\alpha}) = 0, \quad \text{with } h_{10} < 0. \tag{3.2.24}$$

The condition $h_{10} < 0$ reads $-1 < E_3 < 1$, which means $-1 < 1 - \frac{\alpha_1}{c} < 1$. From (3.1.10) and taking into account that $b^* = -c$ at Q_1, the condition $h_{10} < 0$ becomes

$$-c < b < c. \tag{3.2.25}$$

Replacing into (3.2.24) the expressions for $A(\boldsymbol{\alpha})$ from (3.2.20) and $h_{00}(\boldsymbol{\alpha}), h_{10}(\boldsymbol{\alpha})$ from (3.2.19) and taking into account (3.2.4), it follows

$$(x^*)^2 = 1 - \frac{1}{b} \pm \frac{7}{5}\left(-\frac{b}{c^2} + \frac{1}{b}\right). \tag{3.2.26}$$

As x^* is an equilibrium point, from (2.2.1), we obtain

$$a = \frac{b}{3}x^*\left((x^*)^2 - 3 + \frac{3}{b}\right). \tag{3.2.27}$$

Taking into account that in the neighborhood of Q_1 we have $x^* > 0$, $b < 0$, $a > 0$, it follows that in (3.2.26) we must choose the minus sign. In addition we have

$$b \in (b_1, b_2), \text{ where } b_{1,2} = \frac{-5c^2 \pm c\sqrt{25c^2 + 336}}{14}. \tag{3.2.28}$$

Then, from (3.2.26) and (3.2.27), the equation (3.2.1) of the curve BT_1 is obtained. As (3.2.28) was deduced for $b < 0$, the intersection between the intervals (3.2.25) and (3.2.28) shows that BT_1 is defined for $b \in (-c, 0)$. This proves Theorem 3.2.1.

The curve BT_1 defined by (3.2.1) for $b \in (-c, 0)$ and $c = 5$ is given in Figure 3.2.1.

However, the interval $(-c, 0)$ is related to the normal form method and not to the domain of validity of the asymptotic expansions involved into (3.2.24).

For instance, to neglect the higher order terms in the transformations and in the equations is not justified far from Q_1. This is why the theoretical curve BT_1 approximates the exact curve of homoclinic bifurcation points which emerge from Q_1 only in the neihgborhood of Q_1. This fact will be put into evidence in Section 3.3.1 by comparison with numerical computations.

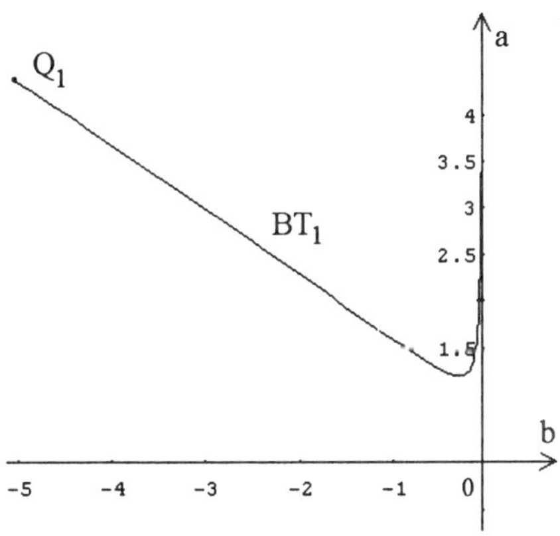

Fig. 3.2.1. The curve BT_1 for $c = 5$.

Bifurcation diagram near Q_1.

In the neighborhood of Q_1, the points (b, a) to whom dynamical systems possessing one limit cycle correspond are situated between the curve H_1 of Hopf bifurcation values and the curve BT_1.

The bifurcation diagram in the neighborhood of Q_1 is given in Figure 3.2.2. Domain 1 is above S_1, domain 2 is between S_1 and H_1, domain 3 is between H_1 and BT_1 and domain 4 is below BT_1, for $b > -c$ and below S_1, for $b \leq -c$.

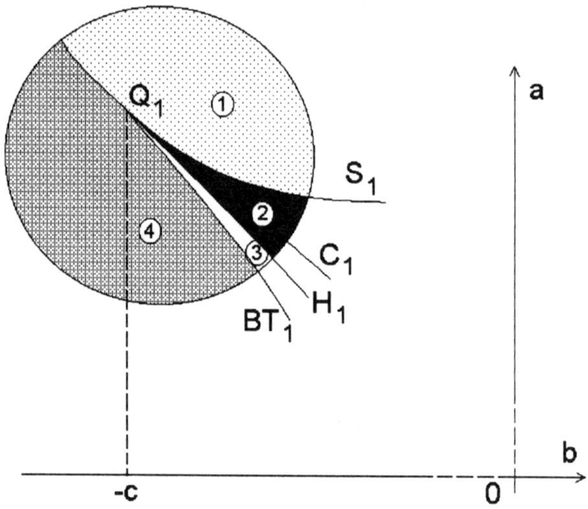

Fig. 3.2.2. The bifurcation diagram in the neighborhood of Q_1.

For points (b, a) from domain 1 in Figure 3.2.2 there exists a unique equilibrium point, and it is a saddle.

For points $(b, a) \in S_1$, in addition a saddle–node equilibrium exists.

If (b, a) is situated in domain 2, the saddle–node equilibium splits into two equilibria: the saddle (x_3, y_3) and the attractive node (x_1, y_1). They coexist with the saddle (x_2, y_2). Recall that $x_2 < x_1 < x_3$.

Consider also the curve C_1 separating the domains where (x_1, y_1)has real or complex-conjugate eigenvalues. On this curve the eigenvalues of (x_1, y_1) are real and equal (Section 2.3.3 and Figure 2.3.5).

If (b, a) crosses the curve C_1 coming from above C_1, the node becomes an attractive focus. For (b, a) on H_1, the attractive focus becomes a weakly repulsive focus. An attractive limit cycle develops in domain 3 for $a < a_{H_1}$, until (b, a) reaches the BT_1 curve, when the saddle (x_3, y_3) and the limit cycle collide. As a result a homoclinic orbit appears.

If (b, a) is situated in domain 4, the limit cycle no longer exists, so no oscillatory regimes are present.

The phase portait of (1.1.17) for $(b, a) = (-4, 3.75)$ and $c = 5$, that is a point from domain 1 of Figure 3.2.2, is represented in Figure 3.2.3a. Region A, from Figure 3.2.3a, is represented at a large scale in Figure 3.2.3b. Keeping $b = -4$, let us follow the phase portaits for points $(-4, a)$ from domains 2, 3 and 4 of Figure 3.2.2. The corresponding phase portraits will change, but their aspects at the scale of Figure 3.2.3a seems unchanged, so we shall not represent them any longer.

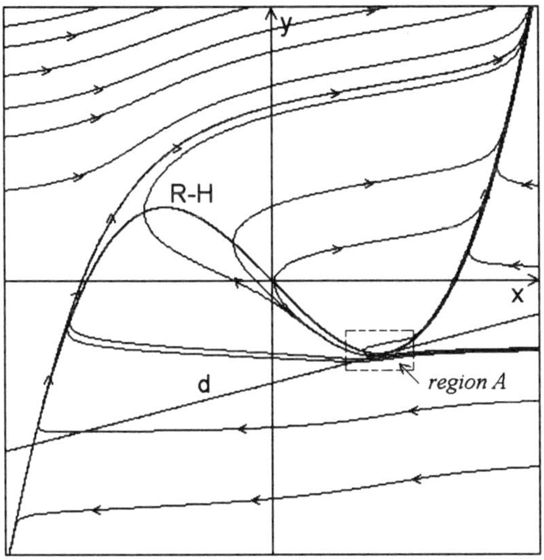

Fig. 3.2.3a. The phase portrait corresponding to the point $(b, a) = (-4, 3.75)$ situated in domain 1 from Figure 3.2.2, for $x \in (-2.5, 2.5)$, $y \in (-2.5, 2.5)$ and for $c = 5$.

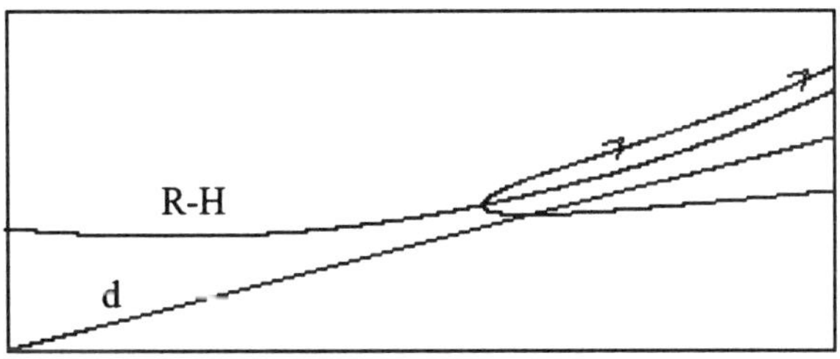

Fig. 3.2.3b. The two nullclines and the trajectory through $(x^0, y^0) = (1.1, -0.66)$ for $x \in (0.95, 1.20)$ and $y \in (-0.7, -0.6)$ corresponding to the point $(b, a) = (-4, 3.75)$ situated in domain 1 from Figure 3.2.2.

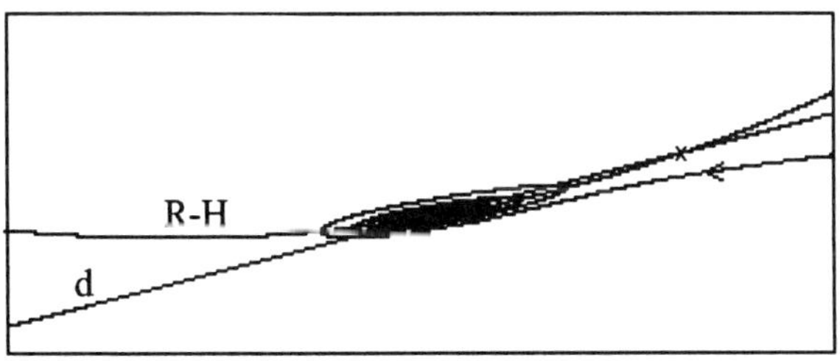

Fig. 3.2.4. The two nullclines, the attractive focus (x_1, y_1), the saddle equilibrium point (x_3, y_3), and the trajectory through $(x^0, y^0) = (1.1, -0.66)$ for $x \in (0.95, 1.20)$ and $y \in (-0.7, -0.6)$, corresponding to the point $(b, a) = (-4, 3.72)$ situated in domain 2 from Figure 3.2.2.

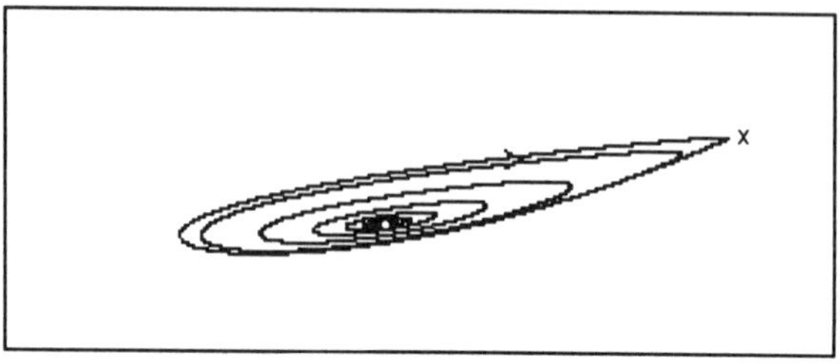

Fig. 3.2.5. The repulsive focus (x_1, y_1), the saddle point (x_3, y_3), the attractive limit cycle, and the trajectory through $(x^0, y^0) = (1.1, -0.66)$ for $x \in (0.95, 1.20)$ and $y \in (-0.7, -0.6)$, corresponding to the point $(b, a) = (-4, 3.7138)$ situated in domain 3 from Figure 3.2.2.

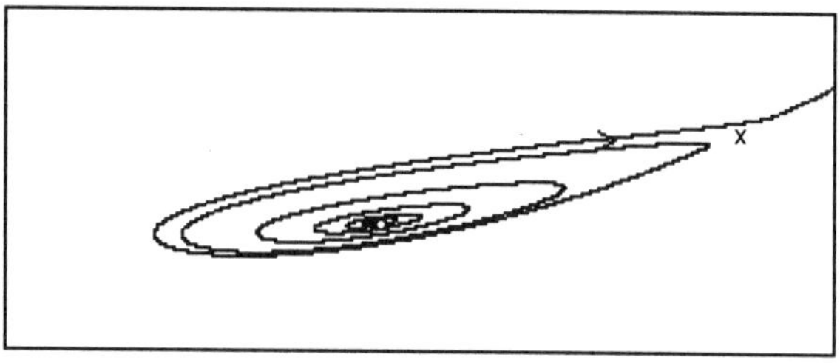

Fig. 3.2.6. The repulsive focus (x_1, y_1), the saddle equilibrium point (x_3, y_3), and the trajectory through $(x^0, y^0) = (1.1, -0.66)$ for $x \in (0.95, 1.20)$ and $y \in (-0.7, -0.6)$, corresponding to the point $(b, a) = (-4, 3.7129)$ situated in domain 4 from Figure 3.2.2.

Yet, choosing a region A as it was done for the point $(-4, 3.75)$ from domain 1, we see that the dynamics at a large scale is different and it is represented in Figures 3.2.4, 3.2.5 and 3.2.6, for the parameters (b, a) situated in domains 2, 3 and 4 respectively from Figure 3.2.2.

By symmetry reasons, in the neighborhood of Q_4, the curve of homoclinic bifurcation is symmetric to BT_1 with respect to the Ob-axis. In the following it will be denoted by BT_2.

3.2.2. BOGDANOV–TAKENS BIFURCATION AT Q_3

A result simillar to that from Section 3.2.1 holds for Q_3.

Theorem 3.2.2 [124] *The system* (1.1.17) *has for Q_3 a codimension-two bifurcation of Bogdanov–Takens type. In the parameter plane, from Q_3, a curve of homoclinic bifurcation values emerges. In the neighborhood of Q_3, it is approximated by the curve BT_3 of equation*

$$a = \frac{7b^2 + 10b\,c^2 - 17c^2}{15c^3}\sqrt{\frac{-7b^2 + 5b\,c^2 + 2c^2}{5b}}, \quad b \in (0, c). \qquad (3.2.29)$$

The investigation of the dynamics in the neighborhood of $Q_3 = \left(c, \frac{2(c-1)}{3}\sqrt{1-\frac{1}{c}}\right)$ is similar to that for Q_1. So, the matrix $\mathbf{A}_0 = \begin{pmatrix} 1 & c \\ -\frac{1}{c} & -1 \end{pmatrix}$ of the linearized system around the double equilibrium point, which has a zero eigenvalue, has the eigenvector $\mathbf{v}_0 = (-c\,r_1, r_1)^{\mathrm{T}}$ and the generalized eigenvector $\mathbf{v}_1 = (-c\,r_3 - c\,r_1, r_3)^{\mathrm{T}}$. The matrix $\mathbf{A}_0^{\mathrm{T}}$ has the eigenvector $\mathbf{w}_1 = \left(\frac{r_2}{c}, r_2\right)^{\mathrm{T}}$ and the generalized eigenvector $\mathbf{w}_0 = \left(\frac{r_4 + r_2}{c}, r_4\right)^{\mathrm{T}}$. These vectors are among the simplest possible for $r_1 = -1$, $r_2 = 1$, $r_3 = 1$, $r_4 = 0$, when they become

$$\begin{aligned} \mathbf{v}_0 &= (c, -1)^{\mathrm{T}}, \quad \mathbf{v}_1 = (0, 1)^{\mathrm{T}}, \\ \mathbf{w}_0 &= \left(\frac{1}{c}, 0\right)^{\mathrm{T}}, \quad \mathbf{w}_1 = \left(\frac{1}{c}, 1\right)^{\mathrm{T}}. \end{aligned}$$

From (3.1.11) we have

$$\begin{aligned} E_1 &= c\left(1 - \left(x^*(\boldsymbol{\alpha}, Q_3)\right)^2\right), \\ E_2 &= c\,x^*(\boldsymbol{\alpha}, Q_3), \\ E_3 &= -1 - \frac{\alpha_1}{c}, \end{aligned} \qquad (3.2.30)$$

and the corresponding equations in $\dot{\psi}_1$and $\dot{\psi}_2$ have the same form (3.1.12). We have $x^*(\mathbf{0}, Q_3) = -\sqrt{1-\frac{1}{c}}$, where x^* is the abscissa of the double equilibrium point corresponding to Q_3.

Similarly to Section 3.2.1, we have succesively (eventually up to third order terms)

$$\begin{aligned} \gamma_1 &= \frac{\psi_1}{c}, \\ \gamma_2 &= \frac{\psi_1}{c} + \psi_2, \end{aligned} \qquad (3.2.31)$$

$$\begin{aligned} \dot{\gamma}_1 &= (E_1 - 1)\,\gamma_1 + \gamma_2 - cE_2\gamma_1^2, \\ \dot{\gamma}_2 &= (E_1 - E_3 - 2)\,\gamma_1 + (E_3 + 1)\,\gamma_2 - cE_2\gamma_1^2, \end{aligned} \qquad (3.2.32)$$

$$u_1 = \gamma_1,$$
$$u_2 = (E_1 - 1)\,\gamma_1 + \gamma_2 - cE_2\gamma_1^2, \tag{3.2.33}$$

$$\gamma_1 = u_1,$$
$$\gamma_2 = (1 - E_1)\,u_1 + u_2 + cE_2u_1^2, \tag{3.2.34}$$

$$\dot{u}_1 = u_2,$$
$$\dot{u}_2 = (-E_1E_3 - 1)\,u_1 + (E_1 + E_3)\,u_2 + cE_2E_3u_1^2 - 2cE_2u_1u_2. \tag{3.2.35}$$

Since the equations (3.2.35)are the same as (3.2.14), formally we deduce that equations (3.2.17)–(3.2.23) take place also for Q_3. In addition, h_{00}, h_{10}, h_{20}, h_{11} and, thus A and B, have the same expressions in terms of E_1, E_2, E_3, as in Section 3.2.1. Of course, here E_1, E_2, E_3 correspond to Q_3 and it follows immediately that $s = +1$.

Thus, the normal form of equation (3.1.12) at Q_3 is (3.2.22) with $s = +1$, whilst for Q_1 we had $s = -1$.

The case from this section can be reduced to that of Section 3.2.1 if $\eta_2 \to -\eta_2$ and $t \to -t$. This means that in these two cases the properties of attraction are converse to each other.

Taking into account that in the neighborhood of Q_3 we have $x^* < 0$, $b > 0$, $a > 0$, it follows that in (3.2.26) the solution with the plus sign must be chosen and we have

$$b \in \left(\frac{5c^2 - c\sqrt{25c^2 + 56}}{14}, \frac{5c^2 + c\sqrt{25c^2 + 56}}{14} \right). \tag{3.2.36}$$

From (3.2.26) and (3.2.27) the equation (3.2.29) of the BT_3 curve is obtained.

From the conditions $h_{10} < 0$, (3.2.36) and taking into account that $b > 0$ and $b^* = c$ at Q_3, it follows $b \in (0, c)$. This proves Theorem 3.2.2.

The graph of BT_3 is given in Figure 3.2.7 for $c = 5$.

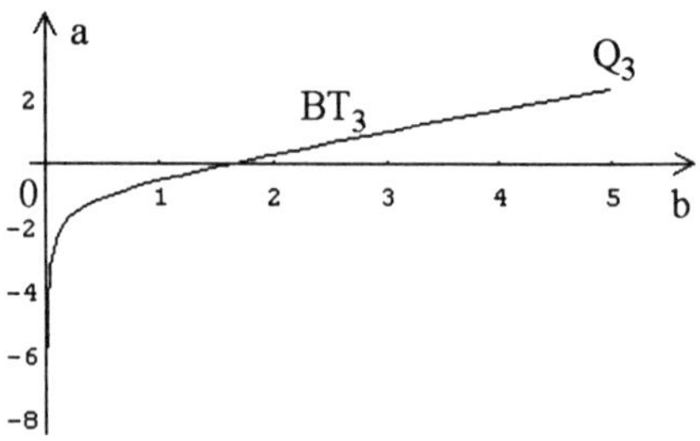

Fig. 3.2.7. The BT_3 curve for $c = 5$.

For the same reasons concerning the validity of the asymptotic expansions involved in Section 3.2.1, the approximation of the curve of homoclinic bifurcation values by the curve BT_3 is good only in the neighborhood of Q_3.

The curve symmetric to BT_3, with respect to the Ob-axis, will be denoted by BT_4. It approximates the curve of homoclinic bifurcation values in the neighborhood of Q_2.

3.3. HOMOCLINIC BIFURCATION

3.3.1. CURVES OF HOMOCLINIC BIFURCATION VALUES

The curves BT_i, $i = \overline{1,4}$ obtained theoretically represent good approximations of the exact curves of homoclinic bifurcation only in the nighborhood of the points Q_i, $i = \overline{1,4}$. Indeed, Tables 3.3.1–3.3.3 show a good agreement between the curves BT_i, $i = \overline{1,4}$ obtained in Sections 3.2.1 and 3.2.2 and those denoted by BTn_i, $i = 1,2,3,4$, obtained by numerical computations, using methods from [56], [57], [58].

b	a_{Hopf1}	a_{BT1}	a_{BTn1}
−5.	4.3817	4.3817	4.3817
−4.5	4.0518	4.0501	4.0501
−4.	3.7194	3.7130	3.71
−3.5	3.3846	3.3711	3.37
−3.	3.0479	3.0251	3.03
−2.5	2.7094	2.6756	2.68
−2.	2.3694	2.3235	2.34
−1.5	2.0282	1.9700	2.00 and 0.40
−1.	1.6860	1.6189	1.67 and 0.97
−0.9	1.6175	1.5500	1.60 and 1.09
−0.8	1.5490	1.4821	1.53 and 1.20
−0.7	1.4804	1.4158	1.46 and 1.32
−0.6	1.4118	1.3521	− − −

Table 3.3.1. Values of the parameters (b, a) of the second quadrant of the (b, a)-plane, situated on the curve of Hopf bifurcation values H_1 and on the curves of homoclinic bifurcation values obtained theoretically (BT_1) and numerically (BT_{n1}), for $c = 5$.

b	a_{Hopf2}	a_{BT3}	a_{BTn3}
5.	2.3851	2.3851	2.3851
4.5	2.0555	2.0534	2.0534
4.	1.7230	1.7146	1.716
3.5	1.3879	1.3690	1.374
3.	1.0506	1.0165	1.03
2.5	0.7115	0.6564	0.69
2.	0..3708	0.2864	0.35
1.5	0.0290	−0.0993	0.01

Table 3.3.2. Values of the parameters (b, a) of the first quadrant of the (b, a)-plane, situated on the curve of Hopf bifurcation values H_2 and on the curves of homoclinic bifurcation values obtained theoretically (BT_3) and numerically (BT_{n3}), for $c = 5$.

b	a_{Hopf1}	a_{BT4}	a_{BTn4}
0.002	1.4545	1.6234	1.468183
0.05	1.3844	1.5622	1.3971435
0.1	1.7230	1.4991	1.323152

Table 3.3.3. Values of the parameters (b, a) of the first quadrant of the (b, a)-plane, situated on the curve of Hopf bifurcation values H_1 and on the curves of homoclinic bifurcation values obtained theoretically (BT_4) and numerically (BT_{n4}), for $c = 5$.

In Figure 3.3.1, for $b \in [-c, c]$ and $a \geq 0$, the curves of static bifurcation S_1, Hopf bifurcation $H_{1,2}$, homoclinic bifurcation BT_1, BT_3, BT_4 and the curves of homoclinic bifurcation obtained numerically, BTn_1, BTn_3, BTn_4 are represented.

Figure 3.3.1 is qualitative, owing to the very small gap between H_1 and BT_1, H_2 and BT_3, H_1 and BT_4 and owing to the small distance between the curves BT_i, $i = \overline{1,4}$ and the homoclinic bifurcation curves BTn_i, $i = \overline{1,4}$ obtained numerically.

Two exceptional points are Q_5 and Q_6, situated at the intersection of the curves of homoclinic bifurcation (obtained numerically) and the Ob-axis. The dynamics in the neighborhood of Q_5 will be treated in Section 3.4, and that around Q_6 is given in the following.

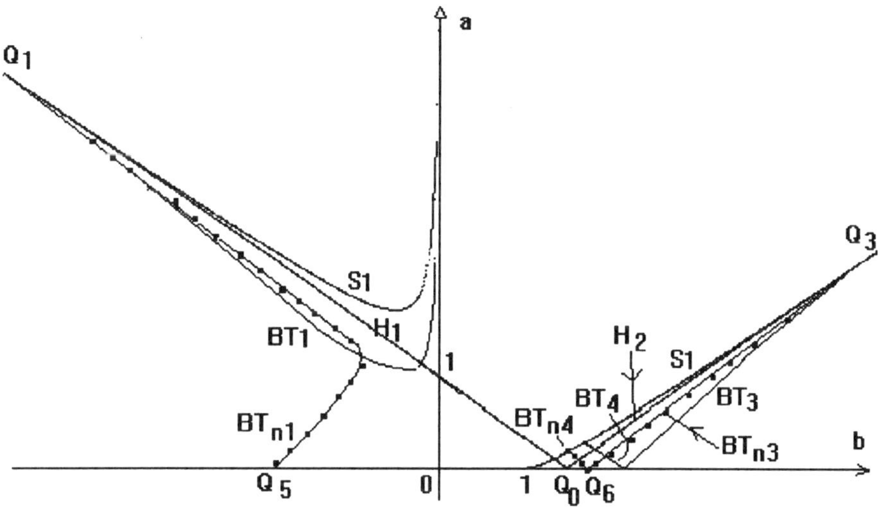

Fig. 3.3.1. The curves S_1, $H_{1,2}$, BT_1,BT_3, BT_4 and the homoclinic bifurcation curves obtained numerically (dashed).

3.3.2. DOUBLE HOMOCLINIC BIFURCATION

To the exceptional point $Q_6 = BTn_3 \cap BTn_4 \cap Ob$, a double homoclinic bifurcation corresponds. Although the unfolding of a codimension-two vector field presenting a double homoclinic bifurcation does not appear in the literature, heuristic arguments indicates that its bifurcation diagram is that from Figure 3.3.2 [64] where the equilibria represented by $\bullet$ are attractors, those represented by $\times$ are saddles, the limit cycles represented by a continous line are attractive and those represented dashed are repulsive.

Thus, except for the two curves of homoclinic bifurcation values passing through Q_6, two extra curves of homoclinic bifurcation values exist, corresponding to some 'big' homoclinic orbits. In our case these curves are denoted by D_1 (in the first quadrant) and D_2 (in the fourth quadrant).

The numerical detection of domain 3 from Figure 3.3.2 was made using 14 decimals for the parameters. Specific phase portraits for (b, a) in domains 1, 2, 3 and 4 are given in Figure 3.3.3.

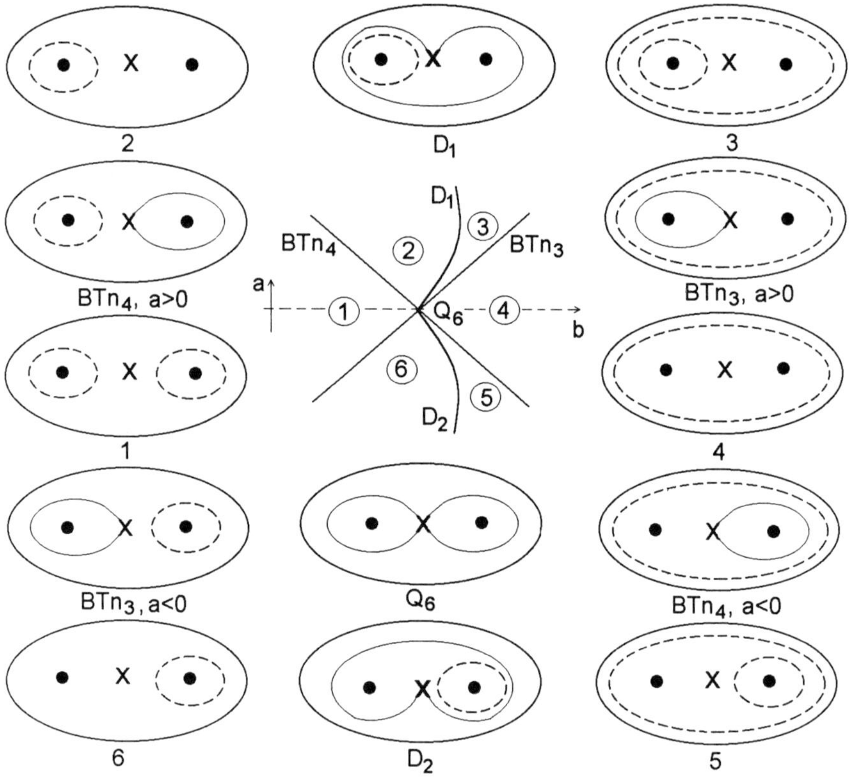

Fig. 3.3.2. Bifurcation diagram in the neighborhood of Q_6.

Thus, for (b, a) in domain 1 of Figure 3.3.2, the trajectories through $(1, 0.2)$ and $(-1, -0.2)$ are represented in Figure 3.3.3a. They have as α-limit set the repulsive limit cycles and as ω-limit set the attractive limit cycle.

For $(b, a) = Q_6$, the repulsive limit cycles are colliding simultaneously at the origin, that is a saddle point, thus forming two homoclinic orbits in a figure eight.

For (b, a) situated in domain 4 of Figure 3.3.2 a repulsive limit cycle surrounds all the three equilibrium points and is situated inside the attractive limit cycle. In Figure 3.3.3b the trajectory through $(1, 0.2)$ has as α-limit set the repulsive limit cycle and as ω-limit set the attractive limit cycle.

In Figure 3.3.3c a partial phase portrait for (b, a) in domain 2 of Figure 3.3.2 is represented. The trajectory through $(-1, -0.1)$ has as α-limit set the repulsive limit cycle and as ω-limit set the attractive limit cycle.

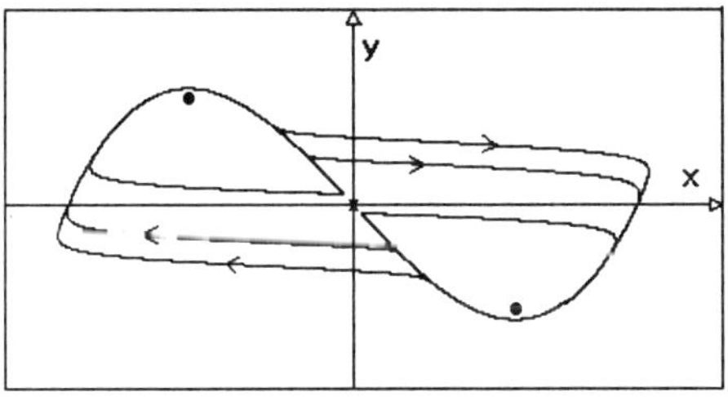

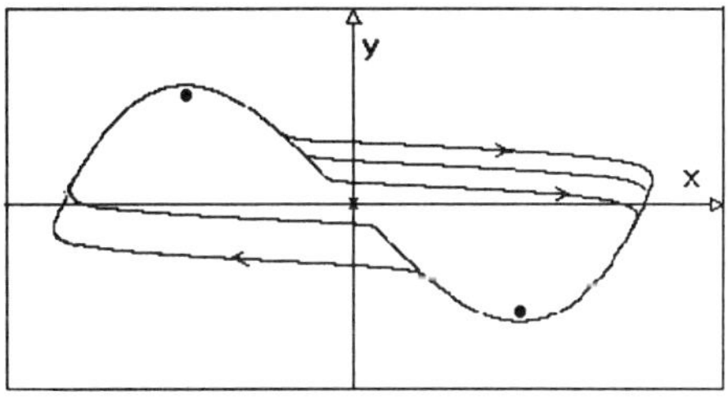

a) $a = 0, b = 1.47114317$ b) $a = 0, b = 1.47114317029$

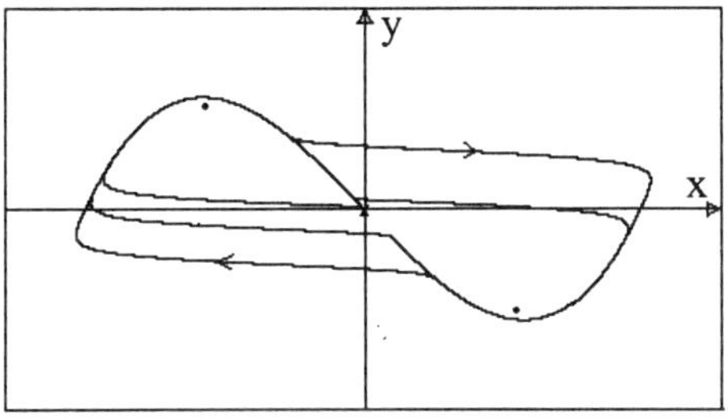

c) $a = 10^{-10}, b = 1.4711431701$

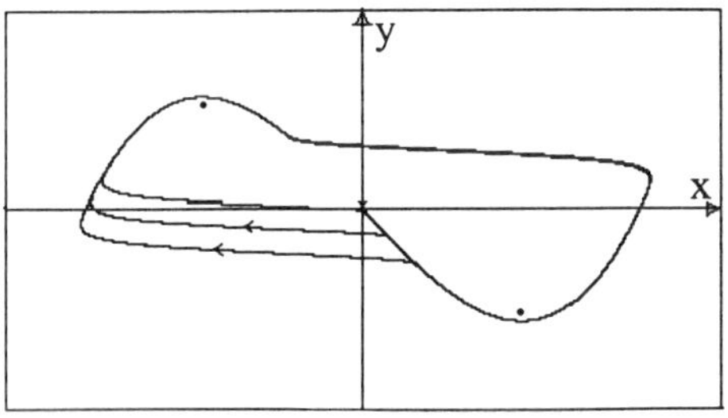

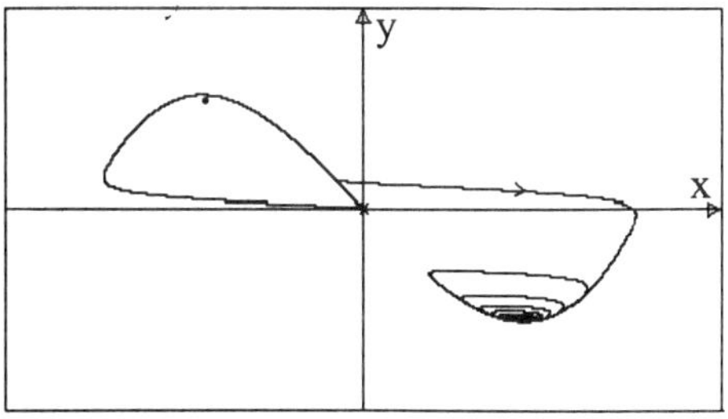

d) $a = 10^{-10}$, $b = 1.47114317015172$

Fig. 3.3.3. Phase portraits for $c = 5$ and some (b, a) situated in the domains 1, 4, 2 and 3 of Figure 3.3.2.

The two Figures 3.3.3d correspond to a point (b, a) situated in domain 3 of Figure 3.3.2. In this case the trajectory through $(-1, -0.1)$ has as α-limit set the new repulsive limit cycle that occured on D_1, while its ω-limit set is the attractive limit cycle. The small repulsive limit cycle continue to exist. It was put into evidence by representing the trajectory through $(1, 0.1)$, which has as α-limit set this limit cycle and as ω-limit set the attractive equilibrium point (x_3, y_3).

3.3.3. SADDLE–NODE HOMOCLINIC BIFURCATION. SADDLE–NODE SEPARATRIX LOOP BIFURCATION

Let $Q_7 = BTn_4 \cap S_1, Q_8 = BTn_3 \cap S_2$, $Q_9 = D_1 \cap S_1$ and $Q_{10} = D_2 \cap S_2$ (Figure 3.3.4). The representation is qualitative, owing to the small gap between BTn_4 and D_1 and BTn_3 and D_2.

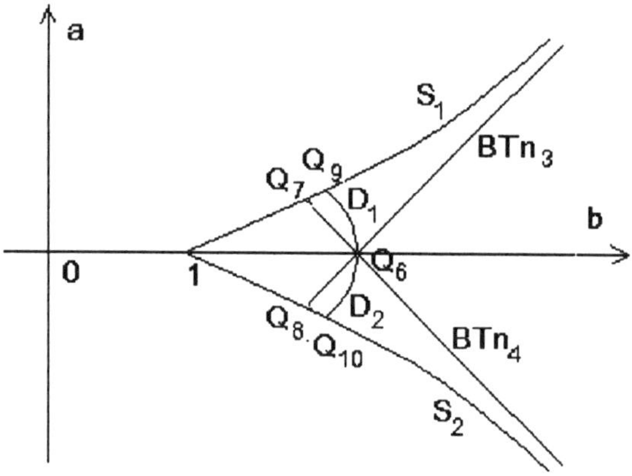

Fig. 3.3.4. The saddle–node homoclinic bifurcation curves Q_7Q_9 and Q_8Q_{10}.

The points of S_1, situated between Q_7 and Q_9 are values of saddle–node homoclinic bifurcation. According to Section 1.3.4, BTn_4 is tangent to S_1 at Q_7 and D_1 is tangent to S_1 at Q_9 (the same results hold for Q_8 and Q_{10}), Q_{7-10} correspond to values of saddle–node separatrix loop bifurcation.

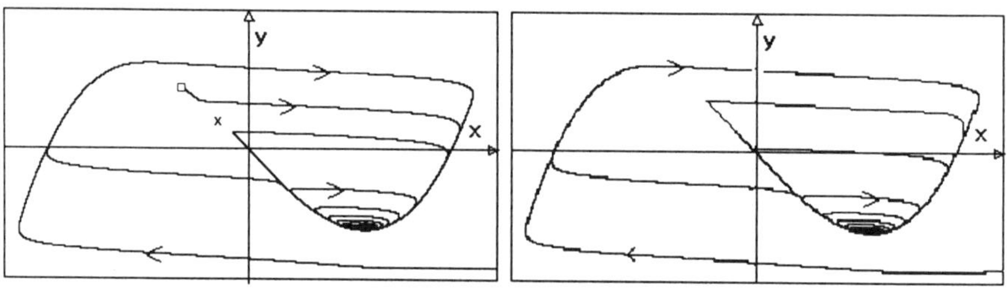

a) $a = 0.1$, $b = 1.323152069045012$ b) $a = 0.10367$, $b = 1.317425464$

Fig. 3.3.5. Phase portraits ilustrating the saddle–node homoclinic bifurcation.

In Figure 3.3.5, two phase portraits of the F–N system are represented. One corresponds to (b, a) situated in the region surrounded by BTn_4, D_1 and S_1, beneath the curve S_1 (a) and the other for (b, a) situated above S_1 (b). In case (a), three equilibrium points and only one attractive limit cycle exist, while in case (b) only an equilibrium point, which is an attractor, exists, but a new repulsive limit cycle occured. When (b, a) is situated on

S_1, the saddle and the node coincide, forming a saddle–node equilibrium with a homoclinic orbit. It will form the new repulsive limit cycle when the saddle–node equilibrium will disappear.

3.4. BREAKING SADDLE CONNECTION BIFURCATION

3.4.1. LOCALLY STABLE AND UNSTABLE MANIFOLDS OF SADDLES

According to [67], for a system of the form

$$\begin{aligned} \dot{u}_1 &= \lambda_1 u_1 + g_1(u_1, u_2), \\ \dot{u}_2 &= \lambda_2 u_2 + g_2(u_1, u_2), \end{aligned} \tag{3.4.1}$$

possessing the saddle equilibrium point $(0,0)$, such that $\mathbf{g}(\mathbf{0}) = \mathbf{0}$ and $D\,\mathbf{g}(\mathbf{0}) = \mathbf{0}$, the stable and unstable local manifolds of the saddle-point are defined by the following equations

$$\begin{aligned} W^s_{\text{loc}} &= \{u_2 = h_s(u_1), \quad |u_1| < \delta\}, \\ W^u_{\text{loc}} &= \{u_1 = h_u(u_2), \quad |u_2| < \delta\}. \end{aligned} \tag{3.4.2}$$

Here h_s and h_u are of the same class as g_1 and g_2 and satisfy the initial conditions $h_s(0) = 0$, $h_u(0) = 0$, $\frac{d\,h_s}{d\,u_1}(0) = \frac{d\,h_u}{d\,u_2}(0) = 0$ and the approximate equations

$$\frac{dh_s}{du_1}\left[\lambda_1 u_1 + g_1(u_1, h_s)\right] = \lambda_2 h_s + g_2(u_1, h_s), \tag{3.4.3}$$

$$\frac{dh_u}{du_2}\left[\lambda_2 u_2 + g_2(h_u, u_2)\right] = \lambda_1 h_u + g_1(h_u, u_2). \tag{3.4.4}$$

Consider the system (1.1.17) and its saddle point (x^*, y^*).

In order to move this point at the origin, like in Section 3.1.2, we use the translation $\psi_1 = x - x^*$ and $\psi_2 = y - y^*$. Then the system (1.1.17) becomes (3.1.12), where

$$\begin{aligned} E_1 &= c\left(1 - x^{*2}\right), \\ E_2 &= c\,x^*, \\ E_3 &= -\frac{b}{c}. \end{aligned} \tag{3.4.5}$$

In this way, the new system (3.1.12) has a saddle point at $(0,0)$.

The linearized system of (3.1.12) around $(0,0)$ is defined by the matrix

$$\mathbf{A} = \begin{pmatrix} E_1 & c \\ -\frac{1}{c} & E_3 \end{pmatrix}.$$

Consider the eigenvectors of $\mathbf{A}$ corresponding to the eigenvalues λ_1 and λ_2

$$p_1 = ((E_3 - \lambda_1)\, c, 1)^{\mathrm{T}}, \qquad p_2 = ((E_3 - \lambda_2)\, c, 1)^{\mathrm{T}} \tag{3.4.6}$$

and let $P = (p_1, p_2)$ be the fundamental matrix. Then

$$P^{-1} = \frac{1}{\Delta}\begin{pmatrix} 1 & -(E_3 - \lambda_2)\, c \\ -1 & (E_3 - \lambda_1)\, c \end{pmatrix} \tag{3.4.7}$$

where $\Delta = c\,(\lambda_2 - \lambda_1)$. Using the transformation $\boldsymbol{\psi} = P\,\mathbf{u}$, $\mathbf{u} = P^{-1}\boldsymbol{\psi}$, system (3.1.12) becomes

$$\dot{u}_1 = \lambda_1 u_1 + g_1(u_1, u_2), \qquad \dot{u}_2 = \lambda_2 u_2 - g_1(u_1, u_2), \tag{3.4.8}$$

where

$$g_1(u_1, u_2) = -E_2 c^2 \frac{[(E_3 - \lambda_1)\, u_1 + (E_3 - \lambda_2)\, u_2]^2}{\Delta} - c^4 \frac{[(E_3 - \lambda_1)\, u_1 + (E_3 - \lambda_2)\, u_2]^3}{3\Delta},$$

so it has the form (3.4.1).

Owing to the differentiability properties of the functions h_s and h_u and taking into account (3.4.2), we obtain the asymptotic expansions for small u_1 and u_2

$$h_s(u_1) \sim \frac{d^2 h_s}{d\,u_1^2}(0)\frac{u_1^2}{2} + \frac{d^3 h_s}{d\,u_1^3}(0)\frac{u_1^3}{6} + \dots, \tag{3.4.9}$$

$$h_u(u_2) \sim \frac{d^2 h_u}{d\,u_2^2}(0)\frac{u_2^2}{2} + \frac{d^3 h_u}{d\,u_2^3}(0)\frac{u_2^3}{6} + \dots, \tag{3.4.10}$$

$$\frac{dh_s}{du_1}(u_1) \sim \frac{d^2 h_s}{d\,u_1^2}(0)\, u_1 + \frac{d^3 h_s}{d\,u_1^3}(0)\frac{u_1^2}{2} + \dots, \tag{3.4.11}$$

$$\frac{dh_u}{du_2}(u_2) \sim \frac{d^2 h_u}{d\,u_2^2}(0)\, u_2 + \frac{d^3 h_u}{d\,u_2^3}(0)\frac{u_2^2}{2} + \dots. \tag{3.4.12}$$

Substituting (3.4.9) and (3.4.11) into (3.4.3) and identifying the coefficient of u_1^2, it follows

$$\frac{d^2 h_s}{du_1^2}(0) = \frac{2cE_2\,(E_3 - \lambda_1)^2}{(2\lambda_1 - \lambda_2)(\lambda_2 - \lambda_1)}. \tag{3.4.13}$$

Then, from (3.4.2), the approximate equation of the stable local manifold W^s_{loc}

$$u_2 = \frac{d^2 h_s}{du_1^2}(0)\,\frac{u_1^2}{2},$$

follows. Taking into account (3.4.13), in the ψ_1, ψ_2 variables, it becomes

$$-\frac{\psi_1}{\Delta} + \frac{(E_3-\lambda_1)\,c}{\Delta}\psi_2 = \frac{cE_2\,(E_3-\lambda_1)^2}{(2\lambda_1-\lambda_2)\,(\lambda_2-\lambda_1)}\left[\frac{\psi_1}{\Delta} - \frac{(E_3-\lambda_2)\,c}{\Delta}\psi_2\right]^2.$$

Coming back to the system (1.1.17) and the variables x and y, the equation

$$\begin{aligned} &-(x-x^*) + (E_3-\lambda_1)\,c\,(y-y^*) \\ &= \frac{E_2\,(E_3-\lambda_1)^2}{(2\lambda_1-\lambda_2)\,(\lambda_2-\lambda_1)^2}\left[x - x^* - (E_3-\lambda_2)\,c\,(y-y^*)\right]^2 \end{aligned} \tag{3.4.14}$$

is found as an approximate equation for W^s_{loc}.

Similarly, substituting (3.4.10) and (3.4.12) into (3.4.4) there follows

$$\frac{d^2 h_u}{du_2^2}(0) = \frac{2cE_2\,(E_3-\lambda_2)^2}{(\lambda_1-2\lambda_2)\,(\lambda_2-\lambda_1)}, \tag{3.4.15}$$

such that the local unstable manifold W^u_{loc} of the saddle-point (x^*, y^*) is approximated by

$$\begin{aligned} &(x-x^*) + (E_3-\lambda_2)\,c\,(y-y^*) \\ &= \frac{E_2\,(E_3-\lambda_2)^2}{(\lambda_1-2\lambda_2)\,(\lambda_2-\lambda_1)^2}\left[-(x-x^*) - (E_3-\lambda_1)\,c\,(y-y^*)\right]^2. \end{aligned} \tag{3.4.16}$$

The local manifolds obtained analytically are in good agreement with those obtained numerically in the neighborhood of saddles.

3.4.2. CURVES OF BREAKING SADDLE CONNECTION BIFURCATION VALUES

Global manifolds of saddles.

The global manifolds of saddle equilibria were obtained numerically [56]. Some of them are presented in Figure 3.4.1.

Recall that, according to Section 2.2, $x_2 < x_1 < x_3$.

In Figure 3.4.1a the saddle equilibrium point (x_2, y_2) has a stable manifold emerging from the repulsor (x_1, y_1), whilst in Figure 3.4.1b both stable manifolds of (x_2, y_2) have as α-limit set the infinity. In addition, the unstable manifolds of the saddle equilibrium point (x_3, y_3) have as ω-limit set infinity, but one of them is situated under the stable manifold of (x_2, y_2) in Figure 3.4.1a and above it in Figure 3.4.1b. This shows that, for $a = 1.418$ and $c = 5$, there exist one value $b \in (-3.2, -3.1)$ such that the two saddles are connected, that is one unstable manifold of (x_3, y_3) is a stable manifold for (x_2, y_2). Figures 3.4.1a,b are represented for $x, y \in (-3, 3)$.

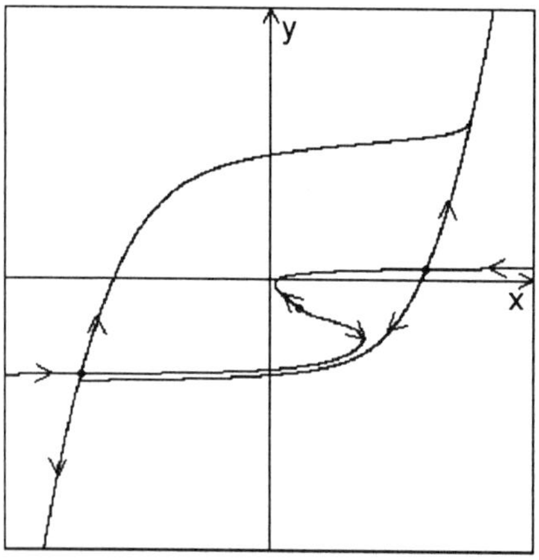

a) $a = 1.418$, $b = -3.2$

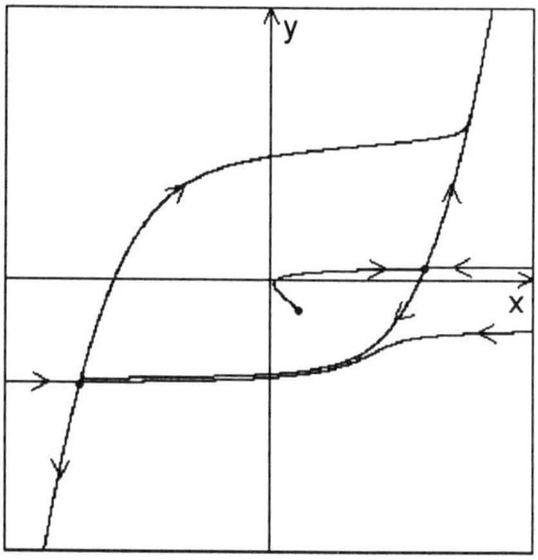

b) $a = 1.418$, $b = -3.1$

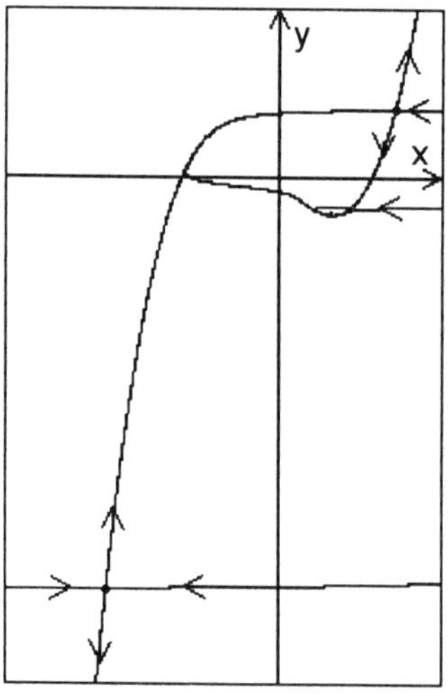

c) $a = 1.418$, $b = -0.62437$

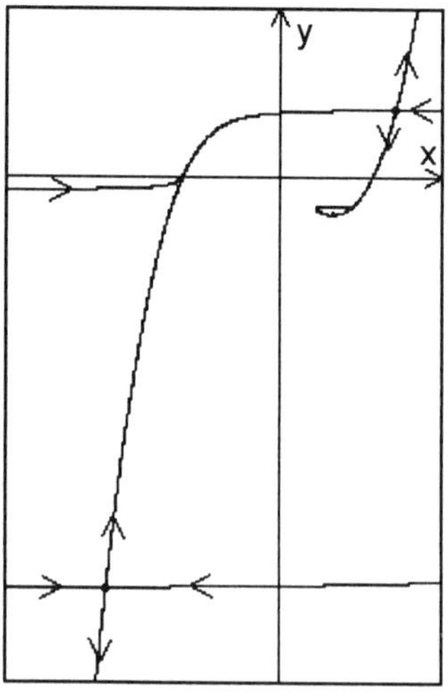

d) $a = 1.418$, $b = -0.6243675$

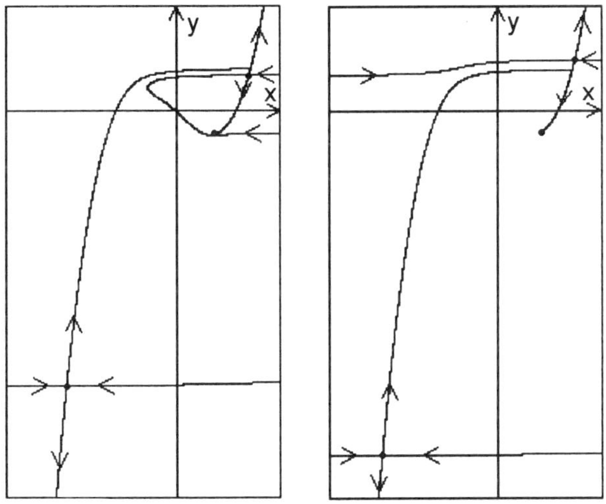

e) $a = 1.5$, $b = -0.6$ f) $a = 1.5$, $b = -0.5$

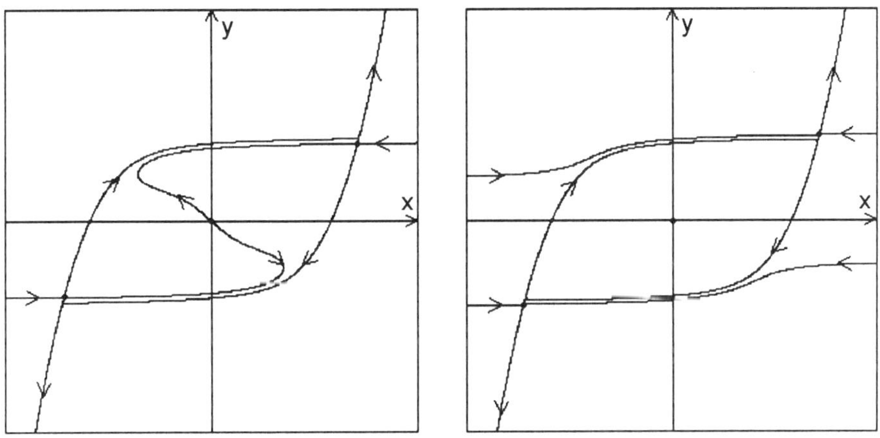

g) $a = 0, b = -1.95$ h) $a = 0, b = -1.8$

Fig. 3.4.1. Stable and unstable manifolds of the saddle-points for $c = 5$.

Similarly, in Figure 3.4.1c the saddle point (x_2, y_2) has both unstable manifolds going to infinity, whilst in Figure 3.4.1d one of these manifolds has as its ω-limit set the attractive limit cycle. The other saddle point, (x_3, y_3), possesses a stable manifold coming from infinity, but situated under the unstable manifold of (x_2, y_2) in Figure 3.4.1c and above this manifold in Figure 3.4.1d. This is why for $a = 1.418$ and $c = 5$ there exists a value $b \in (-0.62437, -0.6243675)$ such that the saddles are connected by a heteroclinic orbit, that is one of the unstable manifolds of (x_2, y_2) is a stable manifold for (x_3, y_3). Figures 3.4.1c,d are represented for $x \in (-5, 3)$ and $y \in (-9, 3)$.

Figures 3.4.1e and f also emphasize a breaking saddle connection. Indeed, in Figure 3.4.1e both unstable manifolds of (x_2, y_2) have infinity as their ω-limit set, and in Figure 3.4.1f one of these manifolds has as its ω-limit set the attractive equilibrium point. In addition, the saddle (x_3, y_3) possesses a stable manifold coming from infinity and situated below the unstable manifold of (x_2, y_2) in Figure 3.4.1e and above this manifold in Figure 3.4.1f. At the bifurcation value, one of the unstable manifolds of (x_2, y_2) is a stable manifold for (x_3, y_3). Figures 3.4.1e and f are represented for $x \in (-5, 3)$ and $y \in (-11, 3)$.

At the point Q_5 the two saddle points (x_2, y_2) and (x_3, y_3) are connected by two heteroclinic orbits. This is revealed by Figures 3.4.1g, h for increasing b and $a = 0$. Recall that for $a = 0$ the F–N system (1.1.17) possesses phase portraits which are symmetric with respect to the origin.

In Figure 3.4.1g no oscillatory regime exists and each of the two saddles possesses one stable manifold having as its α-limit set the repulsive equilibrium point, whilst their unstable manifolds have as their ω-limit set the infinity. In Figure 3.4.1h all the stable manifolds have the point at infinity as α-limit set, whilst each saddle possess an unstable manifold having as ω-limit set the attractive limit cycle. Figures 3.4.1g,h are represented for $x, y \in (-3, 3)$.

Curves of breaking saddle connection bifurcation values.
A numerical study of the phase dynamics taking into account the manifolds of saddle equilibria emphasizes the breaking saddle connection bifurcation. It takes place for (b, a) situated on the curves K_i, $i = \overline{1,4}$ (Figure 3.4.2) passing through the point Q_5, already introduced in Section 3.3 as the point at which the curves $BTn1$ and $BTn3$ intersect the Ob-axis.

The curves K_2, K_4 are symmetric to K_1, K_3 with respect to the Ob-axis.

Thus, the phase portraits from Figures 3.4.1 a, b, c, d, e, f, g, h correspond to the domains 1A, 1B, 2A, 2B, 3A, 3B, 1A, 2B respectively.

Consider now the following important points $Q_{11} = K_1 \cap S_1$, $Q_{13} = K_3 \cap H_1$, $Q_{15} = K_3 \cap S_1$ and denote by Q_{12}, Q_{14}, Q_{16} the symmetric points of Q_{11}, Q_{13}, Q_{15} with respect to the Ob-axis. In this way, the domain of the (b, a)-plane bounded by S_1 for $b \in (-\infty, -c)$, BT_{n1} and Ob-axis, for $b < b_{Q5}$, within which the F–N system possesses two saddles and a repulsive equilibrium point and no oscillatory regimes, is divided by K_1 into two domains, 1A and 1B, with nontopologically equivalent phase dynamics.

Similarly, the domain 2, bounded by H_1, BTn_1, the Oa-axis for $a \in (0, 1)$ and the Ob-axis for $b \in (b_{Q5}, 0)$ is divided by K_3 into the domains 2A and 2B. For (b, a) in 2A or 2B, the F–N system possesses two saddles, a repulsive equilibrium and an attractive limit cycle. The phase portraits of 2A are not topologicaly equivalent to those of 2B.

The domain 3, bounded by S_1 for $b \in (-c, 0)$, H_1 and the Oa-axis for $a > 1$, is divided by K_3 into the domains 3A and 3B, with non-topologically equivalent dynamics. For these domains the F–N system possesses two saddles, an attractive equilibrium and no oscillatory regimes.

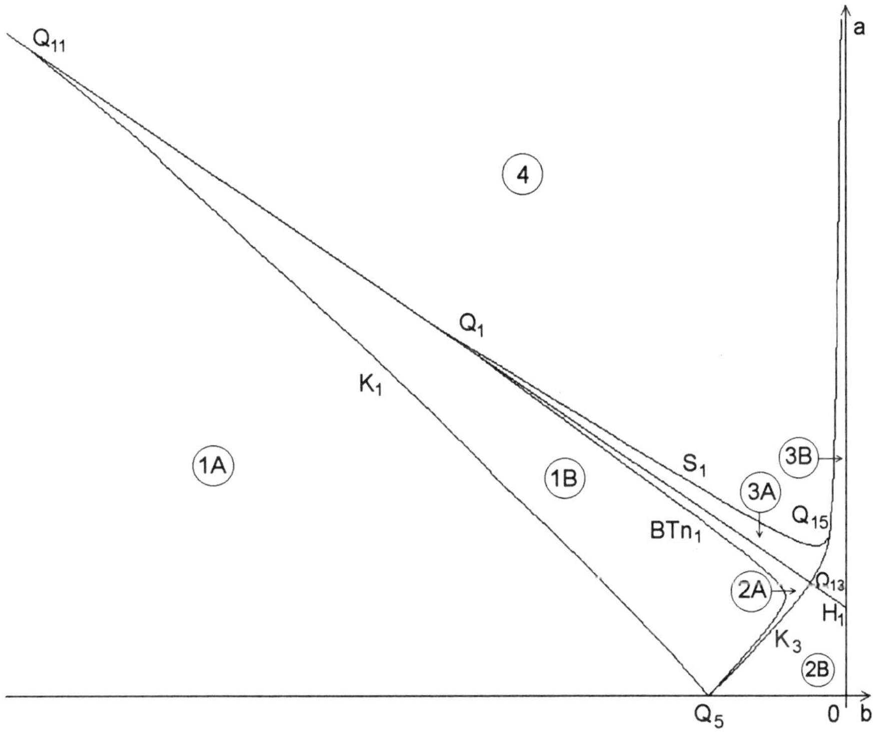

Fig. 3.4.2. The curves $K_{1,3}$.

3.4.3. DOUBLE BREAKING SADDLE CONNECTION BIFURCATION

The analysis of the phase dynamics in a neighborhood of Q_5 revealed that at Q_5 the two saddle equilibria are connected by two heteroclinic orbits. This type of bifurcation is referred to as a **double breaking saddle connection bifurcation**. We present the local bifurcation diagram around such a point in Figure 3.4.3 [120].

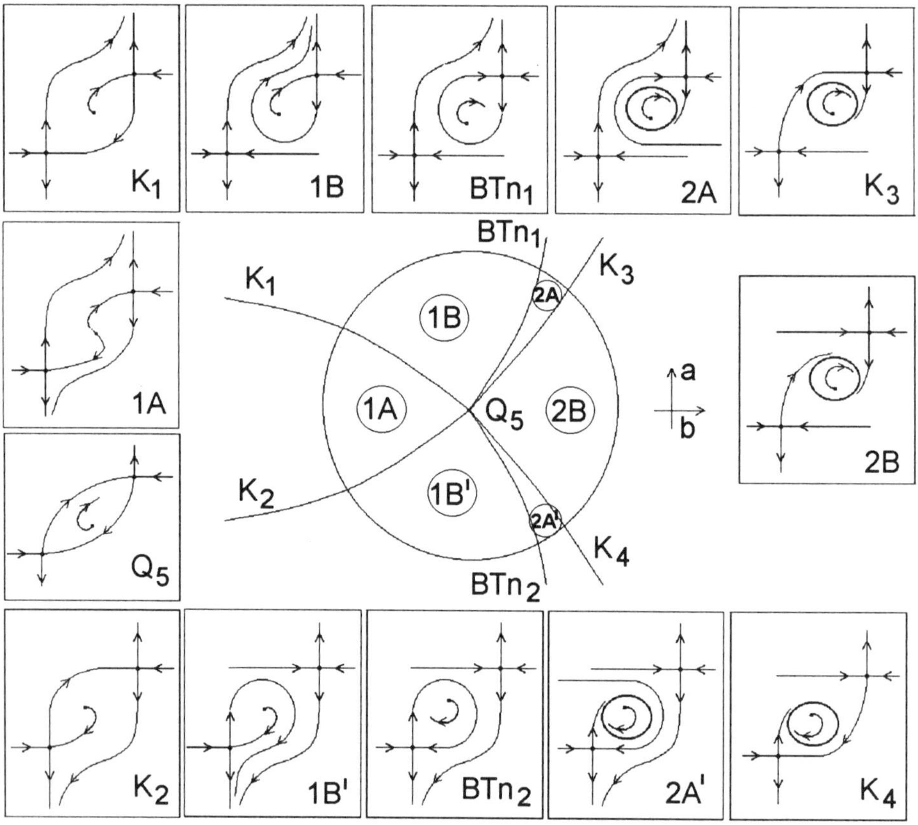

Fig. 3.4.3. Bifurcation diagram for the double breaking saddle connection bifurcation.

3.4.4. SADDLE–NODE–SADDLE CONNECTION BIFURCATION. SADDLE–NODE–SADDLE WITH SEPARATRIX CONNECTION BIFURCATION

Our numerical investigations around the point $Q_{11}(b_{11}, a_{11})$ in Figure 3.4.2 emphasized two new types of bifurcations. One is of codimension-one, corresponding to points of S_1, for $b < b_{11}$ in Figure 3.4.2 and another of codimension-two, corresponding to the point Q_{11}.

This codimension-one bifurcation is referred to as a **saddle–node–saddle connection bifurcation**. This name was suggested by the fact that at the points of S_1, for $b < b_{11}$, a saddle–node is connected with a saddle.

The codimension-two bifurcation, corresponding to Q_{11}, is referred to as a **saddle–node–saddle with separatrix connection bifurcation**.

Its difference from the saddle–node–saddle connection bifurcation consists of the property that in the case of Q_{11} the connection of the saddle–node with the saddle is a separatrix, whilst in the case of the points of S_1, for $b < b_{11}$, this connection is not a separatrix.

Here by **separatrix** it is meant the phase trajectory through (x^0, y^0) such that in any neighborhood of (x^0, y^0) there exist (x^1, y^1), (x^2, y^2) different from (x^0, y^0) such as $\alpha(x^1, y^1) \neq \alpha(x^2, y^2)$ or/and $\omega(x^1, y^1) \neq \omega(x^2, y^2)$.

The local bifurcation diagram around Q_{11} is presented in Figure 3.4.4 [120].

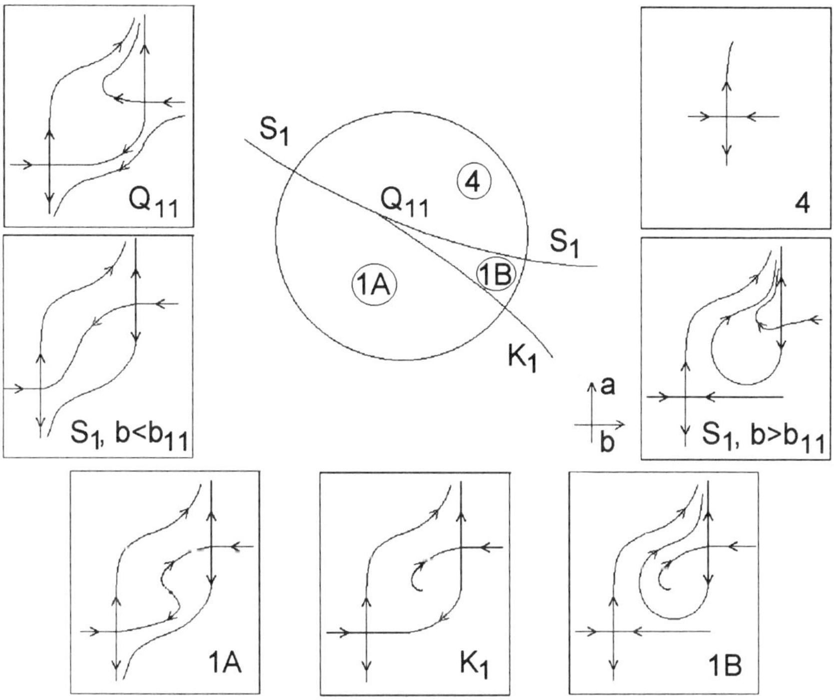

Fig. 3.4.4. Bifurcation diagram for the saddle–node–saddle with separatrix connection bifurcation. Here the saddle–node is partially repulsive.

A similar situation takes place around $Q_{15}(b_{15}, a_{15})$. However in this case there exist a partially attractive saddle–node. The local bifurcation diagram around Q_{15} is presented in Figure 3.4.5 [120].

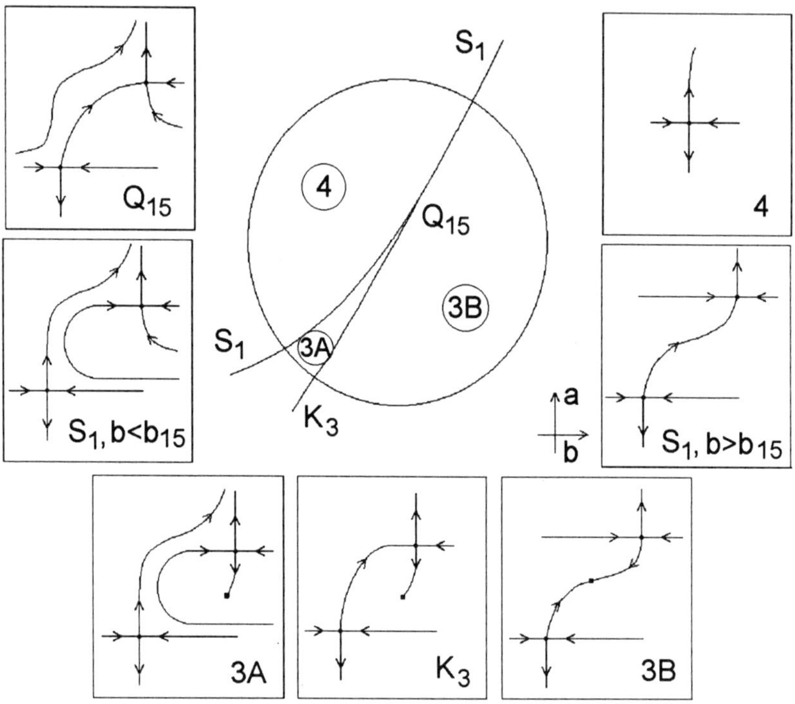

Fig. 3.4.5. Bifurcation diagram for the saddle–node–saddle with separatrix connection bifurcation. Here the saddle–node is partially attractive.

The points of S_1 with $b < b_{11}$ or $b > b_{15}$ correspond to saddle–node–saddle connection bifurcation. A schematic representation of this type of codimension-one bifurcation is given in Figure 3.4.6 for the two situations when the saddle–node is partially repulsive or partially attractive.

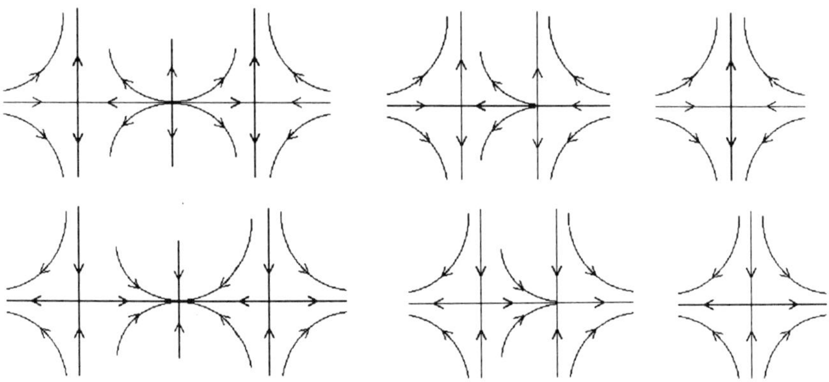

Fig. 3.4.6. Saddle–node–saddle connection bifurcation.

Let us remark a stricking analogy between the feature of bifurcation corresponding to the point Q_6 (double homoclinic bifurcation) and Q_5 (double breaking saddle connection bifurcation) (Figures 3.3.4 and 3.4.2).

For $a > 0$ at Q_6 two curves, BTn_4 and D_1, of homoclinic bifurcation emerge and they cut tangently the curve S_1 at Q_7 and Q_9 (saddle–node separatrix loop bifurcation). Completely similarly, at Q_5 for $a > 0$ two curves K_1 and K_3 of breaking saddle connection bifurcation emerge and they cut tangently the curve S_1 at Q_{11} and Q_{15} (saddle–node–saddle with separatrix connection bifurcation).

Finally, the points of S_1 situated between Q_7 and Q_9 correspond to saddle–node homoclinic bifurcations. Similarly, the points of S_1 for $b < b_{11}$ or $b > b_{15}$ correspond to saddle–node-saddle connection bifurcation.

3.5. BAUTIN BIFURCATION. NON-HYPERBOLIC LIMIT CYCLE BIFURCATION

3.5.1. NORMAL FORM FOR BAUTIN BIFURCATION. LIAPUNOV COEFFICIENTS

The main result of this Section is the following.

Theorem 3.5.1 [126] *For each $c > 1+\sqrt{3}$, the system* (1.1.17) *has two Bautin bifurcation points $Q_{17,18}(b^*_{Ba}, a^*_{Ba})$, where*

$$b^*_{Ba} = c^2 - c\sqrt{c^2-1}, \tag{3.5.1}$$

$$a^*_{Ba} = \pm\frac{4}{3}\left(c\sqrt{c^2-1} - c^2 + 1\right)\sqrt[4]{1-\frac{1}{c^2}}. \tag{3.5.2}$$

The curves of non hyperbolic limit cycles bifurcation values, emerging from the points $Q_{17,18}$, are approximated, near these points, by the curves $Ba_{1,2}$, defined by the equations

$$\left[\mathrm{Re}c_2 - \frac{1}{\omega}\mathrm{Re}c_1\mathrm{Im}c_1 + \mu\left(\frac{1}{\omega^2}(\mathrm{Im}c_1)^2 - \frac{1}{\omega}\mathrm{Im}c_2\right)\right] \tag{3.5.3}$$
$$\times\left(\frac{1}{\omega}\mathrm{Re}\; c_1 - \frac{\nu}{\omega}\mathrm{Im}\; c_1\right)^2 + 4\mu = 0,$$

where Re c_1, Im c_1, Re c_2, Im c_2 *are given in Appendix A*, μ,ω *are given by* (3.1.16) *and*

$$\nu = \frac{\mu}{\omega}. \tag{3.5.4}$$

The largest part of proof of Theorem 3.5.1 is based on the normal form method corresponding to the Hopf bifurcation. In this normal form the Liapunov coefficients ℓ_1 and ℓ_2 of the terms of order 3 and 5, respectively, occur. If $\ell_1(\mathbf{0}) \neq 0$ the Hopf bifurcation is non-degenerate, whilst if $\ell_1(\mathbf{0}) = 0$ but $\ell_2(\mathbf{0}) \neq 0$, a Bautin bifurcation takes place.

The computation of the Liapunov coefficients.
Let us start with the form (3.1.12) of the system (1.1.17). Unlike the investigation from Section 3.2, here the point (b^*, a^*) is unknown. The only information on it is that it belongs to H_1 or H_2.

The system (3.1.12) becomes

$$\dot{\boldsymbol{\psi}} = A\boldsymbol{\psi} + \mathbf{F}(\boldsymbol{\alpha}, \boldsymbol{\psi}) \tag{3.5.5}$$

where the constant matrix $\mathbf{A}$ and the vector $\mathbf{F}$ depend on parameters and have the form

$$\mathbf{A} = \begin{pmatrix} E_1 & c \\ -\frac{1}{c} & E_3 \end{pmatrix} \tag{3.5.6}$$

$$\mathbf{F} = \left(-cx^*\psi_1^2 - c\psi_1^3/3, 0\right)^{\mathrm{T}} \equiv (F_1, F_2)^{\mathrm{T}}. \tag{3.5.7}$$

Thus, for every point (b^*, a^*) of the (b, a)-plane, we have $\mathbf{F}(\boldsymbol{\alpha}, \mathbf{0}) = \mathbf{0}$.

The matrix $\mathbf{A}$ has the eigenvalues $\lambda_{1,2} = \lambda_{1,2}(\boldsymbol{\alpha}) = \mu(\boldsymbol{\alpha}) \pm i\omega(\boldsymbol{\alpha})$ given by (3.1.16).

The point (x^*, y^*) is a Hopf bifurcation point corresponding to the bifurcation value $\boldsymbol{\alpha} = \mathbf{0}$ (i.e., $(b, a) = (b^*, a^*)$), if $\mu(\mathbf{0}) = 0, \omega(\mathbf{0}) \neq 0$ and $\ell_1(\mathbf{0}) \neq 0$, where $\ell_1(\boldsymbol{\alpha})$ will be defined in the following.

If $\mu(\mathbf{0}) = 0$, we have

$$x^*(\mathbf{0}) = \pm\sqrt{1 - \frac{b^*}{c^2}}. \tag{3.5.8}$$

Then from equation (2.2.1) it follows that the Hopf bifurcation values verify

$$a^* = \pm\frac{b^*}{3}\left(-2 + \frac{3}{b^*} - \frac{b^*}{c^2}\right)\sqrt{1 - \frac{b^*}{c^2}}, \quad b^* \in (-c, c), \tag{3.5.9}$$

so in the (b, a)-plane they are situated on the curves $H_{1,2}$. In this way, for the bifurcation value $\boldsymbol{\alpha} = \mathbf{0}$, taking into account (3.1.16) and (3.5.8) we have

$$\omega^2(\mathbf{0}) \equiv \omega_0^2 = 1 - \frac{(b^*)^2}{c^2} > 0. \tag{3.5.10}$$

Using the notation (3.1.11), we obtain

$$\begin{aligned} E_{10} &\equiv E_1(\mathbf{0}) = \frac{b^*}{c}, \\ E_{20} &\equiv E_2(\mathbf{0}) = \pm\sqrt{c^2 - b^*}, \\ E_{30} &\equiv E_3(\mathbf{0}) = -\frac{b^*}{c}. \end{aligned} \tag{3.5.11}$$

Taking into account (3.1.11), (3.1.16) becomes

$$\mu = \frac{E_1 + E_3}{2}, \qquad \omega = \sqrt{1 - \frac{(E_1 - E_3)^2}{4}}. \tag{3.5.12}$$

In the following we shall reduce the system (3.1.12) to its normal form using a sequence of transformations. In the case of Bautin bifurcation the computations will be in the complex field, around the bifurcation point characterized by two purely imaginary and of opposite signs eigenvalues, whilst for the Bogdanov–Takens bifurcation, the two eigenvalues were equal to zero. Correspondingly, it will be more difficult to deduce the first linear transformation. As a result, a more complicate equation will be obtained.

Thus the first transformation is defined by using the eigenvector $\mathbf{q}$ of the matrix $\mathbf{A}$ and the eigenvector $\mathbf{p}$ of $\mathbf{A}^{\mathrm{T}}$. Here $\mathbf{q}$ corresponds to the eigenvalue λ_1 of $\mathbf{A}$ and $\mathbf{p}$ to $\overline{\lambda}_1$. As $\mathbf{A}$ has real elements, the second eigenvalue is $\overline{\lambda}_1$. In general, $\mathbf{q}$, $\mathbf{p} \in \mathbf{C}^2$. We have

$$\mathbf{q} = \left(q_1, (\lambda_1 - E_1)\, q_1/c\right), \quad \mathbf{p} = \left(p_1, c\left(E_1 - \overline{\lambda}_1\right) p_1\right), \tag{3.5.13}$$

where

$$\overline{p}_1 q_1 = \left[1 - (\lambda_1 - E_1)^2\right]^{-1} \tag{3.5.14}$$

and p_1, q_1 are, for the time being, arbitrary complex numbers.

It is easy to check that

$$\mathbf{Aq} = \lambda_1 \mathbf{q}, \quad \mathbf{A}^{\mathrm{T}}\mathbf{p} = \overline{\lambda}_1 \mathbf{p}, \quad \langle \mathbf{p}, \mathbf{q} \rangle = 1, \quad \langle \mathbf{p}, \overline{\mathbf{q}} \rangle = 0. \tag{3.5.15}$$

Here $\langle \mathbf{v}', \mathbf{v} \rangle = \overline{v'}_1 v_1 + \overline{v'}_2 v_2$ is scalar product in $\mathbf{C}^2$.

Let us introduce the new unknown function $\mathbf{z}$ by the relation

$$\boldsymbol{\psi} = (\psi_1, \psi_2) = \mathbf{zq} + \overline{\mathbf{zq}} \tag{3.5.16}$$

with $\mathbf{z}$, $\overline{\mathbf{z}} \in \mathbf{C}^2$.

Unlike the Bogdanov–Takens case, here the matrix $\mathbf{A}$ is used for $\boldsymbol{\alpha} \neq \mathbf{0}$.

Computing the inner product of the form (3.1.12) by $\mathbf{p}$, taking into account $(3.5.15)_{3,4}$ and writting λ instead of λ_1, it follows

$$\dot{\mathbf{z}} = \lambda \mathbf{z} + G(\boldsymbol{\alpha}, \mathbf{z}, \overline{\mathbf{z}}), \tag{3.5.17}$$

where

$$\begin{aligned} G(\boldsymbol{\alpha}, \mathbf{z}, \overline{\mathbf{z}}) &= \langle \mathbf{p}, \mathbf{F}(\boldsymbol{\alpha}, \mathbf{zq} + \overline{\mathbf{zq}}) \rangle \\ &= \overline{p_1} F_1(\boldsymbol{\alpha}, \mathbf{zq} + \overline{\mathbf{zq}}) \end{aligned}$$

$$
\begin{aligned}
&= \sum_{k+j\geq 2} \frac{1}{k!j!} g_{kj}(\boldsymbol{\alpha})\, \mathbf{z}^k \overline{\mathbf{z}}^j \qquad (3.5.18)\\
&= \frac{1}{2} g_{20} z^2 + g_{11} \mathbf{z}\overline{\mathbf{z}} + \frac{1}{2} g_{02} \overline{\mathbf{z}}^2 + \\
&\quad + \frac{1}{6} g_{30} \mathbf{z}^3 + \frac{1}{2} g_{21} \mathbf{z}^2 \overline{\mathbf{z}} + \frac{1}{2} g_{12} \mathbf{z}\overline{\mathbf{z}}^2 + \frac{1}{6} g_{03} \overline{\mathbf{z}}^3 .
\end{aligned}
$$

Remark that the equation (3.5.17) can be alternatively obtained by taking into account that $\mathbf{z} = \langle \mathbf{p}, \boldsymbol{\psi} \rangle$ and, therefore, $\dot{\mathbf{z}} = \langle \mathbf{p}, \dot{\boldsymbol{\psi}} \rangle$, where $\dot{\boldsymbol{\psi}}$ is given by (3.5.5). It is found

$$
\begin{aligned}
g_{20} &= -2\overline{p_1} E_2 q_1^2, & g_{30} &= -2c\overline{p_1} q_1^3, \\
g_{11} &= -2\overline{p_1} E_2 q_1 \overline{q_1}, & g_{21} &= -2c\overline{p_1} q_1^2 \overline{q_1}, \\
g_{02} &= -2\overline{p_1} E_2 \overline{q_1}^2, & g_{12} &= -2c\overline{p_1} q_1 \overline{q_1}^2, \qquad (3.5.19)\\
& & g_{03} &= -2c\overline{p_1}\,\overline{q_1}^3.
\end{aligned}
$$

Choosing $q_1 = 1$ and using the notation $g = -\overline{p_1} E_2$ and $r = \frac{c}{3E_2}$, we get

$$
\begin{aligned}
g_{20} = g_{11} = g_{02} &= -2\overline{p_1} E_2 = 2g, \\
g_{30} = g_{21} = g_{12} = g_{03} &= -2c\overline{p_1} = 6gr
\end{aligned} \qquad (3.5.20)
$$

and, correspondingly, (3.5.17) becomes

$$
\dot{\mathbf{z}} = \lambda \mathbf{z} + g(\mathbf{z} + \overline{\mathbf{z}})^2 + gr(\mathbf{z} + \overline{\mathbf{z}})^3. \qquad (3.5.21)
$$

In the following, a new invertible but nonlinear transformation $(\mathbf{z}, \overline{\mathbf{z}}) \longleftrightarrow (\mathbf{w}, \overline{\mathbf{w}})$ is applied

$$
\mathbf{z} = \mathbf{z}_1 + \mathbf{z}_2 + \mathbf{z}_3 + \mathbf{z}_4 + \mathbf{z}_5 + O(|\mathbf{w}^6|), \qquad (3.5.22)
$$

where

$$
\begin{aligned}
\mathbf{z}_1 &= \mathbf{w}, \\
\mathbf{z}_2 &= \frac{h_{20}}{2}\mathbf{w}^2 + h_{11}\mathbf{w}\overline{\mathbf{w}} + \frac{h_{02}}{2}\overline{\mathbf{w}}^2, \\
\mathbf{z}_3 &= \frac{h_{30}}{6}\mathbf{w}^3 + \frac{h_{21}}{2}\mathbf{w}^2\overline{\mathbf{w}} + \frac{h_{12}}{2}\mathbf{w}\overline{\mathbf{w}}^2 + \frac{h_{03}}{6}\overline{\mathbf{w}}^3, \qquad (3.5.23)\\
\mathbf{z}_4 &= \frac{h_{40}}{24}\mathbf{w}^4 + \frac{h_{31}}{6}\mathbf{w}^3\overline{\mathbf{w}} + \frac{h_{22}}{4}\mathbf{w}^2\overline{\mathbf{w}}^2 + \frac{h_{13}}{6}\mathbf{w}\overline{\mathbf{w}}^3 + \frac{h_{04}}{24}\overline{\mathbf{w}}^4, \\
\mathbf{z}_5 &= \frac{h_{50}}{120}\mathbf{w}^5 + \frac{h_{41}}{24}\mathbf{w}^4\overline{\mathbf{w}} + \frac{h_{32}}{12}\mathbf{w}^3\overline{\mathbf{w}}^2 + \frac{h_{23}}{12}\mathbf{w}^2\overline{\mathbf{w}}^3 + \\
&\quad \frac{h_{14}}{24}\mathbf{w}\overline{\mathbf{w}}^4 + \frac{h_{05}}{120}\overline{\mathbf{w}}^5.
\end{aligned}
$$

Therefore, the new unknown vector function is $\mathbf{w}$. The coefficients h_{ij} are determined such that the ODE for $\mathbf{w}$ have the form

$$\dot{\mathbf{w}}= \lambda \mathbf{w} + c_1 \mathbf{w}^2 \overline{\mathbf{w}} + c_2 \mathbf{w}^3 \overline{\mathbf{w}}^2, \tag{3.5.24}$$

so it has a linear term, a cubic term and another term of the fifth degree.

The terms whose coefficients are c_1 and c_2 are the resonant terms.

Of course, (3.5.24) is exact up to $O(|\mathbf{w}|^6)$ as $\boldsymbol{\alpha} \to \mathbf{0}$.

Introducing (3.5.23) into (3.5.22), then differentiating (3.5.22) it follows that

$$\dot{\mathbf{z}}= \mathbf{D}\ \dot{\mathbf{w}} + \mathbf{K}\ \dot{\overline{\mathbf{w}}}, \tag{3.5.25}$$

where

$$\begin{aligned}\mathbf{D} &= 1 + h_{20}\mathbf{w} + h_{11}\overline{\mathbf{w}} + \frac{h_{30}}{2}\mathbf{w}^2 + \frac{h_{12}}{2}\overline{\mathbf{w}}^2 + \frac{h_{40}}{6}\mathbf{w}^3 \\ &+ \frac{h_{31}}{2}\mathbf{w}^2\overline{\mathbf{w}} + \frac{h_{22}}{2}\mathbf{w}\overline{\mathbf{w}}^2 + \frac{h_{13}}{6}\overline{\mathbf{w}}^3 + \frac{h_{50}}{24}\mathbf{w}^4 \\ &+ \frac{h_{41}}{6}\mathbf{w}^3\overline{\mathbf{w}} + \frac{h_{23}}{6}\mathbf{w}\overline{\mathbf{w}}^3 + \frac{h_{14}}{24}\overline{\mathbf{w}}^4\end{aligned} \tag{3.5.26}$$

and

$$\begin{aligned}\mathbf{K} &= h_{11}\mathbf{w} + h_{02}\mathbf{w}\overline{\mathbf{w}} + \frac{h_{03}}{2}\overline{\mathbf{w}}^2 + \frac{h_{31}}{6}\mathbf{w}^3 \\ &+ \frac{h_{22}}{2}\mathbf{w}^2\overline{\mathbf{w}} + \frac{h_{13}}{2}\mathbf{w}\overline{\mathbf{w}}^2 + \frac{h_{04}}{6}\overline{\mathbf{w}}^3 + \frac{h_{41}}{24}\mathbf{w}^4 \\ &+ \frac{h_{23}}{4}\mathbf{w}^2\overline{\mathbf{w}}^2 + \frac{h_{14}}{6}\mathbf{w}\overline{\mathbf{w}}^3 + \frac{h_{05}}{24}\overline{\mathbf{w}}^4.\end{aligned} \tag{3.5.27}$$

From (3.5.21) and (3.5.25) it follows

$$\mathbf{D}\ \dot{\mathbf{w}} + \mathbf{K}\ \dot{\overline{\mathbf{w}}}= \lambda \mathbf{z} + g\left(\mathbf{z} + \overline{\mathbf{z}}\right)^2 + gr\left(\mathbf{z} + \overline{\mathbf{z}}\right)^3. \tag{3.5.28}$$

On the other hand, $\dot{\mathbf{w}}$ is given by (3.5.24) and $\mathbf{z}$ is given by (3.5.22), (3.5.23). Hence, (3.5.28) is an identity of two polynomials in $\mathbf{w}$ and $\overline{\mathbf{w}}$. From it, a system of algebraic affine equations in h_{ij}, c_1 and c_2 is obtained.

Let us remark that h_{ij} could be determined also from an identity of two polinomals in $\mathbf{z}$ and $\overline{\mathbf{z}}$, but the effort of computation would be much bigger.

Except for the equations which correspond to the indices 21 and 32 (and, so, to the resonant terms), all other equations allow the unique determination of coefficients h_{ij}. Indeed, every equation with $i + j = 2$ contains a single h_{ij}, every equation with $i + j = 3, 4$ or 5 contains a single h_{ij} with $i + j = 3, 4$ or 5 respectively. In addition, such an equation contains some h_{ij} already determined (with $i + j < 3, 4$ or 5 respectively).

The equation for h_{21} contains, in addition, c_1 and the equation for h_{32} contains also c_2. In order for the transformation (3.5.22) to be regular at $\boldsymbol{\alpha} = \mathbf{0}$, it is necessary to choose $h_{32} = h_{21} = 0$. Consequently, for the particular case of system (3.5.5), the coefficients h_{ij}, c_1 and c_2 were determined in [123] using the soft MATHEMATICA [147]. In Appendix A we reproduce only the real and imaginary parts of the coefficients c_1 and c_2. There $g = g_1 + i\ g_2$, $\lambda = L$, the multiplication is denoted by * and the power by ˆ. The coefficients c_1 and c_2 can also be written as

$$\begin{aligned} c_1 &= g\left(3r + 2\ \mathrm{Re}\ h_{11} + h_{20} + \overline{h_{02}}\right), \\ c_2 &= g\left[\frac{h_{12}}{2}\left(h_{20} + \overline{h_{02}} + 3r\right) + \ \mathrm{Re}\ h_{22} + r\frac{h_{30} + \overline{h_{03}}}{2}\right. \\ &+ \frac{h_{31} + \overline{h_{13}}}{3} + 6r\left(\mathrm{Re}\ h_{11}^2 + |h_{11}|^2\right) \\ &+ 6r\left(h_{20} + \overline{h_{02}}\right)\mathrm{Re}\ h_{11} \\ &+ \left(h_{02} + \overline{h_{20}}\right)\frac{h_{30} + \overline{h_{03}}}{6} + 3\frac{r}{2}\left|h_{20} + \overline{h_{02}}\right|^2 \\ &\left. + \overline{h_{12}}\left(2\mathrm{Re}\ h_{11} + 3r\right)\right] \end{aligned} \tag{3.5.29}$$

It follows, in particular, that

$$\begin{aligned} \mathrm{Re}\ c_1 &= \frac{1}{N}\left[12\mu^3 (\mathrm{Re}\ g)^2 - 4\mu (\mathrm{Re}\ g)(\mathrm{Im}\ g) + 3\mu^4 r (\mathrm{Re}\ g)\right. \\ &\left. + 30\mu^2\omega^2 r (\mathrm{Re}\ g) + \mu G_1 + 9\omega^3 G_2 (\mathrm{Re}\ g),\right] \end{aligned} \tag{3.5.30}$$

with

$$\begin{aligned} G_1 &= \omega^2\left[92 (\mathrm{Re}\ g)^2 - 16 (\mathrm{Im}\ g)^2\right], \\ G_2 &= 3r\omega - 4 (\mathrm{Im}\ g), \\ N &= \mu^4 + 10\mu^2\omega^2 + 9\omega^4. \end{aligned}$$

The next transformation concerns the independent variable and it is written as

$$\tau = t\omega(\boldsymbol{\alpha}) \tag{3.5.31}$$

It carries (3.5.24) into

$$\frac{d\mathbf{w}}{d\tau} = (\nu(\boldsymbol{\alpha}) + i)\mathbf{w} + d_1(\boldsymbol{\alpha})\mathbf{w}^2\overline{\mathbf{w}} + d_2(\boldsymbol{\alpha})\mathbf{w}^3\overline{\mathbf{w}}^2 + O(|\mathbf{w}|^6), \tag{3.5.32}$$

where

$$\begin{aligned} \nu(\boldsymbol{\alpha}) &= \frac{\mu(\boldsymbol{\alpha})}{\omega(\boldsymbol{\alpha})}, \\ d_1(\boldsymbol{\alpha}) &= \frac{c_1(\boldsymbol{\alpha})}{\omega(\boldsymbol{\alpha})}, \\ d_2(\boldsymbol{\alpha}) &= \frac{c_2(\boldsymbol{\alpha})}{\omega(\boldsymbol{\alpha})}, \end{aligned} \tag{3.5.33}$$

The next transformation introduces a new time scale and is given in the differential form

$$d\tau = \{1 - \mathrm{Im} d_1(\boldsymbol{\alpha})\,|\mathbf{w}|^2 + \Big((\mathrm{Im} d_1(\boldsymbol{\alpha}))^2 - \mathrm{Im} d_2(\boldsymbol{\alpha})\Big)\,|\mathbf{w}|^4\} d\theta. \tag{3.5.34}$$

In this way, from (3.5.32), it is obtained the equation

$$\frac{d\mathbf{w}}{d\theta} = \nu(\boldsymbol{\alpha}) + i)\mathbf{w} + \ell_1(\boldsymbol{\alpha})\mathbf{w}^2\overline{\mathbf{w}} + \ell_2(\boldsymbol{\alpha})\mathbf{w}^3\overline{\mathbf{w}}^2 + O(|\mathbf{w}|^6), \tag{3.5.35}$$

where

$$\ell_1(\boldsymbol{\alpha}) = \mathrm{Re} d_1(\boldsymbol{\alpha}) - \nu(\boldsymbol{\alpha})\mathrm{Im} d_1(\boldsymbol{\alpha}), \tag{3.5.36}$$

$$\ell_2(\boldsymbol{\alpha}) = \mathrm{Re} d_2(\boldsymbol{\alpha}) - \mathrm{Re} d_1(\boldsymbol{\alpha})\mathrm{Im} d_1(\boldsymbol{\alpha}) + \nu(\boldsymbol{\alpha}) \left[(\mathrm{Im}\, d_1(\boldsymbol{\alpha}))^2 - \mathrm{Im} d_2(\boldsymbol{\alpha})\right].$$

Here $\nu(\boldsymbol{\alpha})$, $\ell_1(\boldsymbol{\alpha})$, $\ell_2(\boldsymbol{\alpha})$ are real functions. The functions $\ell_1(\boldsymbol{\alpha})$ and $\ell_2(\boldsymbol{\alpha})$ are referred to as the **first** and the **second Liapunov coefficients** respectively.

In the case of F–N system (1.1.17), for $\boldsymbol{\alpha} = \mathbf{0}$ we have

$$\begin{aligned} \omega_0 &= \sqrt{1 - \frac{(b^*)^2}{c^2}}, \\ \overline{p_1} &= \frac{1}{2} - i\frac{b^*}{2c\omega_0}, \\ g &= -\frac{\sqrt{c^2 - b^*}}{2} + ib^*\frac{\sqrt{c^2 - b^*}}{2c\omega_0}, \\ r &= \frac{c}{3\sqrt{c^2 - b^*}}, \\ \ell_1(\mathbf{0}) &= \frac{\mathrm{Re}\ c_1(\mathbf{0})}{\omega_0} = \mathrm{Re}\ g\frac{3\omega_0 r - 4\ \mathrm{Im}\ g}{\omega_0^2} \\ &= \frac{-(b^*)^2 + 2b^*c^2 - c^2}{2\omega_0^3 c}. \end{aligned} \tag{3.5.37}$$

The Bautin bifurcation values
There are two situations.

In the first one we have $\ell_1(\mathbf{0}) \neq 0$. For those b_H^* such that $\ell_1(\mathbf{0}) \neq 0$, for $\boldsymbol{\alpha} = \mathbf{0}$ the normal form (3.5.35) characterizes the presence of a Hopf bifurcation point. Therefore these Hopf bifurcation points are

$$P_H^* = \left(x_H^*\left(\mathbf{0}, b_H^*, a_H^*\right), y_H^*\left(\mathbf{0}, b_H^*, a_H^*\right), b_H^*, a_H^*\right),$$

where $x^*(\mathbf{0})$ is given by (3.5.8), $\mu^*(\mathbf{0}) = 0$ represents the degeneracy condition and b_H^* and a_H^* satisfy the nondegeneracy conditions (3.5.10) and $\ell_1(\mathbf{0}) \neq 0$. In addition, the condition (3.5.41) must be satisfied.

The second situation is $\ell_1(\mathbf{0}) = 0$. For those values b_{Ba}^* for which $\ell_1(\mathbf{0}) = 0$, i.e.,

$$\left(b_{Ba}^*\right)^2 - 2b_{Ba}^* c^2 + c^2 = 0, \tag{3.5.38}$$

and $\ell_2(0) \neq 0$, the Hopf bifurcation point degenerates into the Bautin bifurcation point. For each fixed c, this takes place at the point

$$P_{Ba}^* = \left(x_{Ba}^*\left(\mathbf{0}, b_{Ba}^*, a_{Ba}^*\right), y_{Ba}^*\left(\mathbf{0}, b_{Ba}^*, a_{Ba}^*\right), b_{Ba}^*, a_{Ba}^*\right),$$

where $\left(b_{Ba}^*, a_{Ba}^*\right)$ are situated on the curves of Hopf bifurcation values, therefore they depend on c. From (3.5.38) the abscissa of the Bautin bifurcation point, b_{Ba}^*, is obtained. It is given in (3.5.1). The other solution of (3.5.38) does not belong to the interval $(-c, c)$.

Taking into account (3.5.9), the corresponding values a_{Ba}^*, given in (3.5.2), follow. From (3.5.8) it is obtained

$$x_{Ba}^* = \pm\sqrt[4]{1 - \frac{1}{c^2}}. \tag{3.5.39}$$

So far, b_{Ba}^* and a_{Ba}^* must satisfy the degeneracy condition $\ell_1(\mathbf{0}) = 0$ and the nondegeneracy condition $\ell_2(\mathbf{0}) \neq 0$.

The normal form equation for the Hopf and Bautin bifurcation points.
We shall perform a treatment common to the Hopf and Bautin cases. The Hopf case corresponds to $\ell_1(\boldsymbol{\alpha}) \neq 0$ and Bautin case to $\boldsymbol{\alpha} = \mathbf{0}$ and such that $\ell_1(\mathbf{0}) = 0$. In other words let us find another normal form of (3.1.12) in which the Hopf bifurcation values are situated on the $O\beta_2$-axis of the new parameter plane (β_1, β_2). In order to do this consider the invertible parameter transformation

$$\mu_1 = \nu(\boldsymbol{\alpha}), \quad \mu_2 = \ell_1(\boldsymbol{\alpha}), \tag{3.5.40}$$

which is regular at $\boldsymbol{\alpha} = \mathbf{0}$, i.e.,

$$\det \begin{pmatrix} \dfrac{\partial \nu}{\partial \alpha_1} & \dfrac{\partial \nu}{\partial \alpha_2} \\ \dfrac{\partial \ell_1}{\partial \alpha_1} & \dfrac{\partial \ell_1}{\partial \alpha_2} \end{pmatrix}\Bigg|_{\boldsymbol{\alpha}=0} = \frac{1}{\omega_0}\det \begin{pmatrix} \dfrac{\partial \mu}{\partial \alpha_1} & \dfrac{\partial \mu}{\partial \alpha_2} \\ \dfrac{\partial \ell_1}{\partial \alpha_1} & \dfrac{\partial \ell_1}{\partial \alpha_2} \end{pmatrix}\Bigg|_{\boldsymbol{\alpha}=0} \neq 0 \quad (3.5.41)$$

In the concrete case of (3.1.12) the relation (3.5.41) takes place. Indeed, taking into account (3.5.12), it follows

$$\overline{p_1} = \frac{1}{2} + i\frac{\frac{\nu}{2} - c\left(1 - (x^*)^2\right)}{2\omega} \quad (3.5.42)$$

where

$$\begin{aligned} \text{Re } g &= -c\frac{x^*}{2}, \\ \text{Im } g &= E_2\frac{E_1 - E_3}{4\omega} = -c\frac{x^*\nu}{2} + c^2x^*\frac{1-(x^*)^2}{2\omega}, \\ r &= \frac{1}{3x^*}, \quad [r \text{ Re } g](\mathbf{0}) = -\frac{c}{6}, \quad [\text{Im } g](\mathbf{0}) = 3\omega_0\frac{r(\mathbf{0})}{4}. \end{aligned} \quad (3.5.43)$$

Since x^* satisfy (2.2.1), we have

$$(x^*)^3 - 3x^*\left(1 - \frac{1}{b}\right) - 3\frac{a}{b} = 0, \qquad y^* = \frac{a - x^*}{b}, \quad b \neq 0. \quad (3.5.44)$$

Differentiating $(3.5.33)_1$ and (3.5.12) with respect to a and b, we obtain

$$\begin{aligned} \frac{\partial \nu}{\partial \alpha_1}(\mathbf{0}) &= \frac{-5(b^*)^2 - c^2}{6c\omega_0 (b^*)^2}, & \frac{\partial \nu}{\partial \alpha_2}(\mathbf{0}) &= \frac{-c}{2\omega_0 b^* x^*(\mathbf{0})}, \\ \frac{\partial \omega}{\partial \alpha_1}(\mathbf{0}) &= \frac{c^2 - b^*}{3c^2\omega_0} = \frac{(x^*)^2}{3\omega_0}(\mathbf{0}), & \frac{\partial \omega}{\partial \alpha_2}(\mathbf{0}) &= \frac{-c^2x^*}{2\omega_0 b^*}(\mathbf{0}). \end{aligned}$$

Then, from (3.5.30) and (3.5.36), it follows

$$\frac{\partial \text{Re} c_1}{\partial \alpha_1}(\mathbf{0}) = \frac{G_1(\mathbf{0})}{9\omega_0^2}\frac{\partial \nu}{\partial \alpha_1}(\mathbf{0}) - \frac{cx^*}{2\omega_0}(\mathbf{0})\frac{\partial(3\omega r - 4 \text{ Im } g)}{\partial \alpha_1}(\mathbf{0}) \quad (3.5.45)$$

$$\frac{\partial \ell_1}{\partial \alpha_1}(\mathbf{0}) = \frac{1}{\omega_0}\left[\frac{\partial \text{Re} c_1}{\partial \alpha_1}(\mathbf{0}) - \text{Im } c_1(\mathbf{0})\frac{\partial \nu}{\partial \alpha_1}(\mathbf{0})\right], \quad (3.5.46)$$

where we took into account that $[\text{Re } c_1](\mathbf{0}) = 0$ and $\mu(\mathbf{0}) = \nu(\mathbf{0}) = 0$.

In this way, the left-hand side of (3.5.41) reads

$$\left[\frac{\partial \nu}{\partial \alpha_1}\frac{\partial(3\omega r - 4\text{Im}g)}{\partial \alpha_2} - \frac{\partial \nu}{\partial \alpha_2}\frac{\partial(3\omega r - 4\text{Im}g)}{\partial \alpha_1}\right]\Bigg|_{\boldsymbol{\alpha} = \mathbf{0}} \quad (3.5.47)$$

Here, we have

$$3\omega r-4\mathrm{Img}=2cx^{*}\nu+\frac{\omega^{2}-2c^{2}\left(x^{*}\right)^{2}\left(1-\left(x^{*}\right)^{2}\right)}{x^{*}\omega},$$

therefore we must consider only

$$\frac{\partial\left(3\omega r-4\mathrm{Img}\right)}{\partial\alpha_{1}}(\mathbf{0})=\left\{\frac{\partial}{\partial\alpha_{1}}\left[\frac{\omega^{2}-2c^{2}\left(x^{*}\right)^{2}\left(1-\left(x^{*}\right)^{2}\right)}{x^{*}\omega}\right]\right\}_{\boldsymbol{\alpha}=\mathbf{0}}.$$

So (3.5.47) is written as

$$\begin{aligned}&\frac{2}{x^{*}}\left(\frac{\partial\nu}{\partial\alpha_{1}}\frac{\partial\omega}{\partial\alpha_{2}}-\frac{\partial\nu}{\partial\alpha_{2}}\frac{\partial\omega}{\partial\alpha_{1}}\right)\\&-\frac{\omega_{0}^{2}-2c^{2}\left(x^{*}\right)^{2}\left(1-3\left(x^{*}\right)^{2}\right)}{\omega_{0}\left(x^{*}\right)^{2}}\left(\frac{\partial\nu}{\partial\alpha_{1}}\frac{\partial x^{*}}{\partial\alpha_{2}}-\frac{\partial\nu}{\partial\alpha_{2}}\frac{\partial x^{*}}{\partial\alpha_{1}}\right)\\=&\frac{2}{x^{*}}\frac{cx^{*}\left(c^{2}+3b^{*}\right)}{6\omega_{0}^{2}\left(b^{*}\right)^{2}}+\frac{4c^{2}\left(x^{*}\right)^{2}}{\omega_{0}}\frac{1}{4c\omega_{0}b^{*}\left(x^{*}\right)^{2}}=\frac{c^{3}}{3\omega_{0}^{2}b^{*}}\neq0,\end{aligned}$$

that proves (3.5.41).

If the condition (3.5.41) holds, by applying the transformation (3.5.40) to the equation (3.5.35), it follows

$$\dot{\mathbf{w}}=\left(\mu_{1}+i\right)\mathbf{w}+\mu_{2}\mathbf{w}\left|\mathbf{w}\right|^{2}+L_{2}\left(\mu\right)\mathbf{w}\left|\mathbf{w}\right|^{4}+O\left(\left|\mathbf{w}\right|^{6}\right),\qquad(3.5.48)$$

where $L_{2}(\mu)=\ell_{2}(\boldsymbol{\alpha}(\mu))$ is a smooth function of μ, with $L_{2}(\mathbf{0})=\ell_{2}(\mathbf{0})\neq 0$.

Finally, using the transformation

$$\mathbf{w}=\sqrt[4]{\left|L_{2}\left(\mu\right)\right|}\;\mathbf{u},\quad\mathbf{u}\in\mathbf{C}^{2}\qquad(3.5.49)$$

and introducing the new parameters

$$\beta_{1}=\mu_{1},\quad\beta_{2}=\sqrt{\left|L_{2}\left(\mu\right)\right|}\mu_{2},\qquad(3.5.50)$$

we obtain the normal form of the equation

$$\dot{\mathbf{u}}=(\beta_{1}+i)\mathbf{u}+\beta_{2}\mathbf{u}|\mathbf{u}|^{2}+s\;\mathbf{u}|\mathbf{u}|^{4}+O(|\mathbf{u}|^{6})\qquad(3.5.51)$$

presenting one Bautin bifurcation point $(\mathbf{u},\beta_{1},\beta_{2})=(\mathbf{0},0,0)$ and a branch of Hopf bifurcation points $(\mathbf{u},\beta_{1},\beta_{2})=(\mathbf{0},0,\beta_{2})$. In (3.5.51) we have $s=$

sign$\ell_2(0)$. Consequently, the normal form for the ODE whose associated dynamical system has a Hopf bifurcation point reads

$$\dot{\mathbf{u}} = i\ \mathbf{u} + \beta_2 \mathbf{u}|\mathbf{u}|^2 + s\ \mathbf{u}|\mathbf{u}|^4 + O(|\mathbf{u}|^6) \tag{3.5.52}$$

and the normal form for the ODE, whose associated dynamical system has a Bautin bifurcation point, reads

$$\dot{\mathbf{u}} = i\mathbf{u} + s\mathbf{u}\,|\mathbf{u}|^4 + O\left(|\mathbf{u}|^6\right) \tag{3.5.53}$$

as $|\mathbf{u}| \to 0$. These two normal forms exist if in addition to the conditions necessary to write (3.5.35), the condition (3.5.41) holds.

The non-hyperbolic limit cycle bifurcation

In the new parameter plane (β_1, β_2), the curves $\beta_2 = 0,\ \beta_1 > 0$ and $\beta_2 = 0,\ \beta_1 < 0$ are the curves of Hopf bifurcation values.

In addition, according to Section 1.3.4, in the plane (β_1, β_2) there exists a curve Ba such that for its points the equation (3.5.51) presents a non-hyperbolic limit cycle. The curve Ba has the equation (1.3.15).

Comming back to the initial variables, the equation of the curve Ba reads

$$\ell_2 \ell_1^2 + 4\nu = 0 \tag{3.5.54}$$

or, equivalently, (3.5.3) taking into account (3.5.36). This proves Theorem 3.5.1.

In the following, the two branches of the curve Ba corresponding to $a \geq 0$ and $a < 0$ will be denoted by Ba_1 and Ba_2, respectively.

Let us remark, finally, that the normal form for the presence of the Hopf bifurcation point was obtained here for the case of two parameters, whilst in Section 3.1.1 we treated the case of one parameter.

3.5.2. NUMERICAL RESULTS SHOWING THE DEGENERATION OF THE HOPF BIFURCATION INTO THE BAUTIN BIFURCATION

Numerical experiments were done for $c = 5$ and fixed b, namely for $b = 0.5;\ 0.55;\ 0.6;\ 0.7;\ 0.8;\ 0.9;\ 1.0;\ 1.2;\ 1.3$ and variable a. They showed the existence of the Bautin bifurcation values at $(b^*_{Ba\ \exp}, a^*_{Ba\ \exp}) = (0.5; \pm 0.65)$. These values agrees very well with the theoretical values given by (3.5.1) for $c = 5$, namely $b^*_{Ba} = 25 - 5\sqrt{24} \approx 0.505103$ and $a^*_{Ba} = \pm\frac{4}{3}\left(5\sqrt{24} - 24\right)\sqrt[4]{\frac{24}{25}} \approx \pm 0.653163$.

In addition, using the coefficients given in Appendix A, it was found that, for $c = 5$, the second Liapunov coefficient is $\ell_2(0) = -1.42482$.

In the parameter plane, the curves $Ba_i,\ i = \overline{1,2}$ emerging from the points Q_{17} and Q_{18} consist of non-hyperbolic limit cycle bifurcation values.

For the concrete case of the system (1.1.17), in Figure 3.5.1 the curves (3.5.3) asymptotically approximating the exact curves $Ba_{1,2}$ were represented for $c = 5$. The representation is qualitative, owing to the small distance between Ba_i and H_i, $i = \overline{1,2}$.

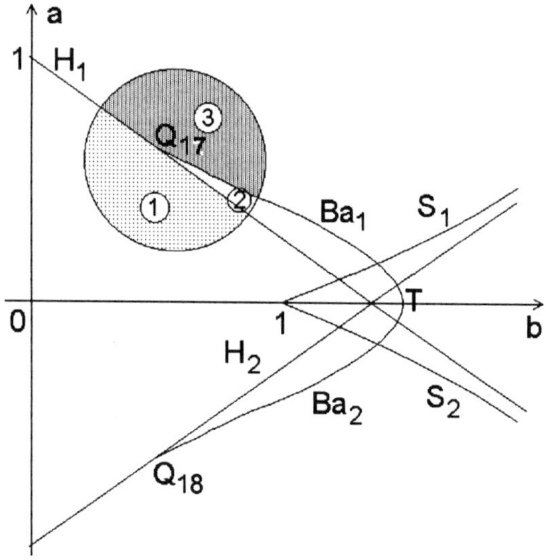

Fig. 3.5.1. The curves $Ba_{1,2}$ of non-hyperbolic limit cycle bifurcation values and the points $Q_{17,18}$ of Bautin bifurcation values, for $c = 5$.

Some values (b, a) situated on H_1 and Ba_1 are given in Table 3.5.1. Remark that the first three decimals of a_{Ba_1} obtained theoretically are the same with those obtained numerically. This is way the approximate curves of non-hyperbolic limit cycle bifurcation values obtained numerically were denoted by $Ba_{1,2}$, too.

b	a_{H1}	a_{Ba1}
0.55	0.6223	0.6227
0.60	0.5880	0.5889
0.70	0.5193	0.521
0.80	0.4507	0.453
0.90	0.3821	0.386
1.00	0.3135	0.318
1.20	0.1764	0.183
1.30	0.1078	0.115
1.3104	0.1034	0.108

Table 3.5.1. Values of (b, a) situated on the Hopf curve H_1 and on the curve Ba_1 obtained theoretically for $c = 5$.

In Figure 3.5.2 phase portraits for the different regions of Figure 3.5.1 around Q_{17} are presented. In Figure 3.5.2a the trajectory through $(1, 0.2)$ is drawn for parameters (b, a) in domain 1 of Figure 3.5.1. Its α-limit point is the only equilibrium point (repulsor) and its ω-limit set is the attractive limit cycle. In Figure 3.5.2b, the trajectory through the same point is represented for (b, a) in domain 2 of Figure 3.5.1. Its α-limit set is the repulsive limit cycle and its ω-limit set is the attractive limit cycle. The only equilibrium point is an attractor, situated inside the repulsive limit cycle. In Figure 3.5.2c, the trajectory through the point $(1, 0.2)$ is represented for parameters in domain 3 of Figure 3.5.1. It emerges from infinity, whilst its ω -limit set is the only equilibrium point, the attractor. Both limit cycles disappeared, due to their collision that took place for (b, a) situated on the curve Ba_1.

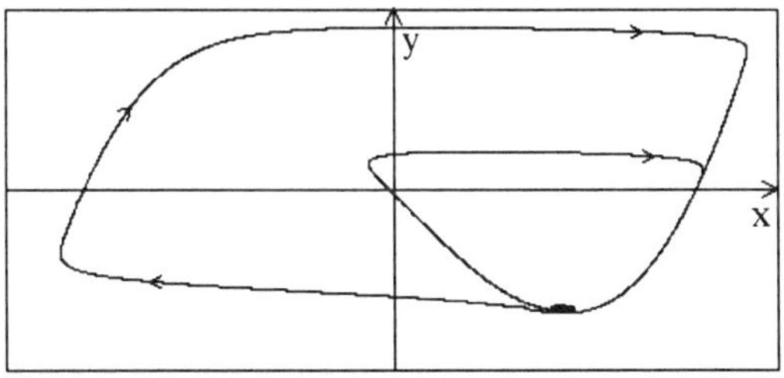

a) $b = 0.55, a = 0.6$

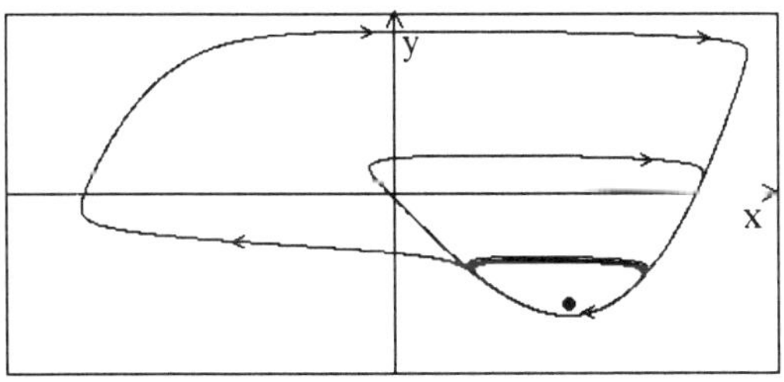

b) $b = 0.55, a = 0.6227481$

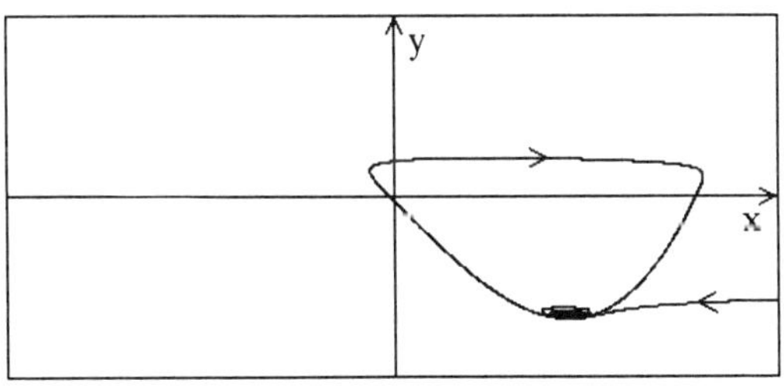

c) $b = 0.55, a = 0.63$

Fig. 3.5.2. Phase portraits for $c = 5$ and (b, a) in domains 1, 2, 3 of Figure 3.5.1.

Numerical computations show that the curves Ba_1 and Ba_2 exist also far from Q_{17} and Q_{18}. They intersect on the Ob-axis at the point T having the abscissae $b_T \approx 1.471143170299$ if $c = 5$. Here the two limit cycles which exist for $a = 0$, $b < b_T$ (represented in Figure 3.3.3b) collide, forming a non-hyperbolic limit cycle.

For $a = 0, b > b_T$, oscillatory regimes do not exist any more.

In Figure 3.5.3a, the three equilibrium points (two attractors and a saddle), the two limit cycles (one attractive, the other repulsive) and the trajectory through the point $(1; 0.2)$ are represented for a value $b < b_T$, whilst in Figure 3.5.3b the three equilibria (two attractors and a saddle) and the trajectory through the same point $(1; 0.2)$ are represented for a value $b > b_T$.

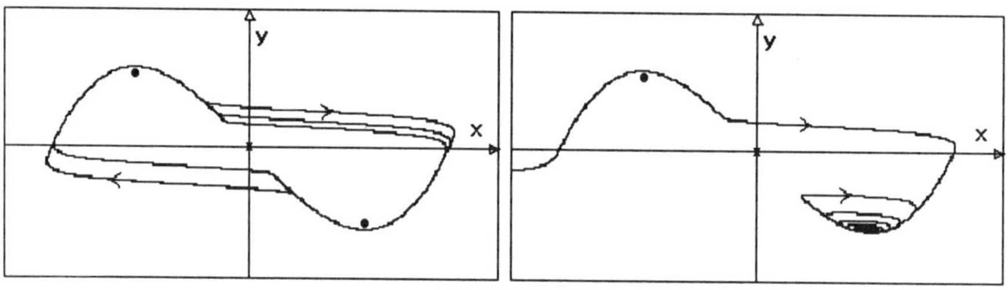

a) $a = 0, b = 1.471143170299$ b) $a = 0, b = 1.4711431702999$

Fig. 3.5.3. Phase portraits for $c = 5$, before and after the collision of the limit cycles, that takes place for $a = 0$ and $b = b_T$.

Finally, let us remark the coexistence of the canard phenomenon and the Hopf and Bautin bifurcations. It will be treated in Section 4.5.

3.5.3. LOCUS OF BAUTIN BIFURCATION VALUES AS c VARIES

Elimination of c between (3.5.1) and (3.5.2) provides, in the (b, a)-plane, the curve locus of Bautin bifurcation values. In order to obtain it, from (3.5.39) and (3.5.44) we find

$$a^*_{Ba} = \pm\frac{4}{3}\left(1 - b^*_{Ba}\right)\sqrt[4]{1 - \frac{1}{c^2}}. \tag{3.5.55}$$

Taking into account in (3.5.55) the expression of c, as a function of b^*_{Ba}, obtained from (3.5.1), it follows that

$$\left(\frac{3a^*_{Ba}}{4\left(1 - b^*_{Ba}\right)}\right)^2 = \frac{1 - b^*_{Ba}}{b^*_{Ba}}. \tag{3.5.56}$$

Hence $b^*_{Ba} \leq 1$.

Then, from (3.5.1), we have $b^*_{Ba} \to \frac{1}{2}$ as $c \to \infty$. Correspondingly, $a^*_H \in (-\frac{2}{3}, \frac{2}{3})$. Finally, (3.5.56) written in the form

$$a^*_{Ba} = \pm\frac{4}{3}\left(1 - b^*_{Ba}\right)\sqrt{\frac{1}{b^*_{Ba}} - 1}, \quad b^*_{Ba} \in (0.5, 1) \tag{3.5.57}$$

represents the curve which is locus of Bautin bifurcation values. In Figure 3.5.4 we ploted this curve for $c \geq 2$, the only case in which our computations for the Bautin bifurcation hold. This is why $b^*_{Ba} \in (0.5, 0.535)$.

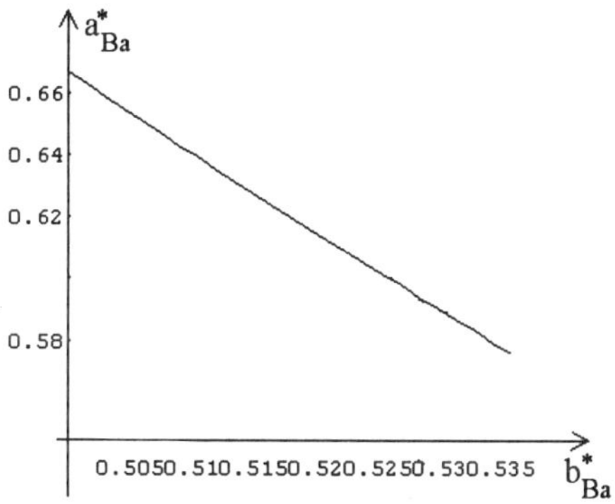

Fig. 3.5.4. The curve locus of Bautin bifurcation values for $a^*_{Ba} > 0$, as c varies.

This curve has a limit point corresponding to $b^*_{Ba} = \frac{1}{2}$. This is why the Bautin phenomenon does not exist for $b^*_{Ba} < \frac{1}{2}$.

The bifurcation value (b^*_{Ba}, a^*_{Ba}) is of Bautin type if the second Liapunov coefficient verifies $\ell_2(\mathbf{0}) \neq 0$. As $\ell_1(\mathbf{0}) = 0, \nu(\mathbf{0}) - 0$, from (3.5.36) it follows $\ell_2(\mathbf{0}) = \dfrac{Rec_2(\mathbf{0})}{\omega_0}$. Taking into account the expression of Re c_2 (Appendix A) and using the expressions (3.5.37) for ω_0, g and r, in which $b^* = b^*_{Ba}$ it is found that

$$\ell_2(\mathbf{0}) = \frac{40 b^*_{Ba}}{9\omega_0^4} \,\mathrm{Re}g\ \mathrm{Im}g \neq 0. \tag{3.5.58}$$

In (3.5.58), b^*_{Ba} is given by (3.5.1). Consequently, all points of the curve in Figure 3.5.4 are Bautin bifurcation values.

3.5.4. THE SYSTEM GENERATING h_{ij} FOR THE CASE OF CUBIC NONLINEARITIES

Consider the case of systems more general than F–N, namely, having the form

$$\begin{aligned} \dot{x} &= f_1(x, y), \\ \dot{y} &= f_2(x, y), \end{aligned} \tag{3.5.59}$$

where f_1 and f_2 are polynomials of the third degree in x and y. From (3.5.25) and (3.5.17) it follows

$$\mathbf{D}\,\dot{\mathbf{w}} + \mathbf{K}\,\dot{\overline{\mathbf{w}}} = \lambda \mathbf{z} + G(\boldsymbol{\alpha}, \mathbf{z}, \overline{\mathbf{z}}) \tag{3.5.60}$$

where $\mathbf{D}$, $\mathbf{K}$ and $\mathbf{w}$ are given by (3.5.26), (3.5.27) and (3.5.24) and G, $\mathbf{z}$ are given by (3.5.18) and (3.5.22) respectively.

Thus, (3.5.60) is an identity of two polynomials in $\mathbf{w}$ and $\overline{\mathbf{w}}$.

The identity (3.5.60) gives the following system of algebraic affine equations in h_{ij}

$$b_{ij} = d_{ij} \tag{3.5.61}$$

where b_{ij} and d_{ij} are the coefficients of $\mathbf{w}^i \overline{\mathbf{w}^j}$ from the left hand side and right hand side of (3.5.60) respectively.

Except for the equations corresponding to indices 21 and 32, all other equations allow unique determination of h_{ij}. Indeed, every equation with $i + j = 2$ contains a single h_{ij} and every equation with $i + j = 3$ or 4 or 5 contains a single h_{ij} with $i+j = 3$, 4 or 5, respectively. In addition, in these last equations, h_{ij} with $i + j < 3$ or 4 or 5 (already determined) occur.

In particular, it follows that

$$h_{20} = \frac{g_{20}}{\lambda}, \qquad h_{11} = \frac{g_{11}}{\overline{\lambda}}, \qquad h_{02} = \frac{g_{02}}{2\overline{\lambda} - \lambda}. \tag{3.5.62}$$

In the equations corresponding to h_{21} and h_{32}, c_1 and c_2 respectively occur. Since the identity (3.5.60) was written with the goal of determining c_1 and c_2, the fact that the equations for h_{21} and h_{32} contain also c_1 and c_2 shows that, in order to determine all these four coefficients, we need supplementary information. This information will be deduced from the condition that the transformation $\mathbf{z} \leftrightarrow \mathbf{w}$ be smooth at $\boldsymbol{\alpha} = \mathbf{0}$ and in the neighborhood of this point in the parameter space.

More precisely, let us recall that equations (3.5.61) followed the requirement that the normal form of the equation (3.5.5) be the simplest possible among all ODEs topologically equivalent to (3.5.5). For the general system (3.5.59) the simplest normal forms are possible, according to various conditions satisfied by f_1 and f_2.

One of these conditions, determining the Hopf bifurcation, is that the linearized system around a Hopf bifurcation point has the eigenvalues $\lambda_1(\boldsymbol{\alpha}) = \overline{\lambda_2(\boldsymbol{\alpha})}$, $\lambda_1 = \mu(\boldsymbol{\alpha}) + i\omega(\boldsymbol{\alpha})$ with the properties $\mu(\mathbf{0}) = 0$, $\omega(\mathbf{0}) > 0$, $\ell_1(\mathbf{0}) \neq 0$, where $\ell_1(\boldsymbol{\alpha})$ is the first Liapunov coefficient.

Similarly, the conditions that determine the Bautin bifurcation read $\mu(\mathbf{0}) = 0$, $\omega(\mathbf{0}) > 0$, $\ell_1(\mathbf{0}) = 0$, $\ell_2(\mathbf{0}) \neq 0$, where $\ell_2(\mathbf{0})$ is the second Liapunov coefficient and

$$\det \begin{pmatrix} \dfrac{\partial \mu}{\partial \alpha_1} & \dfrac{\partial \mu}{\partial \alpha_2} \\ \\ \dfrac{\partial \ell_1}{\partial \alpha_1} & \dfrac{\partial \ell_1}{\partial \alpha_2} \end{pmatrix} \neq 0$$

for $\boldsymbol{\alpha} = \mathbf{0}$.

All these last restrictions allow the reduction of the system (3.5.59) to its normal form presenting the Bautin bifurcation. However, by this reduction, the given system and its normal form must be topologically equivalent. Therefore it is necessary that the transformations of variables, used during the reduction, be, in particular, invertible at the point $\boldsymbol{\alpha} = \mathbf{0}$ and around this point. Let us show that this constraint gives the conditions $h_{21} = h_{32} = 0$.

In the general theory of normal forms these conditions follow from the coefficients of the resonant terms being divided by null eigenvalues of the Lie bracket operators. This is why the terms are not taken into account in the subsequent transformations. Formally, this comes taking $h_{21} = h_{32} = 0$.

In the following we give an equivalent proof for this choice.

In order to do so, consider the equation $b_{21} = d_{21}$. Taking into account the expressions (3.5.62), this equation can be written as

$$\begin{aligned} c_1 &= \frac{g_{21}}{2} + g_{20}g_{11}\frac{2\lambda + \overline{\lambda}}{2\left|\lambda\right|^2} + \frac{\left|g_{11}\right|^2}{\lambda} \\ &\quad + \frac{\left|g_{02}\right|^2}{2\left(2\lambda - \overline{\lambda}\right)} - \frac{\lambda + \overline{\lambda}}{2}h_{21} \\ &\equiv \tilde{G}\left(\boldsymbol{\alpha}\right) - \frac{\lambda + \overline{\lambda}}{2}h_{21}. \end{aligned} \tag{3.5.63}$$

If we would choose $c_1 = 0$, so that the normal form would be simplest, then it will follow that

$$h_{21}\left(\boldsymbol{\alpha}\right) = \frac{2\,\tilde{G}\left(\boldsymbol{\alpha}\right)}{\lambda\left(\boldsymbol{\alpha}\right) + \overline{\lambda}\left(\boldsymbol{\alpha}\right)},$$

such that

$$h_{21}\left(\mathbf{0}\right) = \frac{2\,\tilde{G}\left(\mathbf{0}\right)}{0} \to \infty,$$

and the transformation $\mathbf{z} \leftrightarrow \mathbf{w}$ would not be smooth at $\boldsymbol{\alpha} = \mathbf{0}$. As a consequence, the given system will be not topologically equivalent to its normal form. So c_1 must not be zero. Then

$$h_{21} = \frac{2\left(-c_1 + \widetilde{G}\right)}{\lambda + \overline{\lambda}},$$

so $h_{21} \neq \infty$ iff $c_1 = \widetilde{G}$, in order to have the indeterminable $\frac{0}{0}$. Then (3.5.63) implies $h_{21} = 0$.

Similarly, h_{32} must be equal to zero. This is because its coefficient $\frac{\lambda + \overline{\lambda}}{6}$ is also proportional to $\lambda + \overline{\lambda}$.

It follows that equations (3.5.61) corresponding to indices 21 and 32 will give c_1 and c_2 respectively. The expressions of h_{ij} ($ij \neq 21$ and $ij \neq 32$), c_1 and c_2 for this general case are given in [123].

CHAPTER 4

MODELS OF ASYMPTOTIC APPROXIMATION FOR THE FITZHUGH–NAGUMO SYSTEM AS $c \to \infty$

After presenting all types of asymptotic behaviour of interest for the solutions of the F–N model (Section 4.1), we emphasize only the case $c \to \infty$ and its periodic evolutions. In order to do this the outer and the inner asymptotic expansions of first order (Section 4.2) and higher orders (Section 4.3) are deduced and matched. In addition, in Section 4.4, the time necessary to cover different segments of the limit cycle for some values of c and for some models of asymptotic approximation, as $c \to \infty$, is computed. In Section 4.5, relaxation oscillations, ducks and non-hyperbolic limit cycle bifurcation are discussed. Some results on the asymptotics involved into their investigation are presented. They form a basis for our numerical results on these oscillations and phase trajectories.

4.1. TYPES OF ASYMPTOTIC BEHAVIOUR OF THE SOLUTION OF THE F–N MODEL

We emphasized that the evolution of the electrical potential of the nodal cell membrane in the heart is governed by the following Cauchy problem for the equation (1.1.17)

$$\begin{aligned} \frac{dx}{dt} &= c\left(x + y - \frac{x^3}{3}\right), \\ \frac{dy}{dt} &= -\frac{1}{c}\left(x - a + b\,y\right), \\ x\left(0\right) &= x^0,\ y\left(0\right) = y^0. \end{aligned} \tag{4.1.1}$$

Here the dynamics of (x^0, y^0) depends on the three parameters $a, b, c \in \mathbf{R}$ and, in the phase space, it takes place along three types of topologically nonequivalent trajectories:

— equilibrium points (trajectories reduced to a point, corresponding to particular stationary motions),

— limit cycles (trajectories which are isolated closed curves, corresponding to periodic motions)

— unbounded trajectories, corresponding to transient motions from equilibria or limit cycles with repulsive properties (or from infinity) to equilibria or limit cycles with attractive properties (or to infinity).

This dynamics is very different for different regions of the parameter space. As an example, for $a = b = 0$ the Van der Pol equation is obtained. For this equation, since 1930 it has been known that the origin is the only equilibrium point (repulsor) and it is situated inside a limit cycle (the attractor). This situation holds no longer for all a and b.

The function $x(t)$, the main component of the solution for (4.1.1), must be written as $x(a, b, c, t, x^0, y^0)$, but for the sake of simplicity we prefer to write it as $x(t)$. If a, b are chosen such that there exists a limit cycle for different values of c, then $x(t)$ is periodic with respect to the time for (x^0, y^0) situated on the limit cycle and it has different forms, for different values of c. For small c, $x(t)$ is almost sinusoidal, whilst as c is increasing it becomes almost like a relaxation oscillation. In this case $1/c$ becomes a small parameter, and so an asymptotic study must be done. This is the main aim of this chapter.

Before performing this study let us mention the other types of asymptotic behaviours necessary to investigate if a complete information about the solution x is in view. Thus with respect to the time t the problem of the behaviour of x as $t \to \pm\infty$ can be formulated, but, in fact, this is the problem of the attractivity or repulsivity of the solution, already dealt with in the previous chapters.

Another important asymptotic aspect corresponds to $c \to 0$, the value $c = 0$ being a limit value for the definition of the variables from (4.1.1).

The first two equations (4.1.1) can be written in the form

$$\begin{aligned} \dot{x} &= c\left(x + y - \frac{x^3}{3}\right), \\ c\,\dot{y} &= -x + a - by. \end{aligned} \tag{4.1.2}$$

Therefore, if $c \to 0$, two limit systems are possible

$$\begin{aligned} &\dot{x} = 0, \\ &-x + a - by = 0, \qquad \text{for } c\,\dot{y} = 0 \end{aligned} \tag{4.1.3}$$

or

$$\begin{aligned} &\dot{x} = 0, \\ &c\,\dot{y} = -x + a - by, \qquad \text{for } c\,\dot{y} \neq 0. \end{aligned} \tag{4.1.4}$$

In the first case (4.1.2) degenerated into a system of one ODE and one algebraic equation. The corresponding dynamics of the points (x, y) of the phase space proceeds along straight lines $x = x^0$ as follows:

1^0 if $\| \dot{y} \| = 0$ and $a, b \neq 0$ the initial points (x^0, y^0) tend to the equilibrium $\left(x^*, \frac{a - x^*}{b}\right)$ situated on the straight line

$$-x + a - by = 0. \tag{4.1.5}$$

This is like a sudden fall of the entire phase space onto the straight line (4.1.5);

2^0 if $\| \dot{y} \| = \infty$ and $b = 0$, then the straight line $x = a$ consists of equilibria, all other points of the phase space falling suddenly onto infinity. We must note that $\| \dot{y} \| = \infty$ at all points but the equilibria;

3^0 if $\| \dot{y} \| < \infty$, the cases 1^0 and 2^0 occur. However, the dynamics along the trajectories proceeds at a bounded velocity.

The second case occurs only for points external to the straight line (4.1.5). In this case there are no equilibria and $\| \dot{y} \| = \infty$. Hence, except for the points of (4.1.5), all points of the phase space fall suddenly onto infinity along the straight lines $x = x^0$.

Consequently in the case $c \to 0$ the dynamics is mostly related to the nullcline (4.1.5). As $c \to 0$ we have: $a_{H_{1,2}} \to \pm 1$, if $b = 0$; $a_{H_{1,2}} \to \mp\infty$, if $b \neq 0$.

We shall not treat the case $c \to 0$, because all our considerations concern the case $c > 1 + \sqrt{3}$, the only one of physiological interest.

The case $c = 1$ is not a limit case. However, it is very important because of the coincidence of the following points of intersection of the main curves $H_{1,2}$ and $C_i, i = \overline{1,4}$ with the Ob-axis: $H_1 \cap H_2 = C_1 \cap C_2 = (-3, 0)$; $H_1 \cap H_2 = S_1 \cap S_2 = C_3 \cap C_4 = (1, 0)$.

For the nonlimit case $c = 3$ we have the following coincidence of intersection $H_1 \cap H_2 = C_1 \cap C_2 = (9, 0)$.

The problems of the asymptotic behaviour of x as $a \to \pm\infty$ and $b \to \pm\infty$, and for all possible combinations where two or all the three parameters are very large, can be treated too. These cases are excluded in our study as physiological nonrelevant. Of course, a last asymptotic investigation could concern the cases of large parameters and time. Each of these cases might be the concern of a future study. Let us add that the asymptotic investigation can be useful not only around the limit points at which the problem, mathematical or physiological, has no meaning, but also around some asymptotic points that belong to the domain of definition of the independent variable and of the parameter. This could also be a future project.

In the case $c \to \infty$ the first two equations (4.1.1) can be written in the form

$$\begin{aligned} \tfrac{1}{c}\, \dot{x} &= y + x - \frac{x^3}{3}, \\ \dot{y} &= -\tfrac{1}{c}\,(x - a + by)\,. \end{aligned} \tag{4.1.6}$$

Therefore in the limit case $c \to \infty$ the following two limit systems

$$\begin{aligned} 0 &= y + x - \frac{x^3}{3} \quad \text{for } \frac{1}{c}\,\dot{x} = 0 \\ \dot{y} &= 0, \end{aligned} \tag{4.1.7}$$

and

$$\begin{aligned} \frac{1}{c}\,\dot{x} &= y + x - \frac{x^3}{3}, \\ \dot{y} &= 0, \end{aligned} \tag{4.1.8}$$

are possible. A sudden fall of the phase space on the R–H curve follows similarly to the case $c \to 0$. This is why here we add only the results concerning the intersection of the main curves of interest for the codimension three bifurcation. We also take into account the results from Section 2.6. The curves H_1 and $C_{1,2}$ tend to their asymptotes A_1, and the curves H_2 and $C_{3,4}$ to A_2. They intersect the Ob-axis at the point $\left(\frac{3}{2}, 0\right)$. Therefore this represents the common point of intersection of H_1, H_2, C_1, C_2, C_3 and C_4 with the Ob-axis.

Finally, let us mention that, in spite of x being the state variable of main interest in physiology, the investigation of y may also be useful and we shall carry it out. We could avoid this study by considering the second order equation in x, but no simplification of computations would be obtained.

4.2. FIRST ORDER ASYMPTOTIC APPROXIMATIONS AS $\varepsilon \to 0$

4.2.1. THE OUTER APPROXIMATION

Suppose that the time-scale c is large, i.e., $c \gg 1$, and introduce a new time $t_1 = t/c$, which is asymptotically fixed, i.e., $t_1 = \text{Ord}\,(1)$, $\varepsilon \to 0$. Then the equations from (4.1.1) become

$$\begin{aligned} \varepsilon \frac{dX}{dt_1} &= Y + X - \frac{X^3}{3}, \\ \frac{dY}{dt_1} &= -\left(X - a + bY\right), \end{aligned} \tag{4.2.1}$$

where $\varepsilon = 1/c^2 \ll 1$ is the relaxation parameter. In addition $x\left(t\left(t_1\right)\right)$ and $y\left(t\left(t_1\right)\right)$ were redenoted by $X\left(t_1\right)$ and $Y\left(t_1\right)$ respectively.

Thus our asymptotic analysis begins with the F–N model consisting of equations (4.2.1) and the initial conditions obtained from $(4.1.1)_3$

$$\begin{aligned} X\left(0\right) &= x^0, \\ Y\left(0\right) &= y^0. \end{aligned} \tag{4.2.2}$$

Consider that a phase trajectory of (4.2.1) possesses a segment on which $\dot{X} = O\,(1)$, $\varepsilon \to 0$. In this case we have $\varepsilon\,\dot{X} \ll 1$ and, in the first asymptotic approximation, (4.2.1) becomes

$$\begin{aligned} &Y_1 + X_1 - \frac{X_1^3}{3} = 0, \\ &\dot{Y}_1 = -\,(X_1 - a + b\,Y_1)\,. \end{aligned} \tag{4.2.3}$$

Equation $(4.2.3)_1$ shows that this segment of the trajectory is very close to the R–H curve and, on it, the dynamics takes place according to $(4.2.3)_2$. In addition, $\dot{X}$ is almost moderate since $\dot{X} = \mathrm{Ord}(1)$ as $\varepsilon \to 0$, or it is small, that is $\dot{X} = o\,(1)$ as $\varepsilon \to 0$. These two conditions can be written as a single one: $\dot{X} = O\,(1)$, as $\varepsilon \to 0$.

4.2.2. THE INNER APPROXIMATION

On the part of the trajectory which is not close to the R–H curve, $\dot{X}$ is very large, such that the product $\varepsilon\,\dot{X}$ in (4.2.1) can be neglected no longer. This means that $\dot{X} \gg 1/\varepsilon$, i.e., X varies fast with t_1. It also means that the time within which this variation takes place is very small, therefore its appropriate scale is small. By **appropriate time scale** we mean the time scale such that the ratio between t_1 and the time scale is neither too large, nor too small. In this way, except for ε, all quantities from equations and conditions are asymptotically fixed as $\varepsilon \to 0$. This rescaling allows us to perform a more or less formal and easy asymptotic study.

Suppose that ε is the appropiate time-scale and, correspondingly, introduce the asymptoticaly fixed new time $\eta = t_1/\varepsilon$. Then (4.2.1) becomes

$$\begin{aligned} &\frac{d\hat{x}}{d\eta} = \hat{y} + \hat{x} - \frac{\left(\hat{x}\right)^3}{3}, \\ &\frac{d\hat{y}}{d\eta} = -\varepsilon\left(\hat{x} - a + b\;\hat{y}\right), \end{aligned} \tag{4.2.4}$$

where

$$X\,(t_1, \varepsilon) = X\,(\eta\varepsilon, \varepsilon) \equiv \hat{x}\,(\eta, \varepsilon)$$

and

$$Y\,(t_1, \varepsilon) = Y\,(\eta\varepsilon, \varepsilon) \equiv \hat{y}\,(\eta, \varepsilon)\,.$$

In first approximation, (4.2.4) becomes

$$\begin{aligned} \frac{d\hat{x}_1}{d\eta} &= \hat{y}_1 + \hat{x}_1 - \frac{\left(\hat{x}_1\right)^3}{3}, \\ \frac{d\hat{y}_1}{d\eta} &= 0. \end{aligned} \tag{4.2.5}$$

The last equation shows that $\hat{y}_1 = \widehat{y_1^0}$, hence the segment on the phase trajectory through (x^0, y^0) along which our assumption holds is almost parallel to the X-axis and the dynamics on it proceeds acording to $(4.2.5)_1$.

As a conclusion, in the first asymptotic approximation as $\varepsilon \to 0$, the dynamics along a transient trajectory of (4.2.1) is described by (4.2.3) and (4.2.5). The trajectory has two regions with different dynamics:

— one of them, near the R–H curve, along which the dynamics is slow (described by (4.2.3));

— another one, which is a segment almost parallel to the OX-axis, which is covered within the time $t_1 = \varepsilon\eta = O(\varepsilon)$, as $\varepsilon \to 0$, characterized by a fast dynamics described by (4.2.5).

Let us remark that we considered values of a, b and c such that the Cauchy problem (4.2.2) for (4.2.1) admits, for some (x^0, y^0), a periodic solution, corresponding to a limit cycle from the phase plane. The equilibrium point inside the limit cycle is a repulsor.

Thus, in the first approximation, taking into account only the approximate equations and supposing that the outer branches of the R–H curve contain no equilibrium points, it can be stated that the dynamics, for $t > 0$, of a point $I_0(x^0, y^0)$ along the trajectory, presents a jump along the straight line $Y = y^0$, from $I_0(x^0, y^0)$ until it reaches the R–H curve at A_1'', then the R–H curve is slowly covered until a point close to the minimum point $m = (X_m, Y_m)$ (Figure 4.2.1). Then a new jump to the other branch of the R–H curve takes place, followed by a slow covering of this branch until its maximum point $M = (X_M, Y_M)$ is reached. Afterwards, the jump along the straight line $Y = Y_M$ follows until the first branch of the R–H curve is reached and the dynamics goes on as a cyclic motion in four steps: a slow motion along the lateral branches of the R–H curve and jumps along the parallel straight lines at the OX-axis, passing through the maximum and the minimum points of the R–H curve. The same behaviour for large time could be obtained if the initial point would belong to $AmBM$. In this way the limit cycle, which is the ω-limit set for the phase trajectory through (x^0, y^0), is approximated by the curve $AmBM$, which in Figure 4.2.1 is drawn as discontinuous lines. The straight lines MA and mB are parallel to the OX-axis, while Am and BM are regions of the R–H curve.

Since the dynamics described by (4.2.3) and (4.2.5) are one-dimensional proceeding along some one-dimensional manifolds, they are degenerate. They may be interpreted as: the entire phase space comes with an infinite velocity in a zero time to the outer branches of the R–H curve. This limit situation is reached as a result of an entire process of fastening the dynamics along segments of the phase trajectories which are not close to the R–H curve. This idea is supported by the results in Table 4.4.1. In order to describe this process, higher-order asymptotic approximations are needed. They will model dynamics for large, but not infinte values of c.

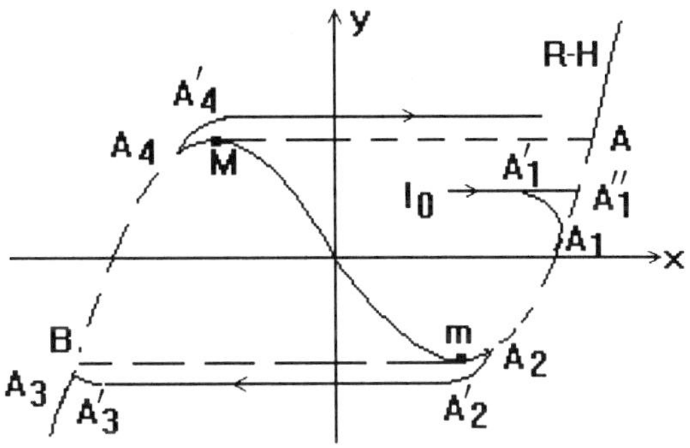

Fig. 4.2.1. The limit cycle (dashed line) and a first order asymptotic approximation transient trajectory, as $c \to \infty$.

It is very important to note that in defining t_1 it was not t which was very small (of order of ε) but the difference $t - t^*$, where t^* is the time at which the trajectory started at (x^0, y^0) passed through the extremum points of the R–H curve. Since all systems of ODEs involved in this investigation are autonomous, we understand that this meaning of t was used. Of course, on the parallel straight line passing thgrough M we have a t^* different from t^* corresponding to the parallel straight line passing through m.

This means that at each return t^* will change. This situation shows the difference between the singular pertubation for Cauchy problems and local problems. In two-point, say, problems, the quick variation in the values of the unknown functions occurs near a few known values of the independent variable whilst in initial value problems this type of variation takes place near an infinity values of time. This will imply difficulties in applying the inner–outer expansion method and will ask for more appropiate perturbation methods, e.g. the boundary layer method [51].

The closed curve $AmBM$ is a first asymptotic approximation of the limit cycle as $c \to \infty$.

The description, in the first approximation, of the dynamics of a point (x^0, y^0) in the phase plane could be also done by eliminating the time between equations (4.2.1). Taking into account that

$$\frac{dY}{dX} = \frac{\dot{Y}}{\dot{X}},$$

the equation

$$\left(Y + X - \frac{X^3}{3}\right) \frac{dY}{dX} = -\varepsilon \left(X - a + b\,Y\right), \tag{4.2.6}$$

would be obtained.

If $X - a + bY$ is not too large, we have either

$$\frac{dY}{dX} = 0,$$

or

$$Y + X - \frac{X^3}{3} = 0,$$

that is, the dynamics proceed either along R–H, or along straight lines parallel to the OX-axis.

4.2.3. INNER–OUTER EXPANSION MATCHING

So far, the initial conditions for (4.2.3) and (4.2.5) were not taken into account, understanding that the initial points for the segments of R–H are M, m, A and B. The Cauchy problem (4.1.1) gives only the two initial conditions $(4.1.1)_3$ that ensure the existence and uniqueness of the solution of (4.2.1), (4.2.2). If $\left(x^0, y^0\right)$ belongs to the curve MAmB, then these conditions will uniquely determine this curve, which is the limit cycle as $c \to \infty$. If $\left(x^0, y^0\right)$ does not belong to this cycle, i.e., $\left(x^0, y^0\right)$ is situated on a transient trajectory, then the initial conditions $(4.1.1)_3$, are not any longer appropriate to ensure the existence and uniqueness of the solution (corresponding to the approximate dynamics) of the Cauchy problem (4.2.2) for (4.2.3) and (4.2.5). Indeed, by associating with (4.2.3) the conditions (4.2.2), an over-determined problem which has no solution is obtained. On the other hand, by imposing only the condition (4.2.2) for Y in the form

$$Y_1\left(0\right) = y^0, \tag{4.2.7}$$

we obtain the Cauchy problem (4.2.3), (4.2.7), which has the solution $(X_1(t), Y_1(t))$ corresponding to the phase trajectory

$$Y_1 = -X_1 + \frac{X_1^3}{3}. \tag{4.2.8}$$

This solution is the parametric form of the curve (4.2.8) obtained by solving $(4.2.3)_1$ with respect to X_1, introducing the expression of X_1 into $(4.2.3)_2$ and solving the obtained ordinary differential equation in Y_1. Since the initial condition

$$X(0) = x^0 \tag{4.2.9}$$

was neglected, we could not expect that the solution (4.2.8) satisfies it. Indeed, in general $\frac{(x^0)^3}{3} - x^0 \neq y^0$ and so $X_1(0) \neq x^0$.

As a conclusion we expect that (4.2.8) approximates the solution of (4.2.1), (4.2.2) on the segment $A_1 m$ where the variation speed with respect to t_1 of Y is the same as that of X or is much higher than it, corresponding to the two situations $\dot{X} = \text{Ord}(1)$ or $\dot{X} \ll 1$ for $\varepsilon \to 0$. Here A_1 is situated on the R–H curve and is different from A_1''.

On the other hand, on the segment $I_0 A_1'$ (where A_1' is situated on this straight line and close to R–H) where the situation is the opposite $\left(\dot{X} \gg Y\right)$, we shall approximate the solution of (4.2.1), (4.2.2) by the solution of (4.2.5), (4.2.2) that is

$$\hat{y}_1 = y^0, \tag{4.2.10}$$

the variation of $\hat{x}_1$ being given by the solution of the problem

$$\begin{aligned} \frac{d\hat{x}_1}{d\eta} &= y^0 + \hat{x}_1 - \frac{\left(\hat{x}_1\right)^3}{3}, \\ \hat{x}_1(0) &= x^0. \end{aligned} \tag{4.2.11}$$

We do not give here the explicit form of $\hat{x}_1(t)$ because it depends on y^0, and, consequently, on the number of real solutions of equation

$$y^0 + \hat{x}_1 - \frac{\left(\hat{x}_1\right)^3}{3} = 0$$

and their values. Anyway, the solution $\left(\hat{x}_1, \hat{y}_1\right)$ with the initial data $\left(x^0, y^0\right)$ is completely determined.

We expect the solution $\left(\hat{x}_1, \hat{y}_1\right)$ to represent a good approximation of (X, Y) only for a short time interval $[0, t_0]$, after which (X, Y) is approximated well by (X_1, Y_1). But asymptotically fixed t_0 means $\eta \to \infty$. This is why the following matching conditions of the two approximations $\hat{x}_1$ and X_1 of X and $\hat{y}_1$ and Y_1 of Y must take place

$$\lim_{\eta\to\infty} \hat{x}_1(\eta) = \lim_{t\to 0} X_1(t) \equiv X_1(0),$$
$$\lim_{\eta\to\infty} \hat{y}_1(\eta) \equiv Y_1(0). \tag{4.2.12}$$

Obviously, the last condition is satisfied. The other one is used to derive $X_1(0)$. Expanding (4.2.11) around $\eta = \infty$, it follows that $X_1(0)$ and $Y_1(0)$ satisfy (4.2.8).

The aim of our asymptotic study was to offer an approximate solution of a transient trajectory situated near a limit cycle of problem (4.2.1), (4.2.2). Until now we have obtained this only for small times: the trajectory through the initial point I_0 consists of a region I_0A_1', parallel to OX-axis, where $A_1'=(X_1(0), Y_1(0))$ is given by (4.2.12) where $\hat{x}_1(\eta)$ is the solution of (4.2.11), while $Y_1(0)$ is solution of (4.2.3) corresponding to the datum

$$y^0 = Y_1(0) = \frac{X_1^3(0)}{3} - X_1(0).$$

For a larger time, the region $A_1'A_1$ of the trajectory is intermediate between the straight line $Y = y^0$ and the branch of the R–H curve and has equations $Y = Y_1(t)$, given by $(4.2.3)_2$ and $X = X_1(t)$, solution of $(4.2.3)_2$. For still larger time the trajectory describes a segment of the R–H curve and reaches a point A_2, analogous to A_1. Then it follows a segment A_2A_2', where A_2 and A_2' are determined from matching conditions. Further, the trajectory follows the segment of the straight line $A_2'A_3$, parallel to the OX-axis and so on.

4.3. HIGHER ORDER ASYMPTOTIC APPROXIMATIONS AS $\varepsilon \to 0$

4.3.1. MODELS OF OUTER ASYMPTOTIC APPROXIMATION AS $\varepsilon \to 0$

Consider now the higher order outer asymptotic approximations, i.e.,

$$X(t_1,\varepsilon) \sim X_1(t_1) + \varepsilon X_2(t_1) + \varepsilon^2 X_3(t_1) + ..., \quad \varepsilon \to 0 \tag{4.3.1}$$
$$Y(t_1,\varepsilon) \sim Y_1(t_1) + \varepsilon Y_2(t_1) + \varepsilon^2 Y_3(t_1) + ..., \quad \varepsilon \to 0 \tag{4.3.2}$$

with respect to the asymptotic sequence $1, \varepsilon, \varepsilon^2, ...$ as $\varepsilon \to 0$.

Introducing now (4.3.1) and (4.3.2) into (4.2.1) it follows that

$$
\begin{aligned}
&X_1 + Y_1 - \frac{X_1^3}{3} = 0,\\
&Y_2 + X_2 - X_1^2 X_2 + \dot{X}_1 - \frac{X_2^3}{3} = 0,\\
&Y_3 + X_3 - X_1 X_2^2 - X_1^2 X_2 + \dot{X}_2 - \frac{X_3^3}{3} = 0,\\
&\ldots\ldots\ldots\ldots\ldots\ldots\ldots\ldots\ldots
\end{aligned}
\tag{4.3.3}
$$

$$
\begin{aligned}
&\dot{Y}_1 = -X_1 + a - bY_1,\\
&\dot{Y}_2 = -X_2 - bY_2,\\
&\dot{Y}_3 = -X_3 - bY_3,\\
&\ldots\ldots\ldots\ldots
\end{aligned}
\tag{4.3.4}
$$

Of course, the first equations (4.3.3) and (4.3.4) are identical to (4.2.3). Once X_1 and Y_1 determined from (4.2.3), then, from $(4.3.3)_2$ and $(4.3.4)_2$, X_2 and Y_2 will follow and the next equations will allow us to determine X_3, Y_3 and so on.

Let us remark (Section 1.4) that the equations for X_2, X_3, ... and Y_2, Y_3, ... have variable coefficients and they are affine. Therefore, they can be immediately solved by the constant variation method. For example, $(4.3.3)_2$ implies the expression

$$X_2 = \frac{-Y_2 + \dot{X}_1}{1 - X_1^2}, \tag{4.3.5}$$

which, introduced in $(4.3.4)_2$, yields the equation for Y_2

$$\dot{Y}_2 = \frac{Y_2\left[1 - b\left(1 - X_1^2\right)\right] + \dot{X}_1}{1 - X_1^2}. \tag{4.3.6}$$

As X_1 is already known from Section 4.2, it follows that (4.3.6) is an affine equation of first order with variable coefficients and its form is

$$\dot{Y}_2 + p(t_1)\, Y_2 + q(t_1) = 0. \tag{4.3.7}$$

This equation has an explicit solution given by

$$Y_2(t_1, Y_2(0)) = e^{-\int_0^{t_1} p(t)\,dt}\left[Y_2(0) - \int_0^{t_1} q(s)\, e^{\int_0^s q(r)\,dr} ds\right]. \tag{4.3.8}$$

Once Y_2 is determined X_2 follows immediately from (4.3.5).

Similarly, X_n, Y_n, for $n \geq 3$ are determined.

In addition, for Y_1 from (4.2.3) we have an explicit solution too, for every explicit solution X_1 as a function of Y_1 of $(4.2.3)_1$. Indeed, by replacing that expression of X_1 into $(4.2.3)_2$ we obtain for Y_1 an equation which has the form (4.3.7) with other coefficients p and q.

Putting together all these results we find that the form of (4.3.1) and (4.3.2) is known. It depends on the initial values $Y_n(0)$, $n = 1, 2, 3, \ldots$.

However, even if the form of the outer expansions for X and Y is known, because these expansions contain the initial data, this form must be particularly determined for every set of initial data.

4.3.2. EQUATIONS OF INNER ASYMPTOTIC APPROXIMATION AS $\varepsilon \to 0$

After introducing the inner independent variable $\eta = t_1/\varepsilon$ and transforming the equations (4.2.1) into (4.2.4), let us perform an inner asymptotic expansion of the solution $\left(\hat{x}, \hat{y}\right)$ with respect to the asymptotic sequence $1, \varepsilon, \varepsilon^2, \ldots, \quad \varepsilon \to 0$. We have

$$\hat{x}(\eta, \varepsilon) \sim \hat{x}_1(\eta) + \varepsilon \hat{x}_2(\eta) + \varepsilon^2 \hat{x}_3(\eta) + \ldots, \quad \varepsilon \to 0, \quad (4.3.9)$$

$$\hat{y}(\eta, \varepsilon) \sim \hat{y}_1(\eta) + \varepsilon \hat{y}_2(\eta) + \varepsilon^2 \hat{y}_3(\eta) + \ldots, \quad \varepsilon \to 0, \quad (4.3.10)$$

Introducing (4.3.9) and (4.3.10) into (4.2.4), we find

$$\begin{aligned}
\dot{\hat{x}}_1 &= \hat{y}_1 + \hat{x}_1 - \frac{\hat{x}_1^3}{3}, \\
\dot{\hat{x}}_2 &= \hat{y}_2 + \hat{x}_2 - \hat{x}_1 \hat{x}_2, \\
\dot{\hat{x}}_3 &= \hat{y}_3 + \hat{x}_3 - \hat{x}_1 \left(\hat{x}_2\right)^2 - \hat{x}_1^2 \hat{x}_2, \\
&\ldots\ldots\ldots\ldots\ldots\ldots
\end{aligned} \quad (4.3.11)$$

$$\begin{aligned}
\dot{\hat{y}}_1 &= 0, \\
\dot{\hat{y}}_2 &= -\hat{x}_1 + a - b\,\hat{y}_1, \\
\dot{\hat{y}}_3 &= -\hat{x}_2 - b\,\hat{y}_2, \\
&\ldots\ldots\ldots\ldots\ldots\ldots
\end{aligned} \quad (4.3.12)$$

The first approximations $(4.3.11)_1$ and $(4.3.12)_1$ coincides with (4.2.5). Using them, from $(4.3.11)_2$ and $(4.3.12)_2$, $\hat{x}_2$ and $\hat{y}_2$ are determined as functions of $\hat{x}_1, \hat{y}_1$and initial data $\hat{x}_2(0)$ and $\hat{y}_2(0)$. Again all $\hat{x}_n$ and $\hat{y}_n$ satisfy first-order affine ODEs with variable coefficients. First, $\hat{y}_n$ will be determined by a simple integration. For instance, we have

$$\hat{y}_2 = \int_0^{\eta} \left(-\hat{x}_1 + a - b\,\hat{y}_1 \right) \left(\eta' \right) d\eta' + \hat{y}_2\,(0)\,. \tag{4.3.13}$$

With $\hat{y}_n$ determined, equations in $\hat{x}_n$ from (4.3.11) are solved. As an example we have

$$\dot{\hat{x}}_2 = \hat{x}_2 - \hat{x}_1 \hat{x}_2 + \int_0^{\eta} \left(-\hat{x}_1 + a - b\,\hat{y}_1 \right) \left(\eta' \right) d\eta' + \hat{y}_2\,(0)\,. \tag{4.3.14}$$

Hence, like in the case of outer asymptotic approximations, to solve the equations for different approximations of order greater or equal to two of the solution of (4.2.1), means to solve some equations of type (4.3.7). The solutions obtained can be written explicitly, but they contain the initial data $\hat{x}_n\,(0)$, $\hat{y}_n\,(0)$.

4.3.3. INNER–OUTER EXPANSION MATCHING

In our case, by a model we understood a Cauchy problem for an ODE. The given model is (4.2.1) and we wish to construct for it models of outer and inner asymptotic approximation of orders $1, 2, 3, ...$ as $\varepsilon \to 0$. The models of the first order approximation are called reduced models and they have been found in Section 4.2. They were not independent. More precisely, from the inner reduced model (4.2.5), the solution $\left(\hat{x}_1, \hat{y}_1 \right)$ is determined as a function of the initial data (x^0, y^0) of problem (4.2.1). Then, with its aid, from (4.2.12), the datum $X_1(0)$ for the outer reduced model (4.2.3) is determined. The datum $Y_1(0)$ for this last model follows from (4.2.8) or $(4.2.3)_1$, in which $X_1(0)$ is given by (4.2.12). As a conclusion, the reduced inner and outer models are

$$\begin{aligned} \frac{d\,\hat{x}_1}{d\eta} &= y^0 + \hat{x}_1 - \frac{\left(\hat{x}_1\right)^3}{3}, \\ \hat{y}_1 &\equiv y^0, \\ \hat{x}_1\,(0) &= x^0 \end{aligned} \tag{4.3.15}$$

and

$$\begin{aligned}
&\frac{dY_1}{dt_1} = -\left(X_1 - a + bY_1\right),\\
&Y_1 + X_1 - \frac{X_1^3}{3} = 0,\\
&X_1(0) = \lim_{\eta\to\infty} \hat{x}_1(\eta),\\
&Y_1(0) = \frac{\left(X_1(0)\right)^3}{3} - X_1(0)
\end{aligned} \tag{4.3.16}$$

respectively.

Let us remark that in (4.2.7) we did not specify $X_1(0)$ because it followed from (4.2.12) and (4.2.8). If we would impose it to be equal to x^0, a contradiction would be obtained.

It follows that the outer reduced model can be determined after the inner one was solved. But if, before reaching m, there exists an equilibrium point A_2, then near it we have $\dot{Y}_1 \ll 1$ and the asymptotic treatment must be investigated in this region and not near m. The matching of the two models proceeds similarly with that from region A_1A_1'.

Let us remark that:

1. $\hat{x}_1$ depends both on x^0 and y^0. This will imply that $X_1(0)$, $Y_1(0)$ and consequently $Y_1(t_1)$ and $X_1(t_1)$ depend also on x^0 and y^0;

2. As we have already remarked in Section 1.4 the choice of the initial conditions attached to equations of various orders of asymptotic approximation is not imposed by any theory, excepting simple particular linear cases;

3. The problem (4.2.5), $(4.1.1)_3$ of first inner asymptotic approximation included all initial conditions of the given problem (4.2.1), whilst problem (4.2.3), (4.2.7) of the first outer asymptotic approximation took into account the conditions obtained from the inner reduced problem;

4. In principle the $n \times m$ matchings with $n, m > 1$ are carried out as for the case $n = m = 1$. Of course, the difference is that now, instead of first approximations, we take the higher approximations.

As an example, let us present the 1×2 matching. The two-terms inner asymptotic expansion for X_1 is $\hat{x}_1(\eta) + \varepsilon\,\hat{x}_2(\eta)$. This function is rewritten as $\hat{x}_1(t_1/\varepsilon) + \varepsilon\,\hat{x}_2(t_1/\varepsilon)$ and it is expanded as the one-term outer asymptotic expansion $\hat{X}_1(t_1) + O(\varepsilon)$.

On the other hand, the one-term outer asymptotic expansion, $X_1(t_1)$, is considered. It is transformed in the inner coordinate as $X_1(\eta\varepsilon)$ and this function is expanded, as the two-terms asymptotic expansion $\hat{x}_{11}(\eta) +$

$\varepsilon \hat{x}_{21}(\eta)$. This sum is written in the variable t_1 as

$$\hat{x}_{11}(t_1/\varepsilon) + \varepsilon\, \hat{x}_{21}(t_1/\varepsilon),$$

which must be the same as $X_1(t_1)$. Taking into account that in these expansions the coefficients $\hat{x}_i$, $\hat{X}_i$ are determined using some limits as $\varepsilon \to 0$ and η or t_1 are fixed, some relationships between the initial conditions of the outer and inner models are determined.

Since for every cycle of X it is necessary to perform four matchings and in spite of the fact that theoretically we know the expressions for $\hat{x}_n$, $\hat{y}_n$, X_n, Y_n, these expressions are very complicated.

A complicated situation occurs if the point of the parameter plane (b, a) is situated in a region for which the canard like limit cycles are present. In this case the trajectory can present a pronounced concavity and we do not expect that the region of trajectory emerging from A_2 will be parallel with the OX-axis any longer. This makes the asymptotic analysis quite different from the case when the concavity is not present. Since the canard like limit cycles occurs at very small variations of the parameters a and b, we do not know the orders of magnitude allowing the initiaton of an asymptotic analysis. The cause is the following. In order to apply an asymptotic approximation method, we must know the order of magnitude of all entities occuring in the model: unknown functions, independent variables, physical parameters, initial conditions. In addition, owing to some computations (differences or divisions), sometimes very small or very large expressions may occur. This results in the presence of secular terms (i.e., $(n+1)$ order terms are much greater than or of the same order as the n order terms).

Moreover, the differentiation of asymptotic expansions may not by allowed. This is why, after deducing the final expressions, we must proof that, indeed, (4.3.1), (4.3.2), (4.3.9) and (4.3.10) are asymptotic expansions. This can be found in [51].

4.4. SOME PARTICULAR CASES

4.4.1. THE CASE OF THE VAN DER POL MODEL

The wide diversity of phase phenomena emphasized in Chapters 2 and 3 make impossible a global treatment, for all parameters. The asymptotic analysis of the F–N model can be useful in some typical special cases. One of them is of the Van der Pol model

$$\begin{aligned} \varepsilon \dot{X} &= X + Y - \frac{X^3}{3}, \\ \dot{Y} &= -X \end{aligned} \tag{4.4.1}$$

corresponding to $a = b = 0$ in (4.2.1).

It will be treated in this section using also the matched inner and outer asymptotic expansions method.

The Cauchy problem (4.2.2) for the system (4.4.1) is a singularly perturbed boundary layer type problem, since in (4.4.1) $\dot{X}$ is multiplied by ε and the reduced system

$$\begin{aligned} X_1 + Y_1 - \frac{X_1^3}{3} &= 0, \\ \dot{Y}_1 &= -X_1 \end{aligned} \tag{4.4.2}$$

does not need a certain initial condition any longer. Indeed, the ODE $(4.4.1)_1$ degenerates into the algebraic equation $(4.4.2)_1$. This is why the approximation realized by (4.4.2) is valid only for those time intervals for which $\frac{dX}{dt_1}$ is (asymptotically) fixed.

In order to find these time intervals, let us compute the time necessary to a point (x^0, y^0) to move along the limit cycle between the points A and B (situated on a straight line almost parallel to the OX-axis), then between B and C (along the R–H curve) in Figure 4.4.1. As $\varepsilon \to 0$, the limit cycle is very close of that from Figure 4.3.1.

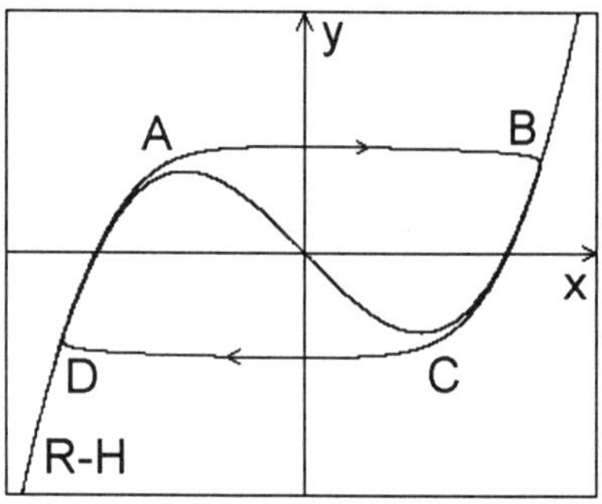

Fig. 4.4.1. The limit cycle for (4.4.1), with $\varepsilon = 1/25$.

Since (4.4.1) is deduced from (4.2.1) for $a = b = 0$, using the results of Chapter 3, (4.4.1) possesses a limit cycle for every $c > 1 + \sqrt{3}$, therefore for every small ε. If, instead of considering the limit cycle from Figure 4.4.1 we should consider another limit cycle close to it, corresponding to a smaller ε, the order of the time intervals during which two of its regions parallel to the Ox-axis are covered would be smaller than for the limit cycle in Figure 4.4.1, while the order of the time intervals during which the other two regions situated close to the R–H curve are covered would be like in the case of the limit cycle from Figure 4.4.1.

In general, the time interval necessary to some point to cover a segment is found by integrating the equation $dt_1 = \frac{dt_1}{dX} dX = \frac{dX}{\dot{X}}$, along AB and $dt_1 = \frac{dt_1}{dY}\frac{dY}{dX} dX = \frac{Y'}{\dot{Y}} dX$, along BC, where $Y' = \frac{dY}{dX}$. We considered that the phase trajectory is given in its explicit form $Y = Y(X)$. So we have

$$\begin{aligned}
\Delta t_1|_{AB} &= \int_{X_A}^{X_B} \frac{dX}{\dot{X}} = \int_{X_A}^{X_B} \frac{\varepsilon}{X + y^0 - \frac{X^3}{3}} dX \qquad (4.4.3)\\
&= O(\varepsilon), \ \varepsilon \to 0, \\
\Delta t_1|_{BC} &= \int_{X_B}^{X_C} \frac{Y'}{\dot{Y}} dX = \int_{X_B}^{X_C} \frac{1 - X^2}{-X} dX = O(1), \quad \varepsilon \to 0 \qquad (4.4.4)
\end{aligned}$$

Similarly $\Delta t_{1CD} = O(\varepsilon)$ and $\Delta t_{1DA} = O(1)$, $\varepsilon \to 0$. As for the first order approximation of the limit cycle, we have $X_B = 2$, $X_C = 1$, the first order approximation of the period is $T_0 = 3 - 2\ln 2$.

Let us remark that if $X \to X_B$, then $X + y^0 - \frac{X^3}{3} \to 0$, so in (4.4.3) $O(\varepsilon)$ does not hold any more. This is why we must understand that, in fact, we consider not X_B but an abscissa close to it. In other words, the passage from segments with constant Y to those from R–H is carried out by portions of curves determinined by the various higher-order asymptotic approximation models. On the segments of the limit cycle for which $\Delta t_1 = O(1)$, $\dot{Y}$ and $\dot{X}$ are $O(1)$. It follows that $\varepsilon \dot{X} = O(\varepsilon)$ and a first approximation of (4.4.1) is (4.4.2).

Equation $(4.4.2)_1$ shows that those regions are, in the first approximation, segments of R–H. The other two segments are covered in $\Delta t_1 = O(\varepsilon)$, $\varepsilon \to 0$, since Y is varying very little, such that $\dot{Y} = O(\varepsilon)$ holds and X is varying very much in a short time-period, such that $\dot{X} \gg 1$ could occur.

Then $\varepsilon \dot{X}$ can be of $O(1)$, such that (4.4.1) can not be approximated by (4.4.2) any longer. In these regions, which are temporal boundary layers, the variation of t_1 is small, so we introduce the new time $\eta = t_1/\varepsilon = O(1)$, $\varepsilon \to 0$. Correspondingly, (4.4.1) becomes, with the notation of Section 4.2.2,

$$\begin{aligned}
\frac{d\hat{x}}{d\eta} &= \hat{y} - \left[\frac{\left(\hat{x}\right)^3}{3} - \hat{x}\right], \\
\frac{d\hat{y}}{d\eta} &= -\varepsilon\, \hat{x},
\end{aligned} \qquad (4.4.5)$$

the Cauchy problem (4.4.1), (4.2.2) having as inner reduced asymptotic approximation model the following Cauchy problem

$$\frac{d\hat{x}_1}{d\eta} = \hat{y}_1 - \left[\frac{\left(\hat{x}_1\right)^3}{3} - \hat{x}_1\right],$$

$$\frac{d\hat{y}_1}{d\eta} = 0, \tag{4.4.6}$$

$$\hat{x}_1(0) = x^0,$$

$$\hat{y}_1(0) = y^0.$$

The solution of $(4.4.6)_2$ is $\hat{y}_1 = y^0$, with y^0 as the initial datum, and $\hat{x}_1 = X_1(\eta, x^0, y^0)$, whose expression can be explicitly obtained. Yet, even in this very simple case of (4.4.1), equation $(4.4.6)_1$ where $\hat{y}_1 = y^0 =$constant has different types of solutions depending on the solutions of the equation $\frac{\left(\hat{x}_1\right)^3}{3} - \hat{x}_1 = y^0$, so they are functions of y^0. The matching of the solutions of the two reduced models (4.4.2) and (4.4.6) is done like in Section 4.2.3 and the higher order models are deduced like in Section 4.3.

4.4.2. EXPANSIONS MATCHING AND THE RUNNING TIME ALONG THE LIMIT CYCLE

Let us consider a particular case of (4.2.1), namely the case corresponding to $a = b = 0.5$

$$\varepsilon\dot{X} = Y + X - \frac{X^3}{3},$$
$$\dot{Y} = -(X - 0.5 + 0.5Y). \tag{4.4.7}$$

For $c = 5, 10, 20$ and correspondingly, $\varepsilon = 1/25,\ 1/100,\ 1/400$, the limit cycles in Figure 4.4.2 are obtained. As ε is decreased, they become closer and closer to the asymptotic cycle $AmBM$ (Figure 4.2.1). In addition, in Table 4.4.1 the time intervals necessary for a point to cover the segments on the limit cycles are given in these three cases. The point I is the initial point. During the integration process, for $t = t_k$, $k = 1, 2, ...$, the points $M_k(x_k, y_k)$ situated on the trajectory through I are obtained.

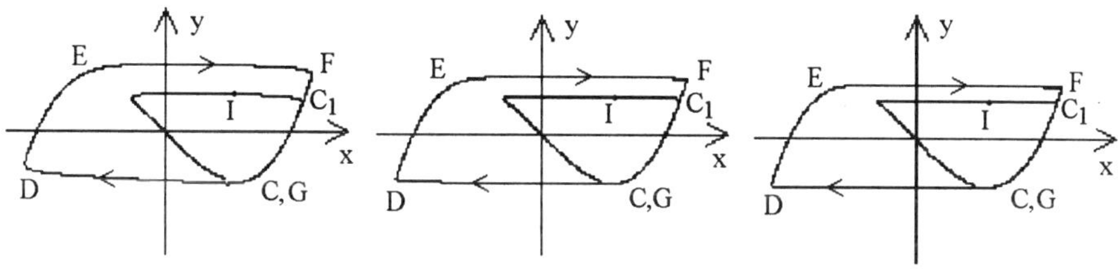

a) b) c)

Fig. 4.4.2. Limit cycle of (4.4.9) with $\varepsilon = 1/25$ (a); $1/100$ (b); $1/400$ (c).

From	to	No steps	Time
I	C1	1120	1.11543
C1	C	3686	16.83007
C	D	3034	0.83007
D	E	3273	7.85302
E	F	3367	0.87315
F	G	4321	18.18908

a)

From	to	No steps	Time
I	C1	1141	0.96064
C1	C	3755	28.40292
C	D	3060	0.41464
D	E	3527	15.29199
E	F	3202	0.42315
F	G	4186	31.29355

b)

From	to	No steps	Time
I	C1	1147	0.88066
C1	C	3788	53.32841
C	D	3060	0.20688
D	E	3631	26.87207
E	F	3119	0.20860
F	G	4131	57.9862

c)

Table 4.4.1. The running time along a trajectory close to the limit cycles for: a) $c = 5$; b) $c = 10$; c) $c = 20$.

Consider also the points $(\text{R–H})_k(x_k, x_k^3/3 - x_k)$ of the R–H nullcline, d_k the distance between the points M_k and $(\text{R–H})_k$ and $\delta > 0$ a small error. For the first steps, $d_k \geq \delta$. We define C_1 as the first point M_k for which $d_k < \delta$. For the next steps, $d_k < \delta$, i.e., the trajectory is close to R–H. Then again is obtained a point M_k such that $d_k \geq \delta$. This point is defined as C. Similarly are obtained the points D, E, F and so on.

The difference of order of magnitude for the time along these segments and the fact that the time corresponding to (almost) orizontal regions tends to zero are manifest.

The asymptotic analysis cannot be carried out in the absence of numerical and theoretical information, concerning the order of magnitude of quantities occuring in the model. We used combined information: those from asymptotic analysis were used in the numerical and theoretical case and conversely. It was the only way to obtain information about the complicate phase dynamics of the F–N model. Other, more precise, information can be determined for particular cases corresponding to fixed values of a and b.

Finally, we mention that the treatment of Cauchy problems which are of singular perturbations is more difficult than the two-point problems. This is owed to the dependence of every trajectory on the initial data, whilst in two-point problems a stationary situation exists. In addition, since the mathematical boundary layer alternates in time with the domain outside the boundary layer, the study of the asymptotic behaviour, as $c \to \infty$, around a limit cycle (or on it) must permanently alternate in time the outer and inner asymptotic approximation models.

We conclude that, even with these difficulties and even using only the first two asymptotic approximations, the asymptotic results realize an important completion to the standard study of the phase dynamics.

4.5. ASYMPTOTIC RESULTS ON DUCKS (FRENCH CANARDS) AND RELATED OBJECTS

4.5.1. CANARD PHENOMENON

The Hopf bifurcation of the F–N system is associated with the canard phenomenon, that is the lymit cycle changes fast its dimension and shape as the parameters have very small variations. This phenomenon was investigated in literature using asymptotic analysis method [18], [32], [33], [36] or nonstandard analysis methods [31], [98], [138], [146].

In classifying periodic oscillations we quote the following criteria: the amplitude, the form, the presence of fast or slow variations, the number of 'steps' (if they can be distinguished). Correspondingly, the limit cycles can be characterized by: their size, form, running velocity along the various portions of the limit cycles, the number of jumps along them.

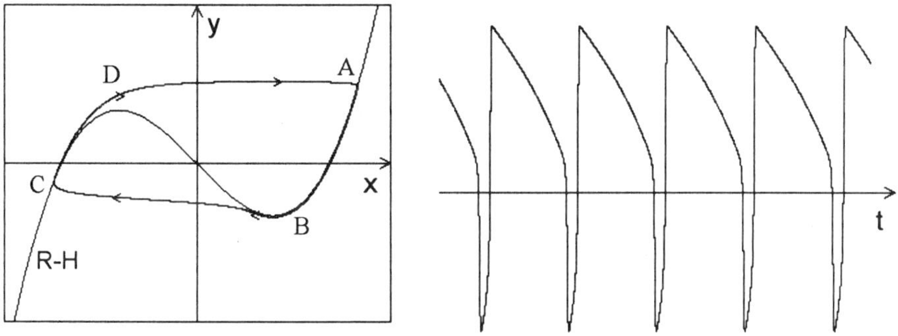

a) $a = 0.994$, $b = 0$

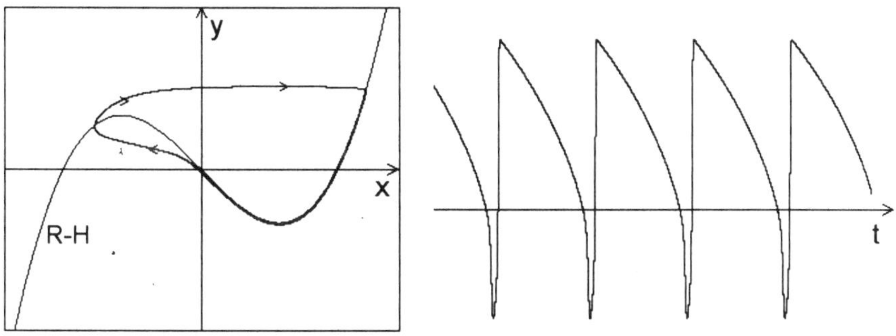

b) $a = 0.994837792331$, $b = 0$

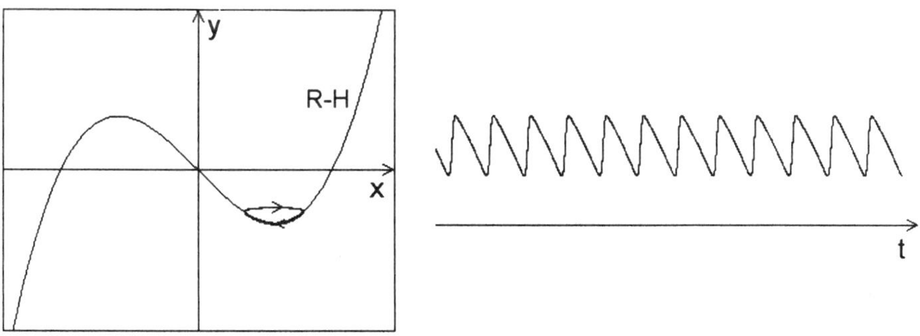

c) $a = 0.9949$, $b = 0$

Fig. 4.5.1. Types of limit cycles and the corresponding oscillations $x(t)$ involved in the canard phenomenon.

For instance, in Figure 4.5.1 there are three typical limit cycles of the F–N model with $b = 0, c = 5$ and the corresponding oscillations occurring in the process of disappearance of some limit cycles as the parameter a is varied.

In Figure 4.5.la we have represented a 'rectangular' limit cycle and the corresponding four 'steps' oscillations.

In Figure 4.5.lb we have drawn a duck-like limit cycle with the corresponding oscillations in three 'steps'.

In Figure 4.5.lc we have sketched a small limit cycle and the corresponding small amplitude oscillations in two 'steps'. This limit cycle looks like a body of a duck, or as a duck without head.

The passage from a) to b) and to c) occurs at a very small change in parameters. This means that the corresponding heart beats undergo a sudden and notable change in the above-quoted characteristics.

The continuous deformation of the limit cycle a) into c) involves an infinity of intermediate forms, more or less similar to b). It is explained in Section 5.4.2 with the aid of the curve locus of the points of inflection of the phase trajectories and in Section 4.5.3 with nonstandard asymptotic arguments [36].

In certain conditions, for the same values of the parameters, two or three nested limit cycles exist. As the parameters are varied, one small limit cycle emerges inside the larger limit cycle, while the outer larger limit cycle diminishes when undergoing the changes presented in Figure 4.5.1. Simultaneously, the smaller cycle develops according to the same scenario, but in the inverse sense. It increases developing the stages of duck body form. In addition, its increase is faster than the decrease of the outer limit cycle (Figure 3.5.2b). At some values of the parameters, the limit cycles collide producing a non-hyperbolic limit cycle.

In Section 3.5 we referred to this phenomenon as non-hyperbolic limit cycle bifurcation and related it to the Bautin bifurcation. The points of the curves $Ba_{1,2}$ of the parameter-plane corresponded to the non-hyperbolic limit cycle bifurcation. This phenomenon is also known as the blue-sky bifurcation [36], [65].

This bifurcation is of global type, like homoclinic bifurcation. In addition, it is related to the local Hopf bifurcation of the inner limit cycle. At the same time the non-hyperbolic limit cycle bifurcation sets in as a result of two processes: the formation and destruction of the outer and inner duck-like limit cycles.

The limit cycle in Figure 4.5.la has two sides situated close to the external branches of the R–H curve. These branches are referred to as the **stable manifolds** [36]. The inner branch of the R–H curve is called the **unstable manifold**.

The limit cycle in Figure 4.5.la has two sides along the stable manifolds of the curve R–H, whilst the limit cycle in Figure 4.5.1c has one side along a stable manifold and another side along the unstable manifold of R–H.

In fact, these limit cycles are very close to these manifolds and they tend to them as some parameter tends to infinity. In the case of the F–N system, this parameter is c.

The tendency of some limit cycles to come closer and closer to the stable and/or unstable manifolds of the R–H curve form the asymptotic basis of the duck formation and of the non-hyperbolic limit cycle bifurcation.

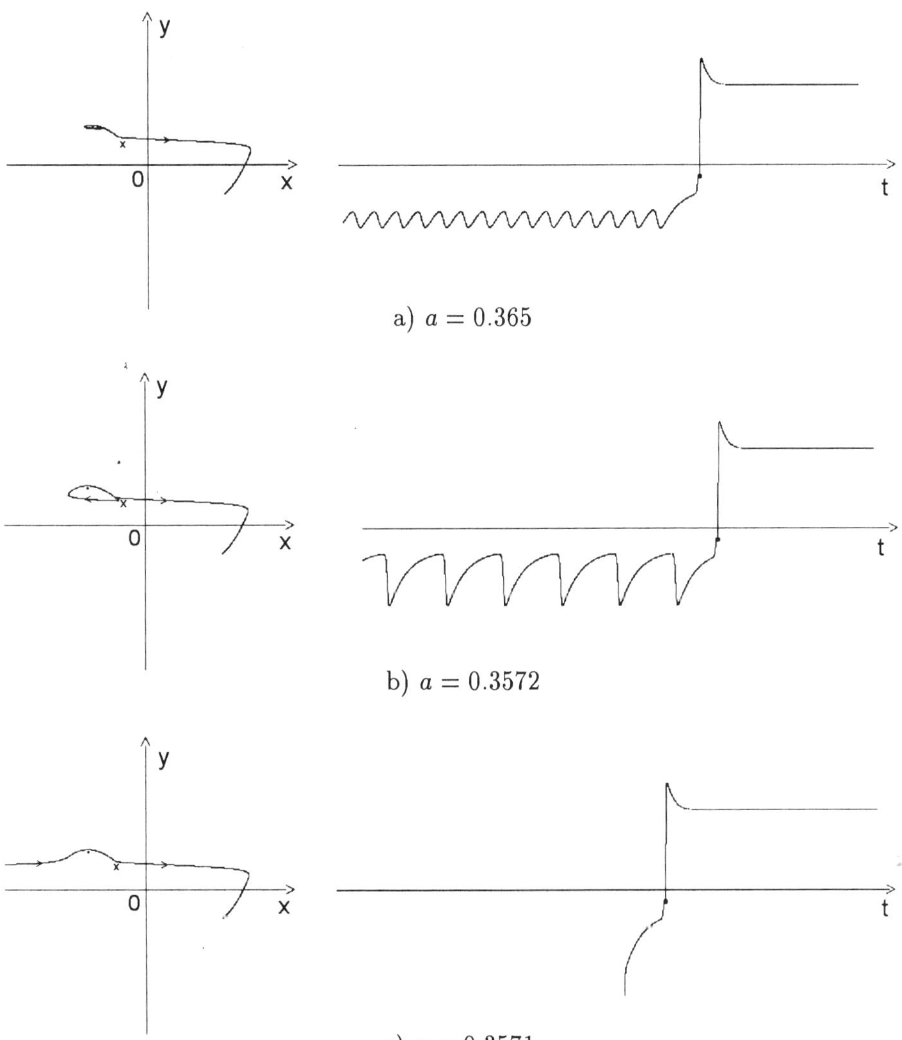

a) $a = 0.365$

b) $a = 0.3572$

c) $a = 0.3571$

Fig. 4.5.2. The canard for $b = 2$, $c = 5$ and variable a and the oscillations corresponding to the trajectory through $(-0.2, 0.45)$.

The canards always accompany the Hopf bifurcation along the curves $H_{1,2}$, but the 'duck' may not develop completely owing to the disappearence of the limit cycle before the 'head' be formed. Such an example was presented in Figures 3.2.5, 3.2.6. In Figure 4.5.2 is given another example for $b = 2$, $c = 5$ and variable a.

In these situations two attractive points and a saddle exist. As a is decreased from the Hopf bifurcation value $a_{H2} \approx 0.3708$ to 0.3572 the repulsive limit cycle becomes bigger and bigger. However, the canard does not develop completely, because for some $a < 0.3572$ it disappears by homoclinic bifurcation. A phase portrait after its disappearance is presented in Figure 4.5.2c.

Some asymptotic arguments explaining the characteristics of the limit cycles which are situated near the stable and unstable manifolds of the R–H curve were given in the previous sections. In the following we supplement them with more precise evaluation of the order of the asymptotic approximation.

4.5.2. RELAXATION OSCILLATIONS

A detailed asymptotic study on relaxation oscillations was done by Eckhaus [36]. He considered a special class of dynamical systems generated by the system of ODEs

$$\begin{aligned} \varepsilon \dot{x} &= y - f(x), \\ \dot{y} &= -(x + \alpha), \end{aligned} \tag{4.5.1}$$

where $\varepsilon > 0$ is a small parameter, α is a parameter, $\alpha \in (0, l)$, $\alpha \neq o(1)$, $\alpha - l \neq o(1)$ as $\varepsilon \to 0$, and $-l$ is the abscissa of the maximum point of the curve $y = f(x)$, $f'(x) > 0$ for $x > 0$; $f'(x) < 0$ for $-l < x < 0$; $f'(x) > 0$ for $x < -l$; near $x = 0$ we have $f'(x) \approx xg(x)$ as $\varepsilon \to 0$, $g(0) > 0$, near $x = -l$ we have $f'(x) \approx (x + l)h(x)$, as $\varepsilon \to 0$, $h(-l) < 0$. Of course, these conditions are related to the points of inflection of the phase trajectories.

Under these assumptions Eckhaus proved the following results on the asymptotic behavior of the phase trajectories close to a limit cycle corresponding to relaxation oscillations and having sides near the stable manifold of the curve $y = f(x)$. In the following we will reffer to this curve as R–H, too. This limit cycle is similar to that from Figure 4.5.la.

Theorem 4.5.1 [36] *Any trajectory that starts outside a singular point reaches $o(1)$ neighborhood of the curve $ABCD$ in finite time and stays in that neighborhood for ever. In particular, the trajectory stays in $O(\varepsilon)$ neighborhood of the open segments AB and CD and in $O(\varepsilon^{2/3})$ neighborhood of the open segments BC and DA. The motion along AB and DC is on time* *intervals of order unity, the motion along BC and DA is on a time scale of order ε.*

Here by 'singular point' is meant a point of equilibrium.

The proof of this theorem is based on suitable change of the time scale, reducing the singular perturbation problem (4.5.1) to a problem of regular perturbation. This is, in fact, the inner problem for fast flow near BC and DA.

Then, near AB or CD, y is looked for as a sum of $f(x)$ and a such perturbation

$$y = f(x) + \sigma(\varepsilon)\Phi, \tag{4.5.2}$$

where $\sigma(\varepsilon) = \mathrm{o}(1)$, $\varepsilon/\sigma(\varepsilon) = \mathrm{o}(1)$ as $\varepsilon \to 0$. Moreover, a time rescaling reduces the given problem to a regular one, for which, under additional hypotheses on various involved quantities, the models of first and second order inner asymptotic approximation are deduced and studied.

Then this asymptotic analysis is extended near the minimum point $(x, y) = (0, 0)$, by assuming for y an asymptotic behaviour as in (4.5.2) and for x taking the following expression

$$x = \delta(\varepsilon)\xi, \tag{4.5.3}$$

where $\delta(\varepsilon) = \mathrm{o}(1)$ as $\varepsilon \to 0$, $\xi \geq d > 0$, d is a number which does not depend on ε.

Further, the neighborhood of y and x are connected by the relation

$$\delta = \sigma^2 \overline{\sigma}(\varepsilon), \tag{4.5.4}$$

where $\overline{\sigma}(\varepsilon) = \mathrm{o}(1)$ as $\varepsilon \to 0$. Next, Eckhaus takes

$$\sigma = \frac{\varepsilon}{\delta} \tag{4.5.5}$$

as valid in an overlapping region of the regions near the curve R–H and the point $x = 0$, in a ε/δ neighborhood of the R–H curve.

It follows an analysis for $\xi = x/\sqrt[3]{\varepsilon}$, i.e., far from the stable manifold, for $y = \eta\sqrt[3]{\varepsilon^2}$ and $\tau = (t - t_0)/\sqrt[3]{\varepsilon^2}$ and the study goes on as the trajectory passes through other and other regions.

In this way, step by step, by suitable rescalings of x, y and t, it is obtained the approximate phase trajectory near the limit cycle corresponding to relaxation oscillations.

Remark that the key point in the Eckhaus investigation is the choice of rescalings. It is validated only by the consistency of computations.

In the case of the F–N model the analysis complicates since a new parameter occurs too and its order as $\varepsilon \to 0$ must be taken into account.

However, the proof of the existence of relaxation oscillations for this model holds at least as far as it can be written in the form, more general than (4.5.1),

$$\begin{aligned} \varepsilon\, \dot{x} &= y - f(x)\,, \\ \dot{y} &= -H(x)\,, \end{aligned} \tag{4.5.6}$$

where f and H must satisfy certain conditions [36].

4.5.3. THE DUCKS: STANDARD VERSUS NONSTANDARD ASYMPTOTIC ANALYSIS

Definition 4.5.1 [36] *A phase trajectory following a stable and then a part of the unstable manifold of the nullcline is called a* **duck trajectory**, *or simply, a* **duck**.

Some authors impose to duck trajectories to be closed [146]. In general, duck trajectories are not closed. Sometimes, only part of the duck trajectories are referred to as ducks. If a duck solution is a limit cycle, it is called a **duck cycle**. The shape of a duck trajectory does not necessarly looks like the eponymous bird.

Therefore we shall refer to the limit cycles in Figures 4.5.lb and 4.5.lc as ducks, despite the head of the duck in Figure 4.5.lc being absent.

The ducks were discovered and intensively studied by the Strasbourg school of nonstandard analysis [14], [15], [16], [31], [8]. For an extensive list of early papers on canards we quote [146]. An equivalent study in terms of the standard asymptotic analysis was performed by Eckhaus [36] by means of a continuous dependence on the data theorem and an extension theorem [37].

Eckhaus dealt with the system

$$\begin{aligned} \varepsilon\, \dot{x} &= y - f(x)\,, \\ \dot{y} &= -x + \alpha(\varepsilon)\,, \end{aligned} \tag{4.5.7}$$

where $\alpha(\varepsilon) = o(1)$ as $\varepsilon \to 0$. Hence the single equilibrium point of (4.5.7) is imposed to be near the minimum point of the curve $y = f(x)$. The proof of the ducks' existence proceeded along the same lines as for the relaxation oscillations (Section 4.5.2). Of course, some changes imposed by the new order of $\alpha(\varepsilon)$ were operated.

For instance, near the curve $y = f(x)$ he took $y = f(x) + \varepsilon\Phi$, where $\Phi = -\dfrac{1}{g(x)} + \varepsilon\Phi_1$. Then he considered the case $\alpha \neq 0$ as a perturbation of the case $\alpha = 0$ and discovered conditions for the existence of bounded solutions for $\alpha = 0$ as $\varepsilon \to 0$. In addition, he first used some particular results of the nonstandard analysis school for parabolic curves $y = f(x)$.

These were suggestive in choosing the order of various involved functions. In particular, the boundedness requirements for the solutions led to $\alpha = \sigma\varepsilon^{\frac{3}{2}}\left(\exp\left(-\frac{1}{2\varepsilon}k^2 g_0^2\right)\right)$, where σ and k are numerical constants and $g(x) = g_0 > 0$. These requirements were necessary in order to avoid exploding (i.e., secular terms) in the asymptotic expansion of the solution as $\varepsilon \to 0$. In the particular case $g(x) = g_0$ and $\alpha = 0$ the solution of the ODE in y as a function of x, obtained from (4.5.7), was found in a closed form. This made easier the task of finding the corresponding perturbed solution for $\alpha \neq 0$. The obtained approximate formulae, led to a duck similar to that in Figure 4.5.1c.

Then the general case $g(x)$ different from a constant was studied in a similar way. This time an asymptotic expansion for α around $\alpha_c(\varepsilon)$

$$\alpha \sim \alpha_c(\varepsilon) + \sigma\varepsilon^{3/2}\exp\left(-\frac{k^2}{\varepsilon}\right), \quad \text{as } \varepsilon \to 0 \tag{4.5.8}$$

is found. Here $\alpha_c \sim \varepsilon\,\dfrac{g'(0)}{g^3(0)} + O(\varepsilon^2)$ as $\varepsilon \to 0$ is a value around which the duck occurs. Moreover, asymptotic expansions are found for the trajectories near the stable and unstable manifolds of the nullcline $y = f(x)$.

According to the sign of σ and the value of k, we can have a duck as in Figure 4.5.lb or 4.5.lc. These situations, referred to by Eckhaus as **subcritical ducks**, correspond to $g'(0) > 0$ and $\alpha > 0$, i.e., the equilibrium point is situated on the unstable manifold of R–H. If $g'(0) < 0$ and the ducks occur for $\alpha < 0$, i.e., the equilibrium point belongs to the stable manifold of R–H, they are referred to as **supercritical ducks** and one of them is represented in Figure 3.5.2b.

This phenomenon is associated with a Hopf bifurcation at $\alpha = 0$. Indeed, as this value of α is crossed, the equilibrium changes its attractivity properties on the account of appearance or disappearance of some limit cycles.

Eckhaus [36] suggested the alternative averaging techniques for studying relaxation and ducks. Their application to the F–N model was carried out in [128] and mentioned in [61].

Benefitting from the old traditions of the famous Russian school of dynamical systems, a lot of studies on ducks were performed.

A systematic presentation in terms of nonstandard analysis of the duck solutions can be found in [146] for the system in the plane ${}^*\mathbf{R}^2$

$$\begin{aligned} \dot{x} &= \tfrac{1}{\varepsilon}\left(y - f(x)\right), \\ \dot{y} &= a - x, \end{aligned} \tag{4.5.9}$$

where $^*\mathbf{R}$ is the set of random real numbers, $\varepsilon \in {}^*\mathbf{R}$, $\varepsilon > 0$, is an infinitely small parameter, $a \in {}^*\mathbf{R}$ is a parameter and f is a standard function of class $\mathbf{C}^\infty$, $f \in \mathbf{C}^\infty(\mathbf{R}, \mathbf{R})$, extended to a map $^*\mathbf{R} \to {}^*\mathbf{R}$. Zvonkin and Shubin stated the following.

Definition 4.5.2 [146] *A solution* $(x(t), y(t))$ *of* (4.5.9) *is called a* **duck solution** *or simply a* **duck**, *if there are standard* $t_1 < t_0 < t_2$ *such that* $o[x(t_0)] = x_0$, *for* $t \in (t_1, t_0)$ *the segment of the trajectory* $(x(t), y(t))$ *is infinitely close to the attracting part and for* $t \in (t_0, t_2)$ *to the repelling part of the slow curve.*

In the above the slow curve is the R–H curve.

For the meaning of the concepts of nonstandard analysis used in equations (4.5.9) and Definition 4.5.2 we recommend [146]. Here we recall only the following. Each finite $x \in {}^*\mathbf{R}$ has a **shadow** (or a standard part), that is the number ${}^0x \in \mathbf{R}$ such that $x - {}^0x$ is infinitely small.

Definition 4.5.2 can be rephrased in standard terms as the following form equivalent, more or less, to Definition 4.5.1.

Definition 4.5.3 [146] *A* **standard duck** *of* $(4.5.9)_{\varepsilon,a}$, *also written as* (4.5.9), *is a sequence* $(x_n(t), y_n(t))$ $(n = 1, 2, ...)$ *consisting of solutions of equations* $(4.5.9)_{\varepsilon_n, a_n}$, *such that:*

a) $\varepsilon_n > 0$ *for all* n; $\lim_{n\to\infty} \varepsilon_n = 0$;

b) $\lim_{n\to\infty} a_n = a_0$ *exists;*

c) the solution $(x_n(t), y_n(t))$ *is defined for* $t \in (c_n, d_n)$ *and there are two closed disjoint subintervals* $\left[c_n', d_n'\right]$ *and* $\left[c_n'', d_n''\right]$ *of* $[c_n, d_n]$, *of which the first lies to the left of second and the restriction of this solution to* $\left[c_n', d_n'\right]$ *(respectively* $\left[c_n'', d_n''\right]$*) defines a curve on the Liénard plane that converges, as* $n \to \infty$, *to the portion of the slow curve* $y = f(x)$ *above a nontrivial interval of the semiaxis* $x > x_0$ *(respectively* $x < x_0$*).*

The extensive study of Zvonkin and Shubin yields a rich information on various classes of ducks. Many of their results agree with our numerical results from Section 5.4.2. However, the non-degenerate F–N case $b \neq 0$ is not included.

By studying ducks with nonstandard analysis methods, Zvonkin and Shubin discovered specific sets of phase trajectories.

Definition 4.5.4 [146] *A* **funnel** *is a bundle of trajectories that at the beginning are at an appreciable distance from one another and then become infinitely close. A* **shower** *is a bundle of trajectories that at the beginning* *are infinitely close and then deviate considerably from one another. A* **tunnel** *is a bundle of trajectories that forms a funnel at the beginning and then a shower.*

The corresponding standard form of this Definition is obvious. In our computations and graphs these sets can be noticed too.

The study of the ducks for planar systems was extended to the three-dimensional dynamical systems generated by systems of two non-autonomous ODEs, among others in [14], [15], [94], [82], [83], [135]. In particular, closed form duck solutions have been constructed for singular perturbation problems of the form

$$\begin{aligned} \dot{x} &= f(x, y, z, a), \\ \dot{y} &= g(x, y, z, a), \\ \varepsilon \dot{z} &= h(x, y, z, a). \end{aligned} \tag{4.5.10}$$

where a is a real parameter and $\varepsilon > 0$ is an infinitely small parameter. In this case, as well as in the two-dimensional case, an important role is played by the constrained system

$$\begin{aligned} \dot{x} &= f(x, y, z, a), \\ \dot{y} &= g(x, y, z, a), \\ 0 &= h(x, y, z, a). \end{aligned} \tag{4.5.11}$$

The equation $(4.5.11)_3$ is the analogous of the R–H curve and the ducks are positioned close to its stable and unstable manifolds.

CHAPTER 5

GLOBAL BIFURCATION DIAGRAM AND PHASE DYNAMICS FOR THE FITZHUGH–NAGUMO MODEL

The partial results obtained in Chapters 2–4 are synthesized in the global bifurcation diagram for the F–N model (Section 5.1). It is obtained by putting together, as in a huge puzzle, all local bifurcation diagrams obtained in the previous chapters. This completes the treatment of the bifurcation for the F–N model.

New results concern: basins of attraction (Section 5.2), transient regimes in the absence of limit cycles or in their presence (Section 5.3). Various types of oscillations, corresponding to different shapes of the limit cycles and the range of the parameter plane corresponding to concave limit cycles are shown in Section 5.4. In Section 5.5, the way in which heart beats can appear is explained. In Section 5.6 concluding remarks of interest to physiologists are presented, whilst in Section 5.7 some mathematical open problems are formulated.

5.1. GLOBAL BIFURCATION DIAGRAM FOR THE FITZHUGH–NAGUMO MODEL

5.1.1. PARAMETRIC PORTRAIT OF THE F–N MODEL

In this section the parametric portrait of the F–N system [123], obtained by using all previous results, is presented. In addition we used the notation $Q_{29} = BTn_3 \cap Ba_1$, $Q_{27} = H_2 \cap Ba_1$, $Q_{25} = S_1 \cap Ba_1$, $Q_{19} = H_2 \cap D_1$, $Q_{21} = H_2 \cap BTn_4$, $Q_{23} = H_1 \cap S_1$, $Q_{31} = H_1 \cap Oa$ and $Q_{30}, Q_{28}, Q_{26}, Q_{20}$, Q_{22}, Q_{24}, Q_{32}, for the points symmetric to them with respect to the Ob-axis. The representation (Figures 5.1.1a and 5.1.1b) is qualitative, owing to the small distance between some of the bifurcation frontiers.

In order to emphasize this fact, we give certain values obtained for $c = 5$. The approximated coordinates for some points Q_i are: $Q_1\,(-5, 4.38)$, $Q_3\,(5, 2.38)$, $Q_5\,(0, -1.85)$, $Q_0\,(0, 1.457)$, $Q_6\,(0, 1.47114317)$, $T\,(0, 1.471143170299)$, $Q_{17}\,(0.50, 0.65)$. The straight line $a = 10^{-10}$ crosses succesively the domains 1A, 1B, 2A, 2B, 5, 8, 9, 11, 12, 13, 16, 18, whilst the straight line $a = 0.1$ crosses the domains 1A, 1B, 2A, 2B, 5, 8, 9, 10, 14, 15, 16, 18.

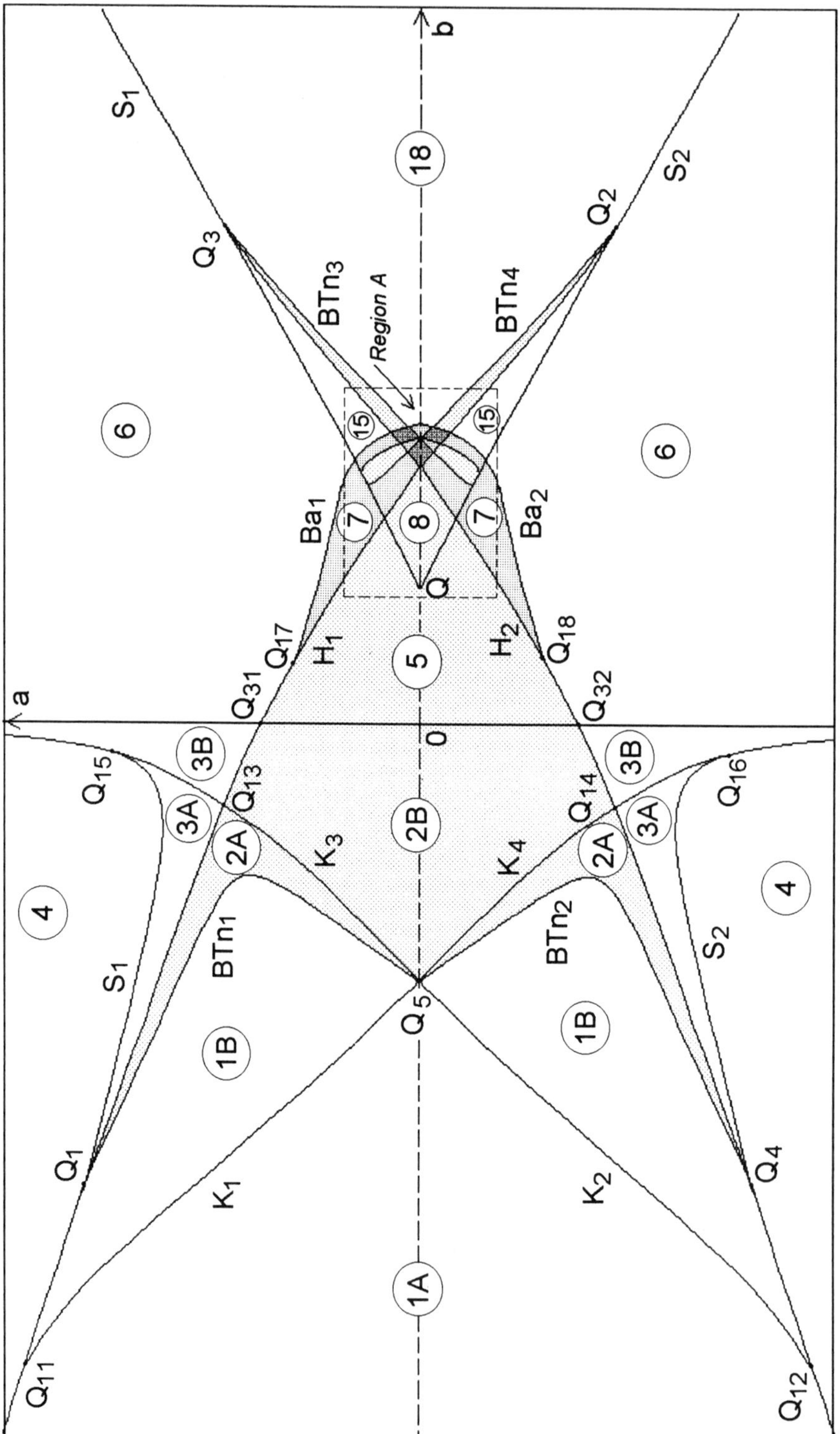

Fig.5.1.1a. Parametric portrait of the FitzHugh-Nagumo system.

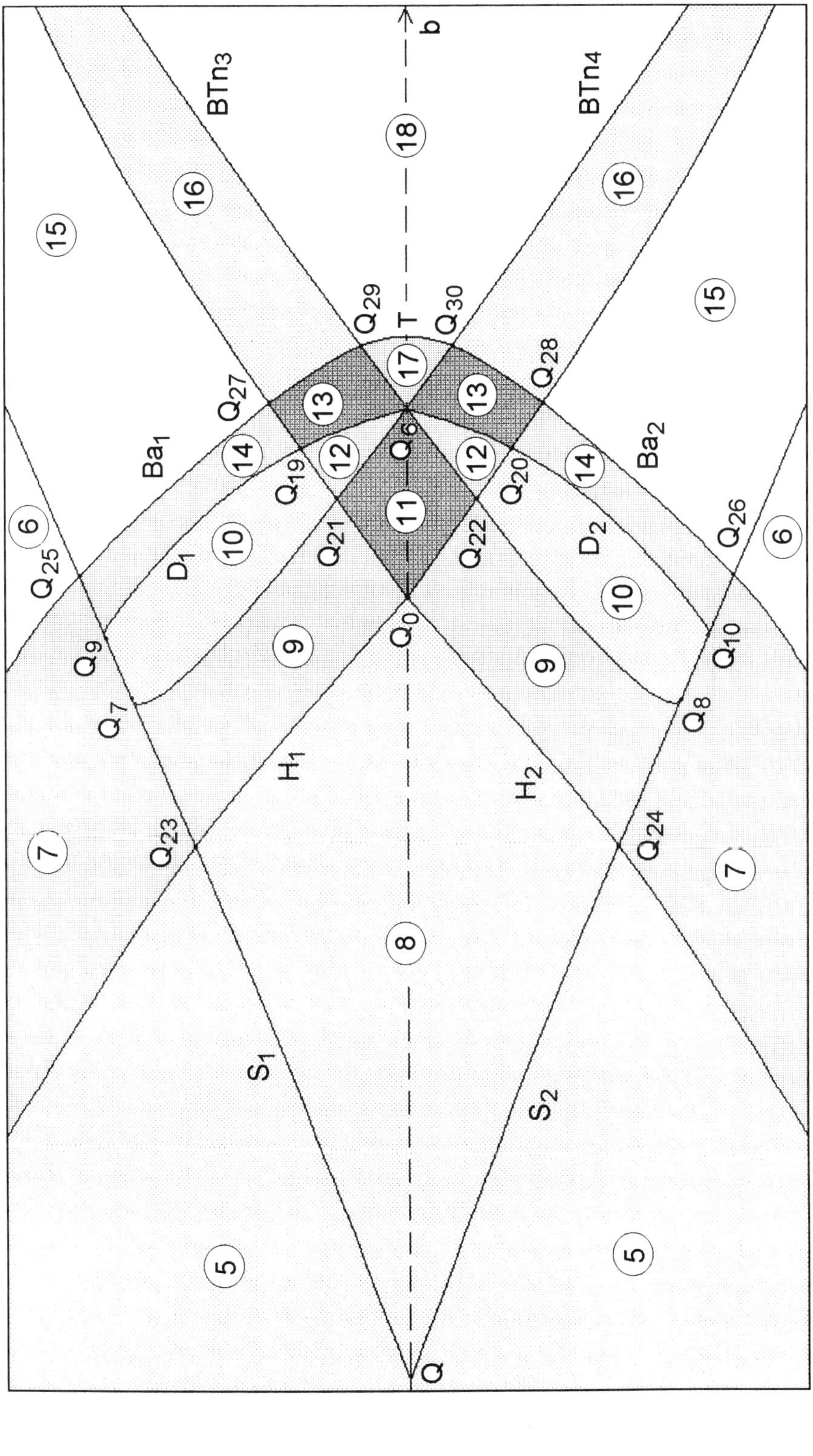

Fig.5.1.1b. Region A from Figure 5.1.1a.

The parametric portraits for all $c > 1+\sqrt{3}$ are topologically equivalent. Hence, it suffices to take $c = 5$. The completion of the parametric portrait with specific dynamics for every layer is given in Section 5.1.2.

Let us remark that all generic codimension-one bifurcations that were described for a planar system (Section 1.3.4) are present for the F–N system: saddle–node bifurcation values are situated on $S_{1,2}$, Hopf bifurcation values are situated on $H_{1,2}$, non-hyperbolic limit cycle bifurcation values are situated on $Ba_{1,2}$, homoclinic bifurcation values are situated on BTn_i, $i = \overline{1,4}$ and $D_{1,2}$, saddle–node homoclinic bifurcation values are situated on Q_7Q_9 and Q_8Q_{10}, whilst breaking saddle connection bifurcation values are situated on K_i, $i = \overline{1,4}$. Another codimension-one bifurcation, namely the saddle–node-saddle connection, takes place for values of the parameters situated on $S_{1,2}$ for $b < b_{Q11}$ or $b \in (b_{Q15}, 0)$. In addition, points of Oa-axis are values of static bifurcation from infinity.

On the other hand, not all points situated on the quoted curves correspond to codimension-one bifurcations. The points Q and Q_i, $i = \overline{0,32}$ correspond to codimension-two bifurcations.

Domains 1–18 are of structural stability.

Note that some of the curves and points of the parametric portrait were obtained analytically, so they are known for every c. Thus, the equations of $S_{1,2}$, $H_{1,2}$ are given by (2.2.5), (2.3.14) respectively, whilst the points whose coordinates are determined exactely are: $Q_1\left(-c, \frac{2}{3}(c+1)\sqrt{1+\frac{1}{c}}\right)$, $Q_3\left(c, \frac{2}{3}(c-1)\sqrt{1-\frac{1}{c}}\right)$, $Q_{17}\left(c^2 - c\sqrt{c^2-1}, \frac{4}{3}\left(c\sqrt{c^2-1} - c^2 + 1\right)\sqrt[4]{1-\frac{1}{c^2}}\right)$, $Q_{23}(b_{23}, a_{23})$, where $a_{23} = \dfrac{\left(\sqrt{9c^4+16c^2} - 32c^2 - 2\right)\sqrt{5c^2 - \sqrt{9c^4+16c^2}}}{6\sqrt{2}c}$ and $b_{23} = \dfrac{-3c^2 + \sqrt{9c^4+16c^2}}{2}$; $Q_{31}(0,1)$, their symmetrics with respect the Ob-axis and the points $Q(1,0)$, $Q_0(-c^2+\sqrt{c^4+3c^2}, 0)$ of the Ob-axis.

The approximate equations of the curves BTn_1 in the neighborhood of Q_1, BTn_3 in the neighborhood of Q_3 and Ba_1 in the neighborhood of Q_{17} were also obtained. They are given by (3.2.1), (3.2.29) and (3.5.3), respectively. The continuation of these curves far from these neighborhoods was done numerically for the case $c = 5$. This is why the points Q_i, $i = \overline{5,30}$, $i \neq 17$, 18, 23, 24 and the curves $K_{1,2}$ and $D_{1,2}$ were obtained by numerical computations for $c = 5$.

In the previous sections we represented the local bifurcation diagrams only for the Bogdanov–Takens (Figure 3.2.2), Bautin (Figure 3.5.1), saddle–node separatrix loop (Figures 1.3.7, 3.3.4), double homoclinic (Figure

3.3.2), double breaking saddle connection (Figure 3.4.3) saddle–node-saddle with separatrix connection (Figures 3.4.4, 3.4.5) bifurcations, since they were of main importance in our study. The rest of local bifurcation diagrams can be deduced easily from the global bifurcation diagram. They are presented, schematically, in the following.

5.1.2. TYPES OF DYNAMICS IN THE F–N MODEL

In order to obtain the global bifurcation diagram of the F–N model, in this section we supplement the parametric portrait from Figure 5.1.1 with schematic phase portraits of every layer (for $a \geq 0$). They are presented in Figure 5.1.2 [50], [123]. For $a < 0$ the phase portraits are deduced using the remarks (Section 1.1.4) concering the symmetry of the phase portrait.

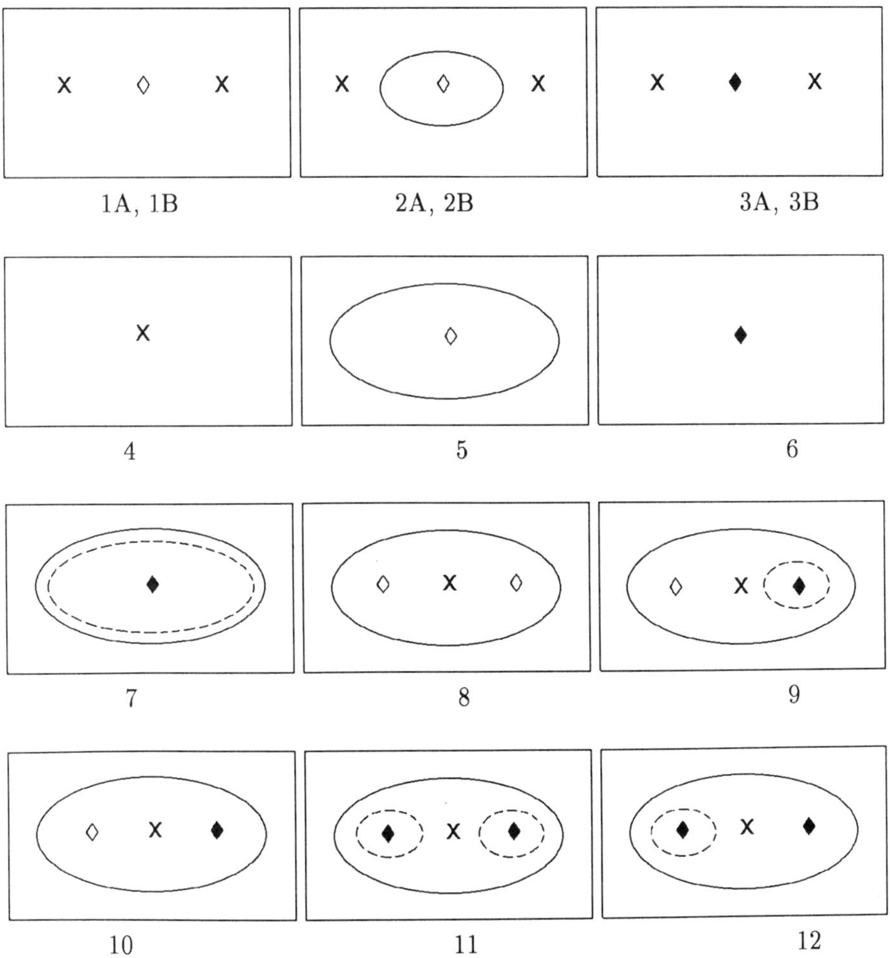

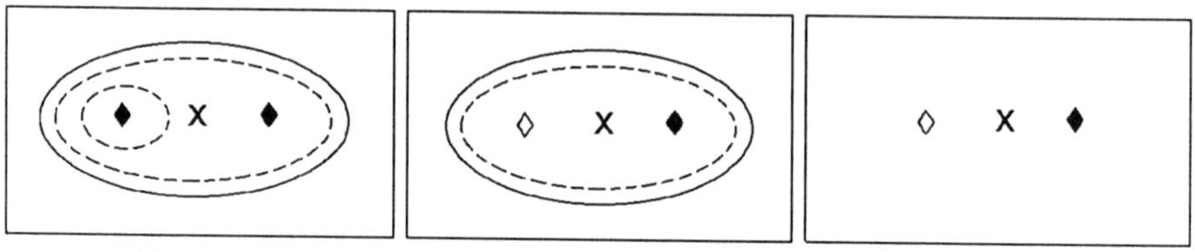

13 14 15

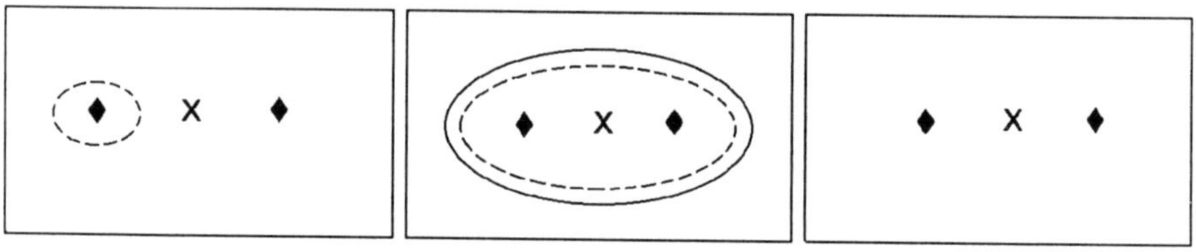

16 17 18

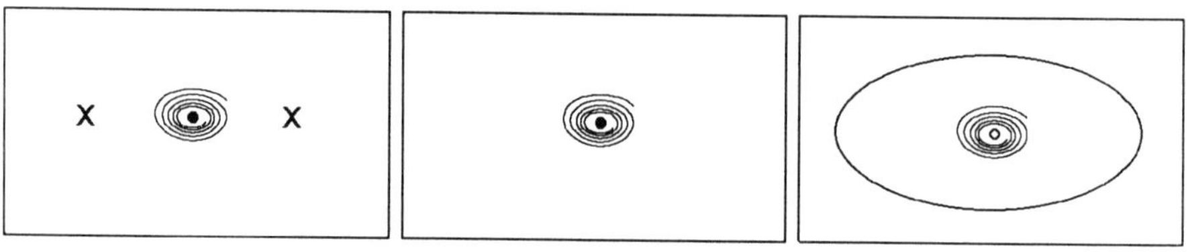

$Q_1Q_{13}, Q_{13}Q_{31}$ $Q_{31}Q_{17}$ $Q_{17}Q_{23}$

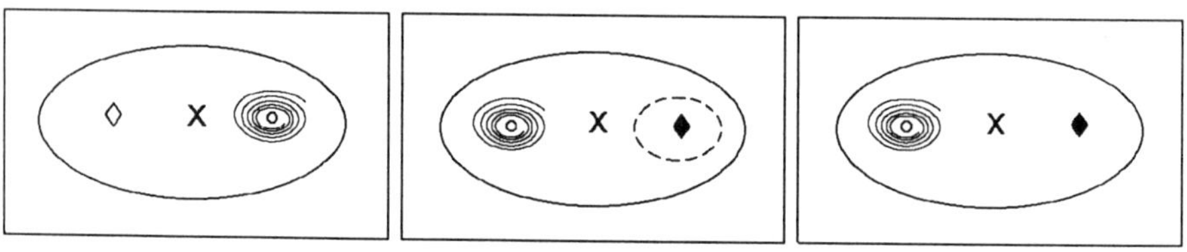

$Q_{23}Q_0$ Q_0Q_{21} $Q_{21}Q_{19}$

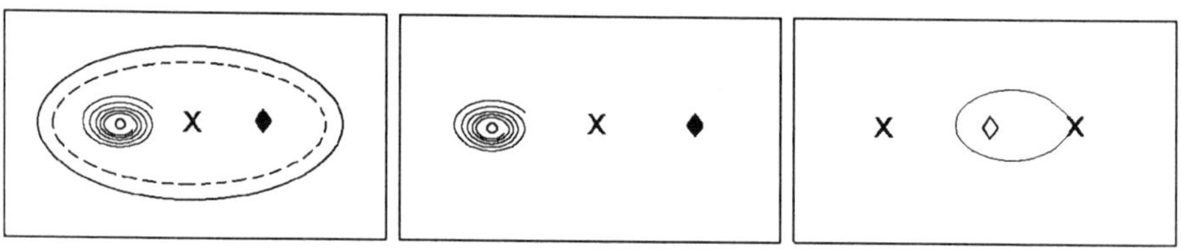

$Q_{19}Q_{27}$ $Q_{27}Q_3$ Q_1Q_5

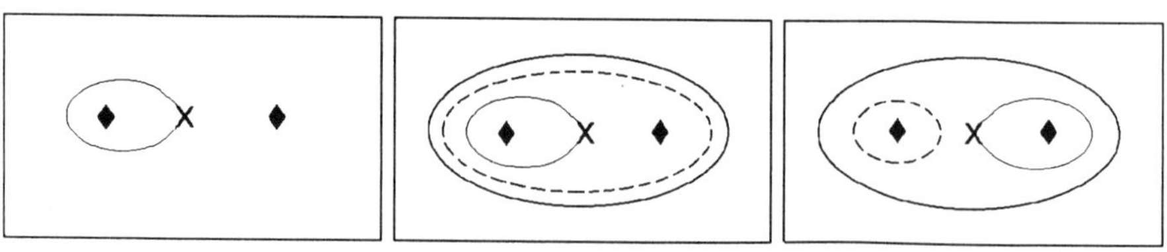

Q_3Q_{29} $Q_{29}Q_6$ Q_6Q_{21}

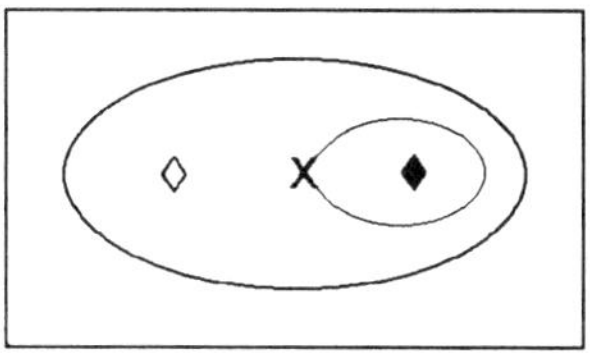

$Q_{21}Q_7$

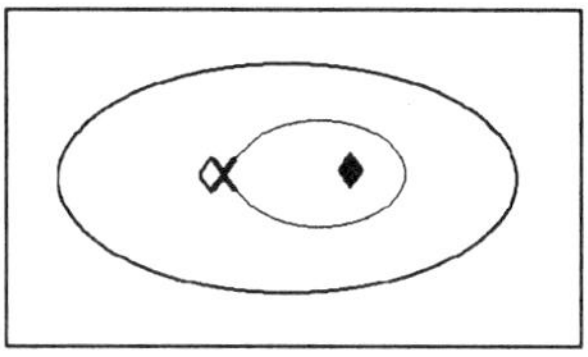

Q_7Q_9

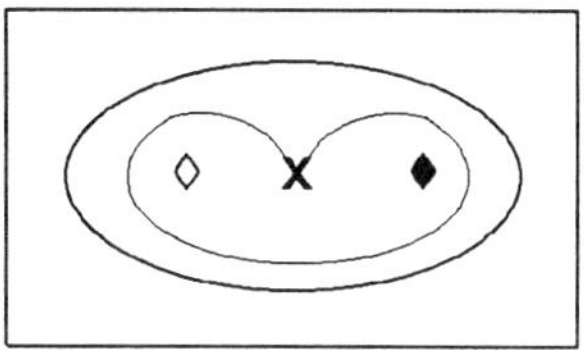

Q_9Q_{19}

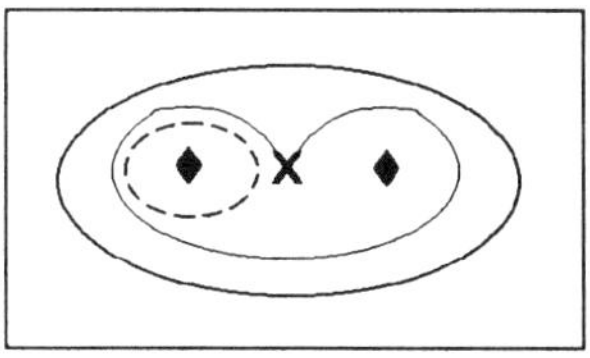

$Q_{19}Q_6$

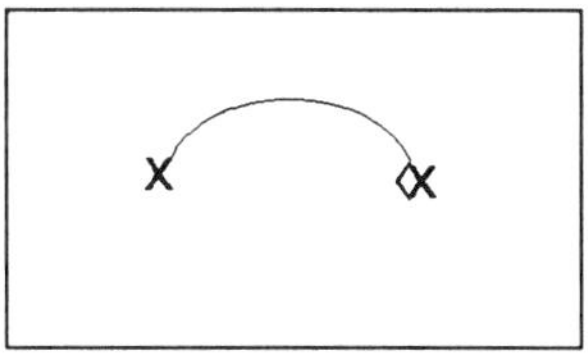

$S_1, b \in (-\infty, b_{11})$

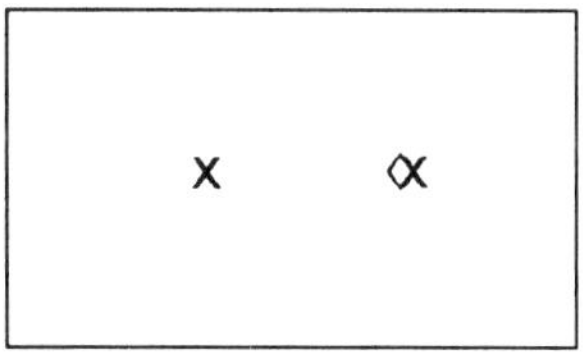

$Q_{11}Q_1$

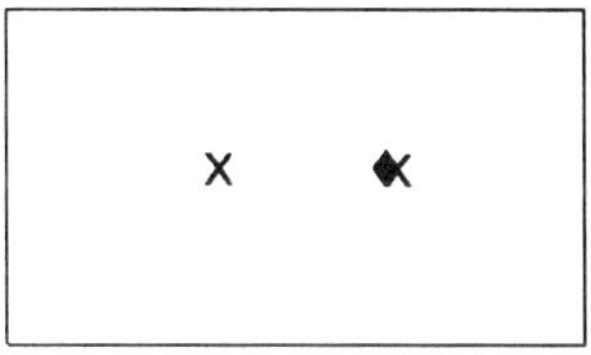

Q_1Q_{15}

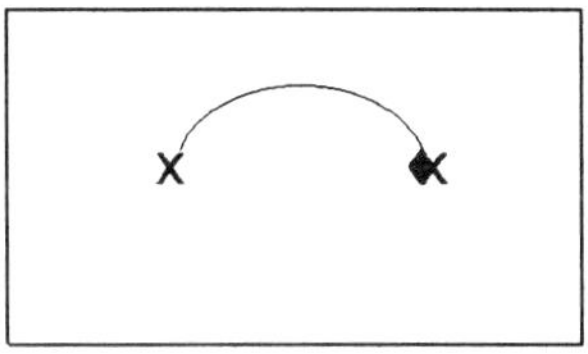

$S_1, b \in (b_{15}, 0)$

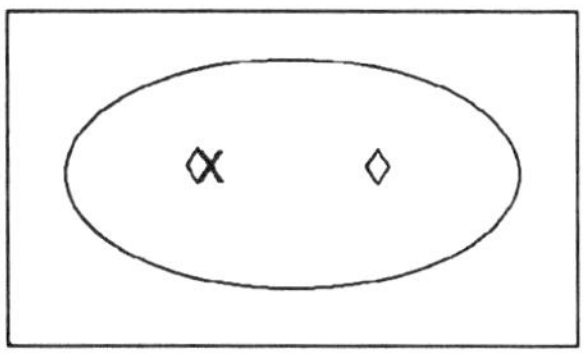

QQ_{23}

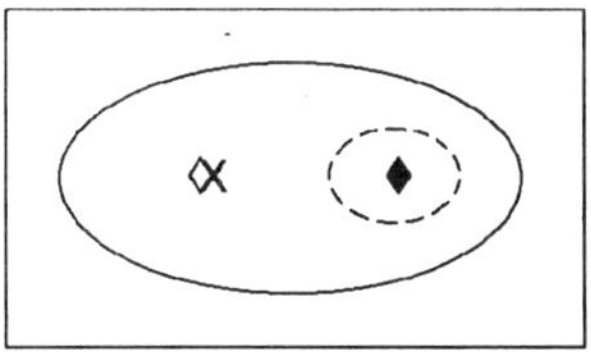

$Q_{23}Q_7$

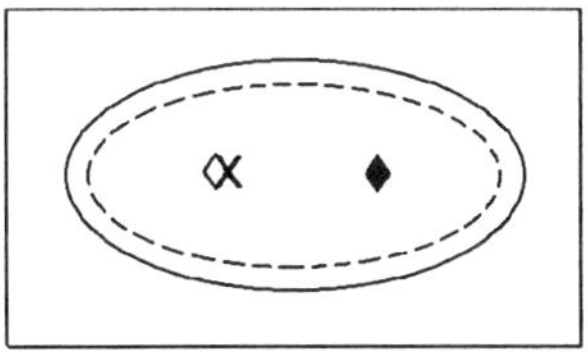

Q_9Q_{25}

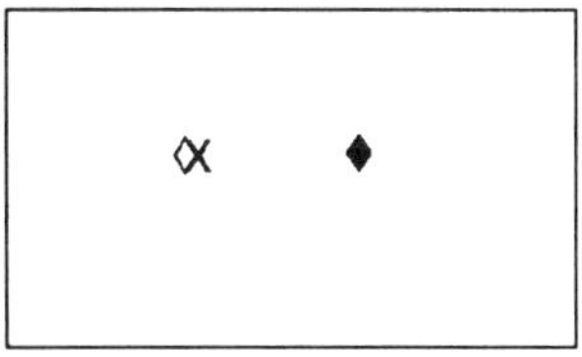

$Q_{25}Q_3$

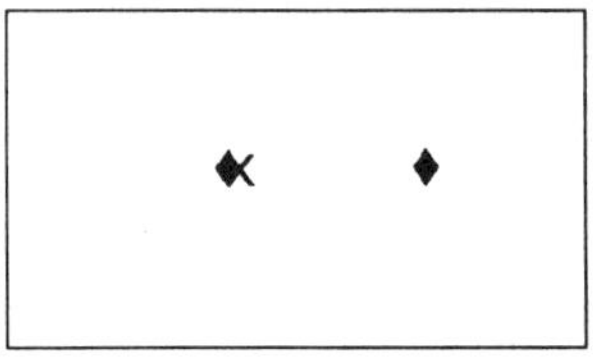

$S_1, b \in (c, \infty)$

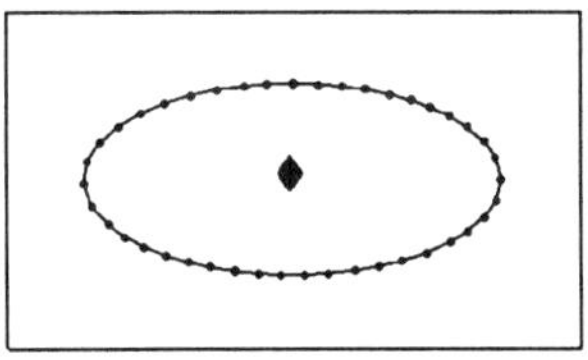

$Q_{17}Q_{25}$

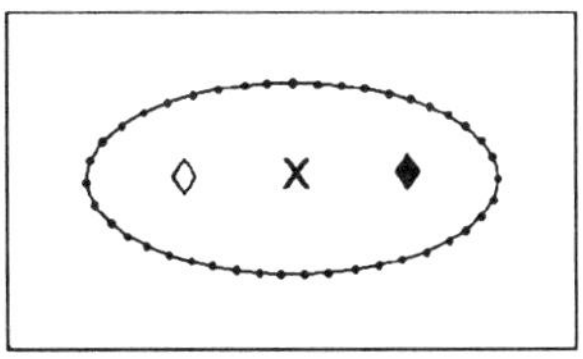

$Q_{25}Q_{27}$

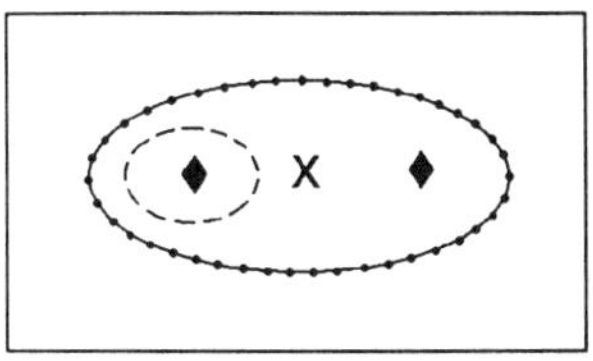

$Q_{27}Q_{29}$

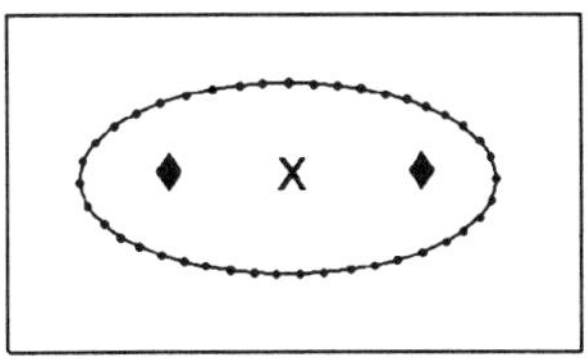

$Q_{29}Q_{30}$

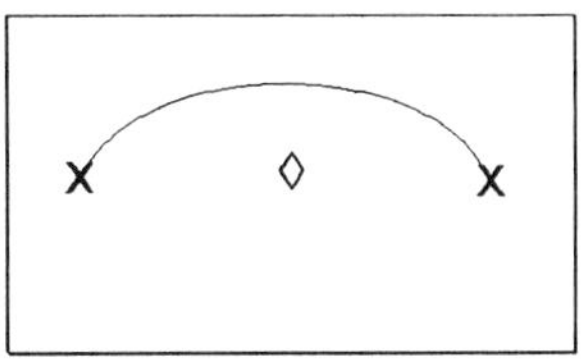

Q_5Q_{11}

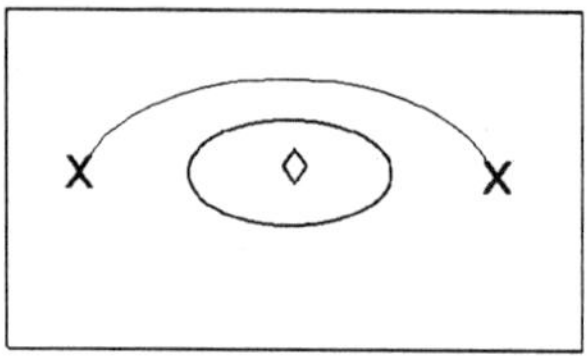

Q_5Q_{13}

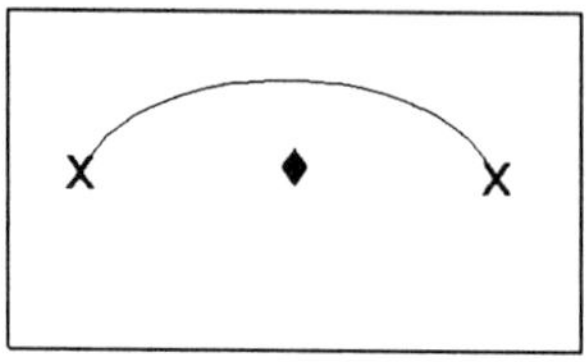

$Q_{13}Q_{15}$

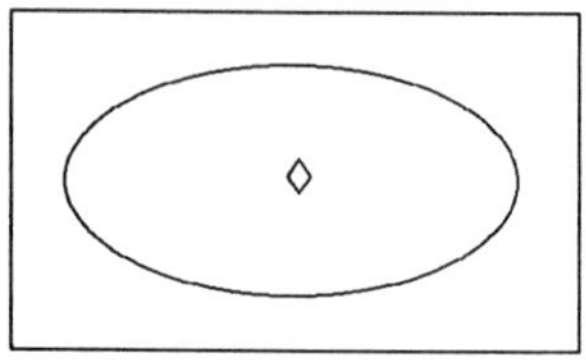

Oa-axis, $a \in [0, 1)$

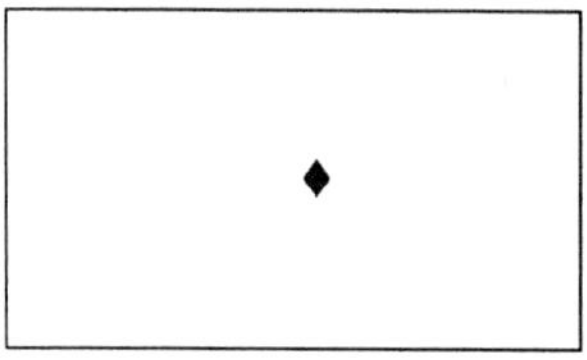

Oa-axis, $a \in (1, \infty)$

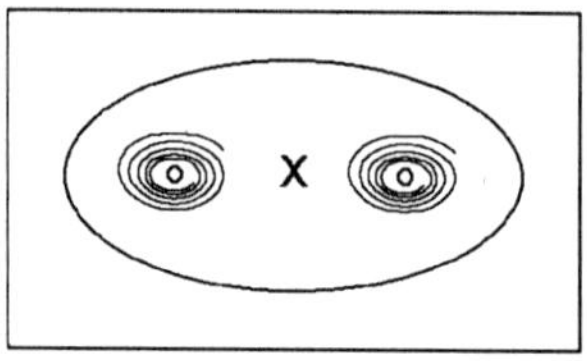

Q_0

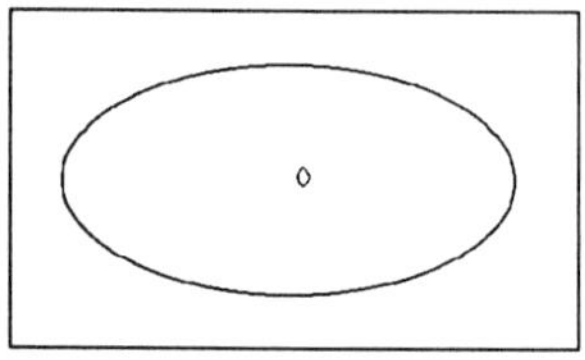

Q

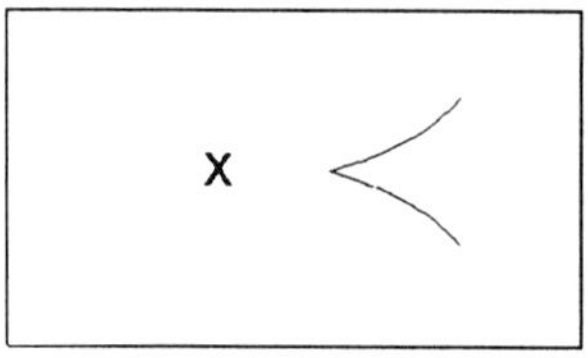

Q_1

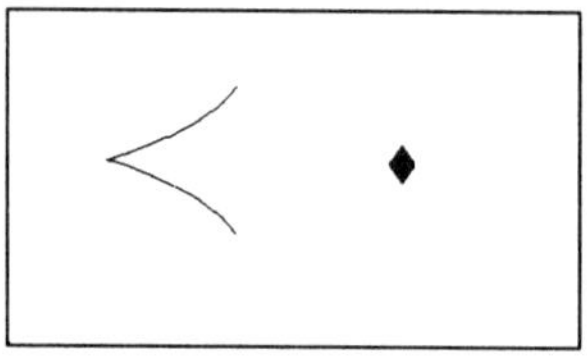

Q_3

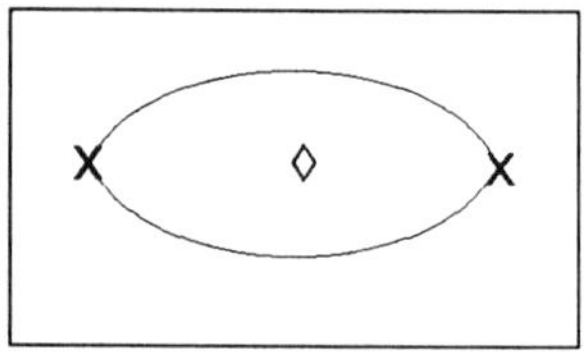

Q_5

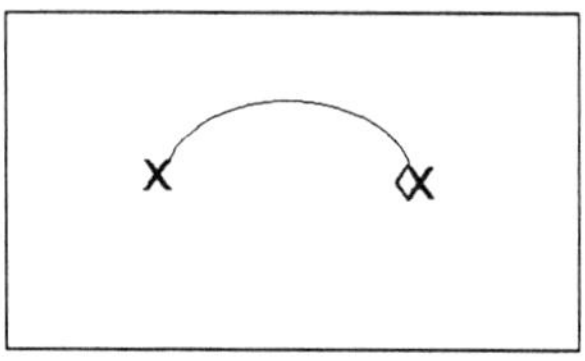

Q_{11}

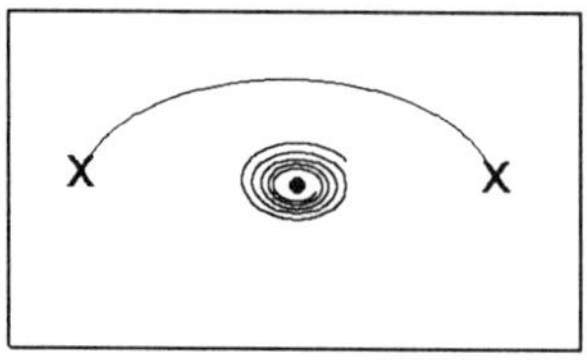

Q_{13}

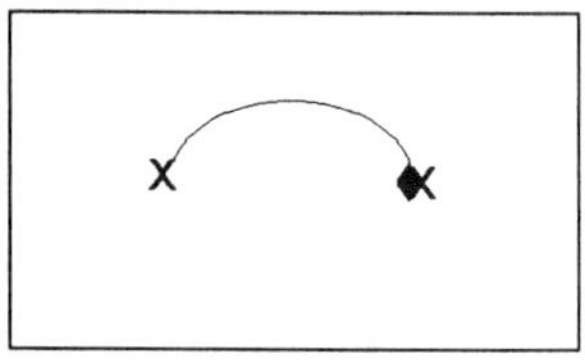

Q_{15}

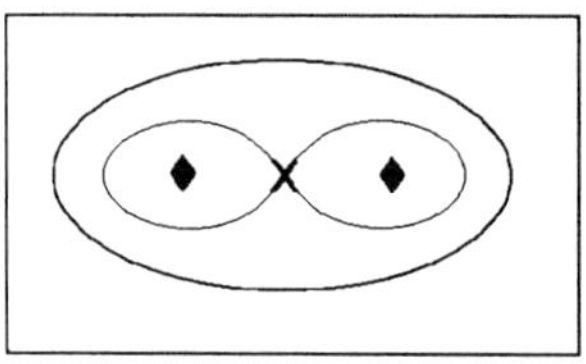

Q_6

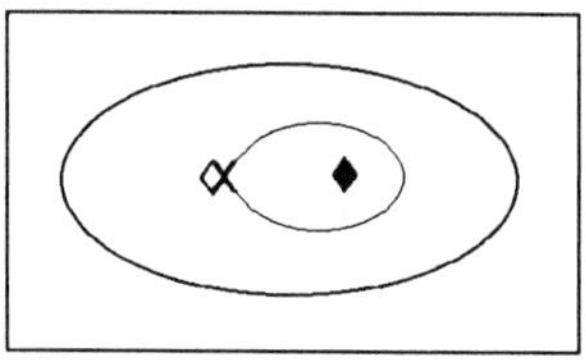

Q_7, Q_9

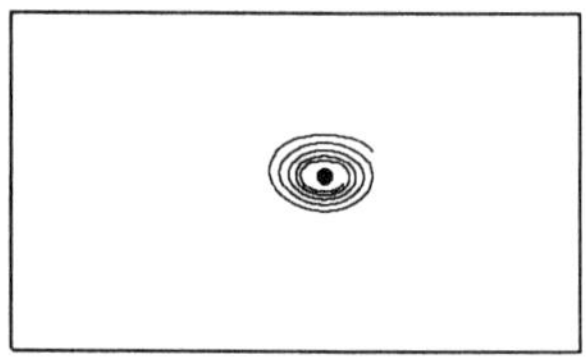

Q_{17}

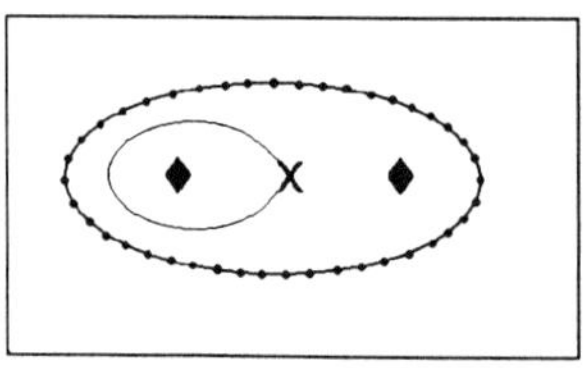

Q_{29}

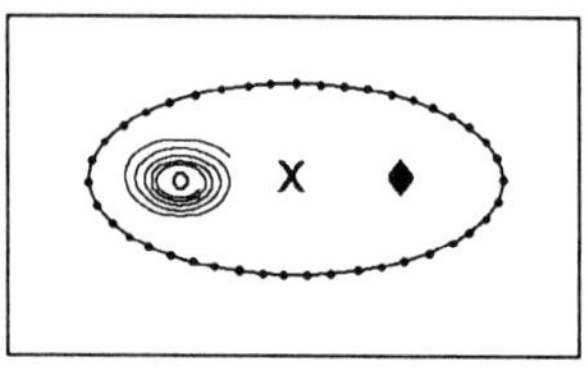

Q_{27}

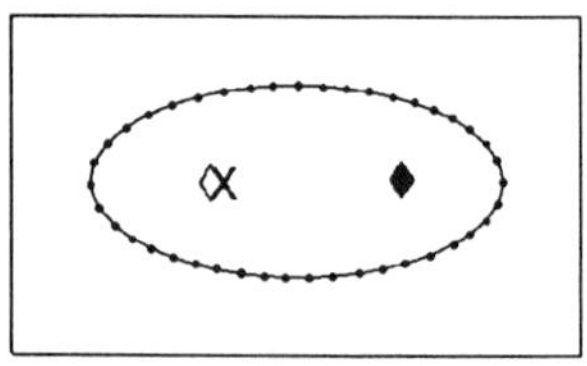

Q_{25}

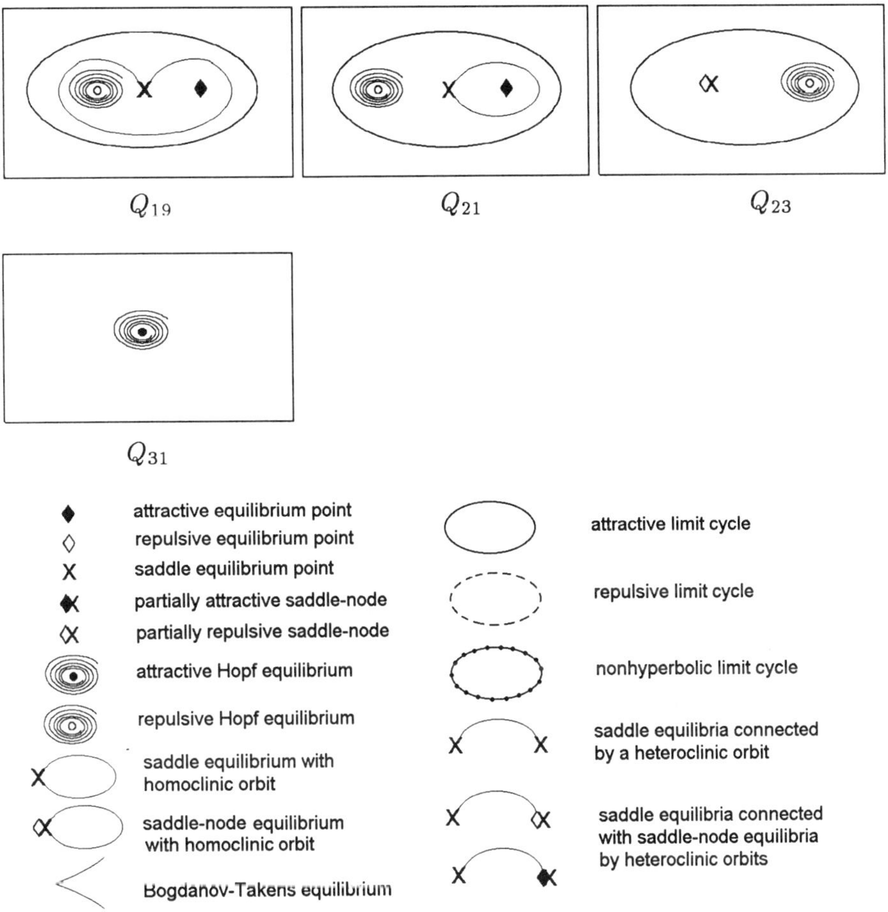

Fig. 5.1.2. Schematic phase portraits corresponding to strata from Figure 5.1.1.

In the following we describe the type of phase portrait for each stratum. The attractive or repulsive equilibria can have either real or complex eigenvalues, according to Section 2.3. This is why they are represented by a void rhomb for the repulsive equilibria and by a heavy rhomb for the attractive equilibria.

Strata of topological dimension equal to two.

These are the domains 1–18:

— for parameters in domains 1A or 1B, the F–N system possesses two saddles and a repulsive equilibrium point. However, the phase portraits from these two domains are not topologically equivalent, owing to the curve K_1 of breaking saddle connection bifurcation, which separates 1A from 1B.

— for parameters in domains 2A or 2B, there are two saddles and an attractive limit cycle surrounding a repulsive equilibrium point. Domains 2A and 2B are separated by the segment Q_5Q_{13} of the curve K_3 of breaking saddle connection bifurcation, therefore their phase dynamics are not topologically equivalent.

— for parameters in domains 3A or 3B, there are two saddles and an attractive equilibrium point. Domains 3A and 3B are separated by the segment $Q_{13}Q_{15}$ of the curve K_3 of breaking saddle connection bifurcation, therefore the phase dynamics from 3A is not topologically equivalent with that from 3B.

The differences of the phase dynamics corresponding to domains 1A, 1B; 2A, 2B or 3A, 3B were presented in Figures 3.4.3, 3.4.4 and 3.4.5.

— for parameters in domain 4, the single equilibrium point is a saddle;

— for parameters in domain 5, an attractive limit cycle surrounds a repulsive equilibrium;

— for parameters in domain 6, the single equilibrium point is an attractor;

— for parameters in domain 7, there are two nested limit cycles, the exterior one is attractive, the interior one is repulsive and inside them there is an attractive equilibrium point;

— for parameters in domain 8, there exists an attractive limit cycle inside which there are two repulsive equilibria and a saddle;

— for parameters in domain 9, there exists a larger attractive limit cycle, inside which there exist a repulsive equilibrium point, a saddle-point and a smaller repulsive limit cycle inside which there is an attractive equilibrium point;

— for parameters in domain 10, there exists an attractive limit cycle inside which there are three equilibria (a saddle, an attractor and a repulsor);

— for parameters in domain 11, there is an attractive limit cycle inside which there is a saddle and two smaller repulsive limit cycles, each of them surrounding an attractive equilibrium point;

— for parameters in domain 12, there is an attractive limit cycle containing a saddle, an attractive equilibrium point and a smaller repulsive limit cycle surrounding an attractive equilibrium point;

— for parameters in domain 13, there exists an attractive limit cycle inside which there exists a smaller repulsive limit cycle inside which there is a saddle, an attractive equilibrium point and a smaller repulsive limit cycle surrounding an attractive equilibrium point;

— for parameters in domain 14, there exists an attractive limit cycle. Inside it there exists a repulsive limit cycle surrounding a saddle, an attractive equilibrium point and a repulsive equilibrium point;

— for parameters in domain 15, there is a saddle, an attractive equilibrium point and a repulsive equilibrium point;

— for parameters in domain 16, there is a saddle, an attractive equilibrium point and a repulsive limit cycle inside which there is an attractive equilibrium point;

— for parameters in domain 17, there exists an attractive limit cycle inside which there exists a repulsive limit cycle surrounding two attractive equilibrium points and a saddle;

— for parameters in domain 18, there are two attractive equilibria and a saddle.

These domains are sets of points (b, a) to which structurally stable dynamical systems correspond. For all points (b, a) of such a domain topologically equivalent dynamical systems correspond.

Strata of topological dimension equal to one.
These are open curvilinear segments. Here is the topological description of the phase portraits.

— Q_1Q_{13}, $Q_{13}Q_{31}$: two saddles and one attractive Hopf bifurcation point. However, the phase portraits corresponding to $Q_{13}Q_{31}$ are not topologically equivalent to those corresponding to Q_1Q_{13}, because at Q_{13} a breaking saddle connection bifurcation takes place.

— $Q_{31}Q_{17}$: an attractive Hopf bifurcation point;

— $Q_{17}Q_{23}$: an attractive limit cycle surrounding a repulsive Hopf bifurcation point;

— $Q_{23}Q_0$: an attractive limit cycle surrounding a repulsive equilibrium point, a saddle and a repulsive Hopf bifurcation point;

— Q_0Q_{21}: an attractive limit cycle inside which there exists a saddle, a repulsive Hopf bifurcation point and a smaller repulsive limit cycle surrounding an attractive equilibrium point;

— $Q_{21}Q_{19}$: an attractive limit cycle surrounding an attractive equilibrium point, a saddle and a repulsive Hopf bifurcation point;

— $Q_{19}Q_{27}$: an attractive limit cycle surrounding a smaller repulsive limit cycle inside which there are a saddle, an attractive equilibrium point and a repulsive Hopf bifurcation point;

— $Q_{27}Q_3$: a saddle, an attractive equilibrium point and a repulsive Hopf equilibrium point;

— Q_1Q_5: two saddles and a repulsive equilibrium point. This point is situated inside the domain delimited by the homoclinic orbit of one among the saddles;

— Q_3Q_{29}: two attractive equilibria and a saddle with a homoclinic orbit surrounding one among the attractive equilibria;

— $Q_{29}Q_6$: an attractive limit cycle surrounding a smaller repulsive limit cycle inside which there exist two attractive equilibrium points and a saddle. The homoclinic orbit of the saddle encloses a domain where there is one among the attractive equilibria;

— Q_6Q_{21}: an attractive limit cycle surrounding a smaller repulsive limit cycle and a saddle with its homoclinic orbit. Each among the small limit cycles and the homoclinic orbit surround an attractive equilibrium point;

— $Q_{21}Q_7$: an attractive limit cycle surrounding a repulsive equilibrium point and a saddle with its homoclinic orbit inside which there exists an attractive equilibrium point;

— Q_7Q_9: an attractive limit cycle surrounds the homoclinic orbit of a partially repulsive saddle–node, inside which there exists an attractive equilibrium point;

— Q_9Q_{19}: an attractive limit cycle surrounds the homoclinic orbit of a saddle inside which there are a repulsive equilibrium point and an attractive equilibrium point;

— $Q_{19}Q_6$: an attractive limit cycle surrounds the homoclinic orbit of a saddle inside which there is an attractive equilibrium point and a small repulsive limit cycle inside which there exists another attractive equilibrium point;

— S_1, for $b \in (-\infty, b_{11})$: a saddle, connected with a partially repulsive saddle–node;

— $Q_{11}Q_1$: a saddle and a partially repulsive saddle–node;

— Q_1Q_{15}: a saddle and a partially attractive saddle–node;

— S_1, for $b \in (b_{15}, 0)$: a saddle, connected with a partially attractive saddle–node;

— QQ_{23}: an attractive limit cycle surrounding a repulsive equilibrium point and a partially repulsive saddle–node;

— $Q_{23}Q_7$: an attractive limit cycle surrounding a partially repulsive saddle–node and a smaller repulsive limit cycle inside which there exists an attractive equilibrium point;

— Q_9Q_{25}: an attractive limit cycle surrounding a smaller repulsive limit cycle inside which there are an attractive equilibrium point and a partially repulsive saddle–node;

— $Q_{25}Q_3$: a partially repulsive saddle–node and an attractive equilibrium point;

— S_1, with $b \in (c, \infty)$: a partially attractive saddle–node and an attractive equilibrium point;

— $Q_{17}Q_{25}$: a non-hyperbolic limit cycle enclosing an attractive equilibrium point;

— $Q_{25}Q_{27}$: a non-hyperbolic limit cycle surrounding a repulsor, a saddle and an attractor;

— $Q_{27}Q_{29}$: a non-hyperbolic limit cycle enclosing a saddle, an attractive equilibrium point and a smaller repulsive limit cycle surrounding an attractive equilibrium point;

— $Q_{29}Q_{30}$: a non-hyperbolic limit cycle surrounding two attractive equilibrium points and a saddle;

— Q_5Q_{11}: a repulsive equilibrium point and two saddles connected by a heteroclinic orbit.

— Q_5Q_{13}: two saddles connected by a heteroclinic orbit and an attractive limit cycle surrounding a repulsive equilibrium point;

— $Q_{13}Q_{15}$: two saddles connected by a heteroclinic orbit and an attractive equilibrium point.

— Oa-axis, $a \in (-1, 1)$: an attractive limit cycle surrounding a repulsive equilibrium point and a static bifurcation at infinity;

— Oa-axis, $a \in (1, \infty)$: one attractive equilibrium point and a static bifurcation at infinity.

These open segments are sets of points (b, a) to which codimension-one bifurcation corresponds. They are part of the boundaries of the domains of structural stability.

We included the Oa-axis among the set of values corresponding to codimension-one bifurcations owing to this static bifurcation at infinity. However, according to Definition 1.2.14 , which ignores the point at infinity, these bifurcations are of codimension-zero, i.e., the corresponding vector fields and dynamical systems are structurally stable.

Strata of topological dimension equal to zero.

The points $Q_i\,(b_i, a_i)$, with $a_i \geq 0$, correspond to phase dynamics characterized as follows:

— at Q_0 : an attractive limit cycle surrounding two repulsive Hopf bifurcation equilibria and a saddle;

— at Q : an attractive limit cycle surrounding a non-hyperbolic repulsive equilibrium point;

— at Q_1 : a Bogdanov–Takens equilibrium point and a saddle;

— at Q_3 : a Bogdanov–Takens equilibrium point and an attractive equilibrium point;

— at Q_5 : a repulsive equilibrium point and two saddles connected by two heteroclinic orbits;

— at Q_{11} : a partially repulsive saddle–node connected with a saddle by a separatrix;

— at Q_{13} : an attractive Hopf equilibrium point and two saddles, connected by a heteroclinic orbit;

— at Q_{15} : a saddle connected with a partially attractive saddle–node, by a separatrix.

Let us remark that the schematic phase portrait for Q_{11} is the same as for S_1, $b < b_{11}$, yet the phase portraits corresponding to $(b, a) \in S_1$, $b < b_{11}$, are not topologically equivalent to that corresponding to Q_{11}, where the heteroclinic orbit connecting the saddle–node and the saddle is a separatrix. A similar conclusion takes place for the phase portraits corresponding to $(b, a) \in S_1$, $b \in (b_{15}, 0)$ and to Q_{15}. More details were given in Section 3.4;

— at Q_6 : an attractive limit cycle surrounding two attractive equilibria and a saddle. The two homoclinic orbits of the saddle encloses two domains, each of them containing one attractive equilibrium point;

— at Q_7 , Q_9 : an attractive limit cycle surrounding an attractive equilibrium point and a partially repulsive saddle–node, with a separatrix loop that surrounds the attractor.

Note that although the schematic phase portraits for Q_7, Q_9 and the open segment Q_7Q_9 were represened in the same way, the phase portraits corresponding to Q_7 or Q_9 contains a separatrix loop and are not topologically equivalent to those corresponding to Q_7Q_9, where the homoclinic orbit of the saddle–node is not a separatrix;

— at Q_{17} : a Bautin bifurcation point;

— at Q_{29} : a non-hyperbolic limit cycle surrounding two attractive equilibria and a saddle having a homoclinic orbit which surrounds one of the attractors;

— at Q_{27} : a non-hyperbolic limit cycle surrounding an attractive equilibrium point, a saddle and a repulsive Hopf bifurcation point;

— at Q_{25} : a non-hyperbolic limit cycle surrounding a partially repulsive saddle–node and an attractive equilibrium point;

— at Q_{19} : an attractive limit cycle surrounding a saddle and its homoclinic orbit, that surrounds a repulsive Hopf bifurcation point and an attractive equilibrium point;

— at Q_{21} : an attractive limit cycle surrounding a repulsive Hopf bifurcation equilibrium point, a saddle and an attractive equilibrium point. The saddle has a homoclinic orbit, surrounding the attractor;

— at Q_{23} : an attractive limit cycle surrounding a partially repulsive saddle–node and a repulsive Hopf bifurcation equilibrium point;

— at Q_{31} : an attractive Hopf bifurcation equilibrium point and a static bifurcation at infinity.

To the points Q and Q_i, $i = \overline{0, 32}$, codimension-two bifurcations correspond. These points are situated on the boundaries of the domains of stucturally stable dynamical systems and of the one-dimensional strata corresponding to codimension-one bifurcations.

5.2. BASINS OF ATTRACTION

In Section 3.4 we were concerned with the stable and unstable manifolds of saddles. They are invariant sets of dimension equal to one.

In this section by **basin of attraction** in the plane we mean the stable set of an attractive equilibrium point or limit cycle. In Chapter 1 (Definition 1.2.17) we defined only the stable set of an equilibrium point. We can also define the stable set of an attractive limit cycle as the set of all points of the phase plane for which the ω-limit set is the cycle. The basin of attraction is a domain of the topological dimension equal to two. We shall determine the basins of attraction numerically for some points of the parameter plane.

If the system possesses a single attractive equilibrium point and no oscillatory regimes exist (domain 6 of the parametric portrait), then the entire (x, y)-plane is the basin of attraction of the attractor. If the attractor coexists with other equilibria (but still no oscillatory regimes exist) its basin of attraction is not the entire (x, y)-plane any more, but some part of it. Such situations are presented in Figures 5.2.1a,b and c.

The parameters from Figure 5.2.1a correspond to domain 3A of the parametric portrait. The basin of the attractive point is delimited by the stable manifolds of the saddle (x_3, y_3), whilst the trajectories through the points (x^0, y^0) which do not belong to this basin (except for the stable manifolds of the saddles) have as ω-limit set the point at infinity.

In Figure 5.2.1b, corresponding to parameters in domain 3B of the parametric portrait, the basin of the attractive point is delimited by the stable manifolds of the two saddle-points, whilst the trajectories through points which do not belong to this basin of attraction (except for the stable manifolds of the saddles) go to infinity, closer and closer to the branches of R–H.

Figure 5.2.1c corresponds to parameters situated in domain 18 of the parametric portrait. There are two attractive equilibrium points, whose basins are delimited by the stable manifolds of the saddle.

If the single equilibrium point of the system is an attractor and it coexists with limit cycles (domain 7 of the parametric portrait), then the basin of the attractive equilibrium is the domain surrounded by the inner limit cycle, whilst the outer limit cycle is the ω-limit set for the trajectories through the points situated outside the inner limit cycle (Figure 5.2.1d).

If three equilibria coexist with limit cycles, the stable and unstable manifolds of the saddles or/and the limit cycles can be boundaries of the basins of attraction. Such examples are presented in Figures 5.2.1e, f, g and h. Recall that $x_2 < x_1 < x_3$.

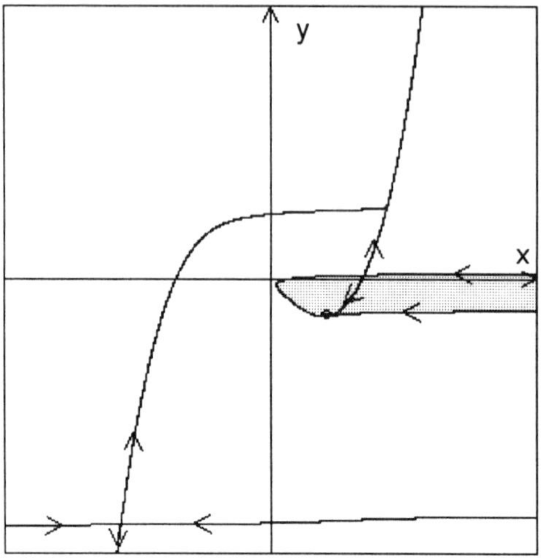

a) $a = 1.7$, $b = -1$

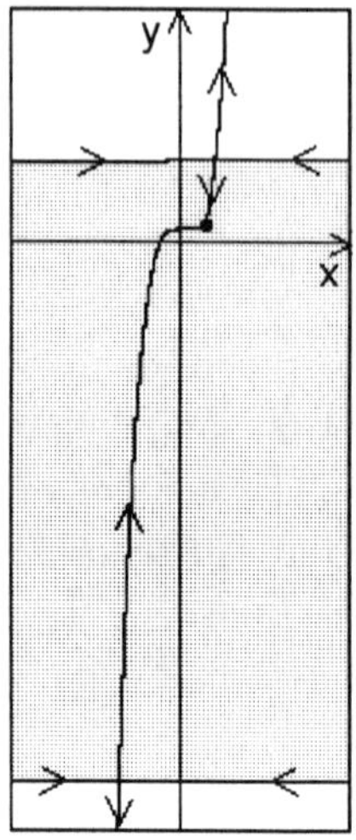

b) $a = 2$, $b = -0.16$

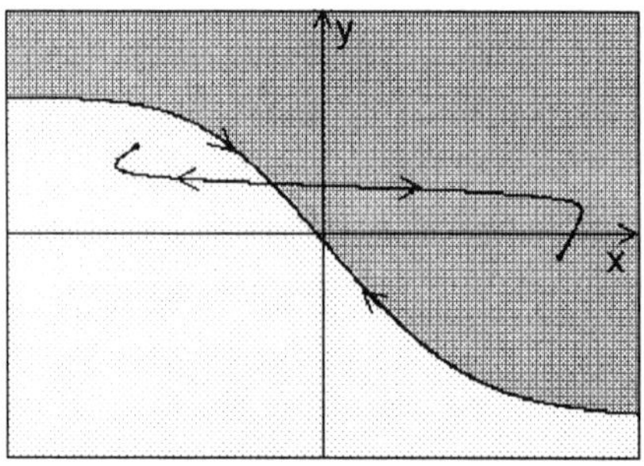

c) $a = 1$, $b = 4$

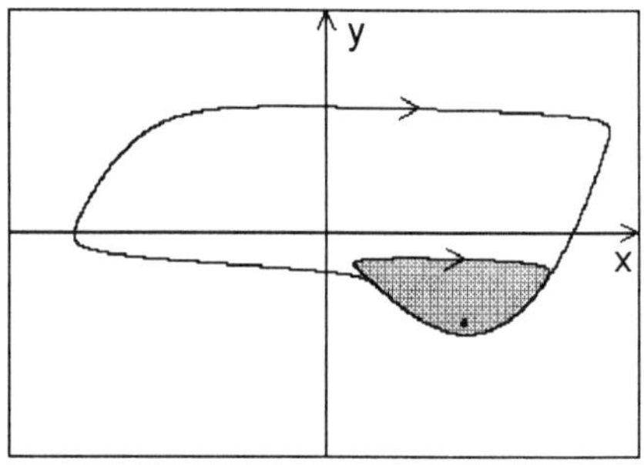

d) $a = 0.4536669535$, $b = 0.8$

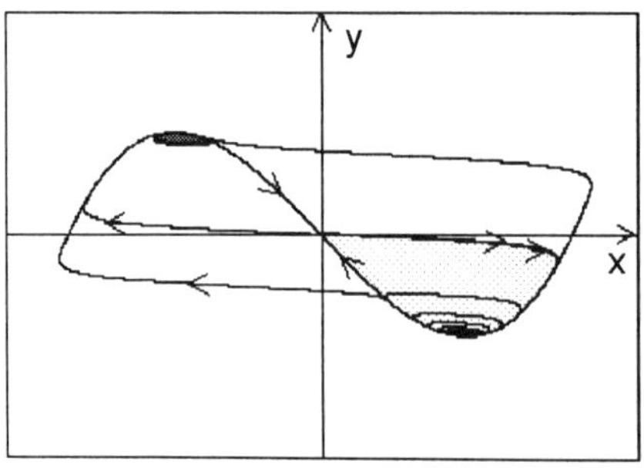

e) $a = 0.002$, $b = 1.4681830098$

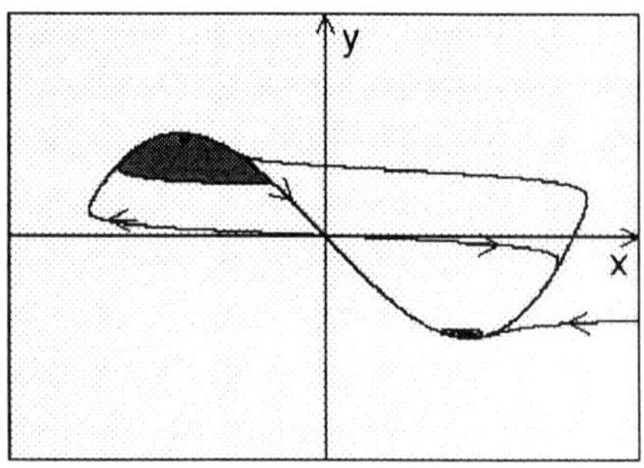

f) $a = 0.002$, $b = 1.4741$

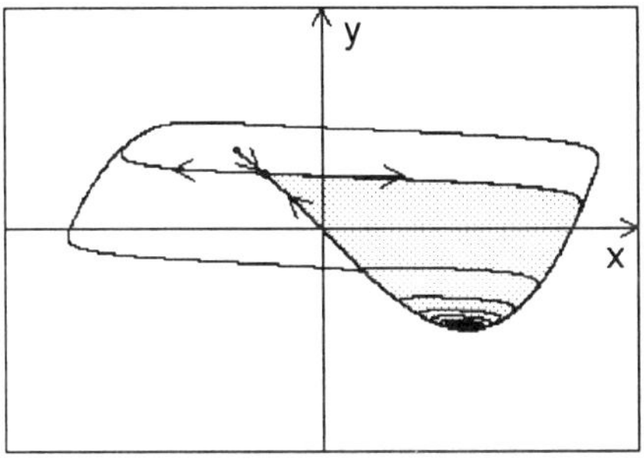

g) $a = 0.1$, $b = 1.323152069045012$

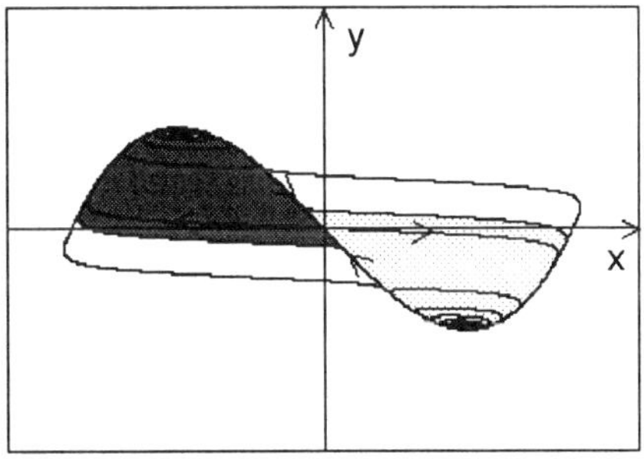

h) $a = 0$, $b = 1.47114317029$

Fig. 5.2.1 Basins of attraction, for $c = 5$ and (b, a) in domains a) 3A; b) 3B; c) 18; d) 7; e) 12; f) 16; g) 10; h) 17 of the parametric portrait.

In Figure 5.2.1e the parameters correspond to domain 12 of the parametric portrait. The stable manifolds of the saddle point (x_1, y_1) have the repulsive limit cycle as α-limit set, whilst one of its unstable manifolds has the attractive limit cycle as the ω-limit set and the other one has the attractor (x_3, y_3) as the ω-limit set. The attractor (x_2, y_2) has as basin of attraction the domain surrounded by the repulsive limit cycle, whilst the basin of the attractor (x_3, y_3) is the domain delimited by the two stable manifolds of the saddle. This is why, in the second quadrant, the last basin becames very narrow and it spirals around the repulsive limit cycle. Of course, such a complicate domain cannot be seen on the figure, owing to the smaller and smaller dimensions of the loops of the spiraling stable manifolds.

In Figure 5.2.1f basins of attraction characteristic to the domain 16 of the parametric portrait are presented. The stable manifolds of the saddle have as α-limit set the repulsive limit cycle and the infinity, whilst both its unstable manifolds have as the ω-limit set the attractor (x_3, y_3). This is why the basins of attraction of the two attractive equilibrium points are separated only by the limit cycle.

In Figure 5.2.1g corresponding to parameters in domain 10 of the parametric portrait, one of the unstable manifolds of the saddle (x_1, y_1) has as the ω-limit set the limit cycle and the other has as the ω-limit set the attractor (x_3, y_3), whilst both stable manifolds have as the α-limit set the repulsive equilibrium point (x_2, y_2). This is why the basin of the attractor is surrounded by the stable manifolds of the saddle, whilst the trajectories through the points from the unhatched region have as ω-limit set the limit cycle.

Figure 5.2.1h is characteristic to domain 17 of the parametric portrait. The unstable manifolds of the saddle point $(x_1, y_1) = (0, 0)$ have as the ω-limit set the attractive equilibria, whilst its stable manifolds have as α-limit set the repulsive limit cycle. This is why the basins of the two attractors (hatched) contain very narrow regions surrounding the repulsive limit cycle, which cannot be distingiushed in the figure.

5.3. TRANSIENT REGIMES AND NON–PERIODIC OSCILLATIONS

5.3.1. TYPES OF TRANSIENT REGIMES IN THE ABSENCE OF LIMIT CYCLES

The local behaviour around an equilibrium point, that does not take into account the presence of other equilibria, must be completed with the nonlocal case, when the aspect of the phase portrait in regions containing several equilibria is of interest.

The presence, simultaneously, of several equilibria, even in the absence of limit cycles (domains 1A, 1B, 3A, 3B, 15 or 18 from the parametric portrait), has as consequence the existence of very complex phase trajectories. They correspond to (nonstationary) transient regimes between partially or totally repulsive equilibria (or the point at infinity) and partially or totally attractive equilibria (or the point at infinity). The non–periodic oscillations can be very complicated and correspond to these regimes. In Figure 5.3.1 some trajectories through points (x^0, y^0) and the corresponding oscillations of $x(t, x^0, y^0)$ and $y(t, x^0, y^0)$ are represented.

They were computed for points of the parameter space corresponding to three equilibria, namely for (b, a) in regions 1A, 18, 15 and 3A of the parametric portrait.

The transient trajectories possess an infinite number of extremum points near a focus and they present regions of monotonicity near saddles and far away from foci.

Some transient trajectories have no maximum or/and minimum points, so the function defining them is monotonous (as the trajectory through $(-1, 1)$ in Figure 5.3.1d). Others oscillate as t runs over a bounded time interval (Figure 5.3.1.a), on a semiaxis (Figure 5.3.1b) or over the real axis (Figure 5.3.1c).

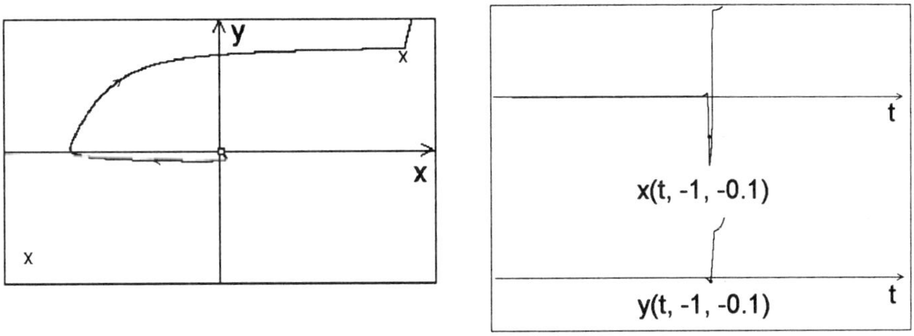

a) $a = 10^{-10}, b = -1.9$

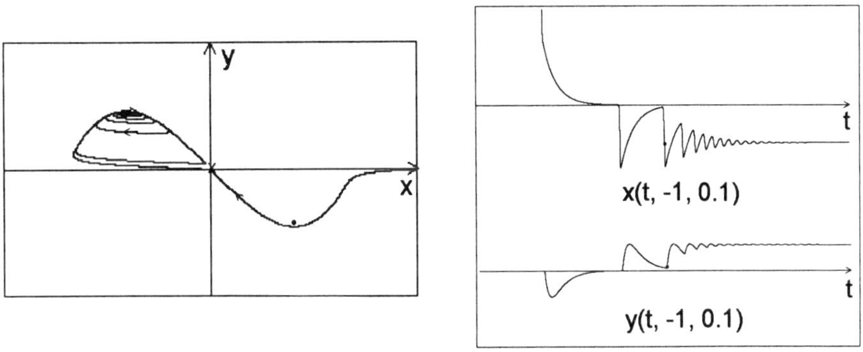

b) $a = 10^{-10}, b = 1.47114317023$

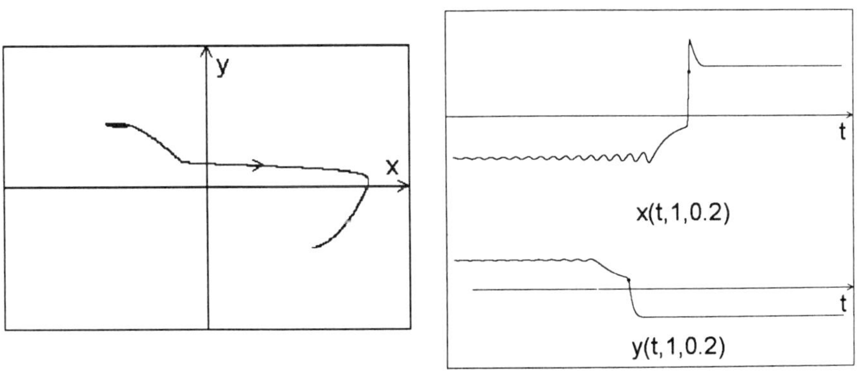

c) $a = 0.1, b = 1.6$

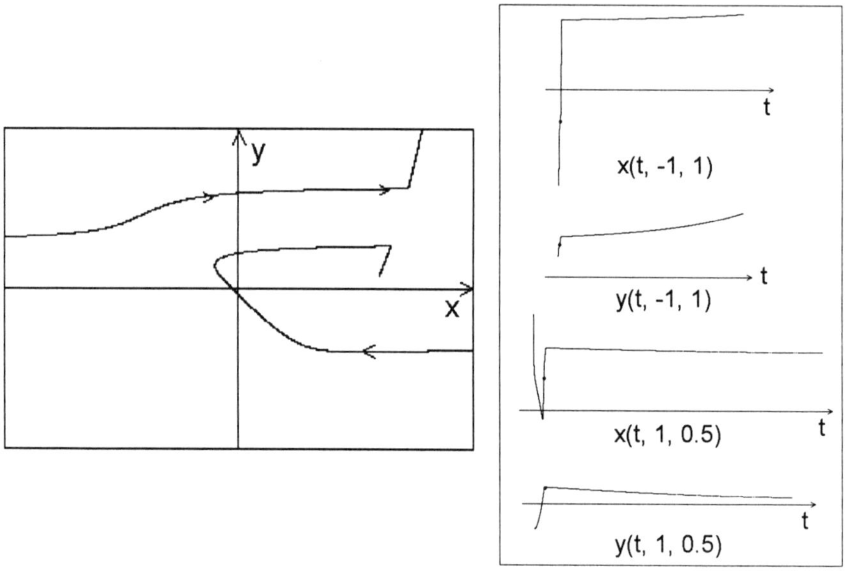

d) $a = 1.75, b = -0.35$

Fig. 5.3.1. Transient trajectories and corresponding oscillations for $c = 5$ and (b, a) situated in domains a) 1A; b) 18; c) 15; d) 3A of the parametric portrait.

In the case where oscillations are present as t runs over a bounded time interval, only a finite number of maximum and/or minimum points exists. Among these oscillations, interesting for applications are the excitation oscillations (Figure 5.3.2).

These oscillations possess only one minimum point, so that, instead of a monotonous decreasing to an equilibrium state, the characteristics of the corresponding physical system first decrease below the equilibrium state and then increase to this state.

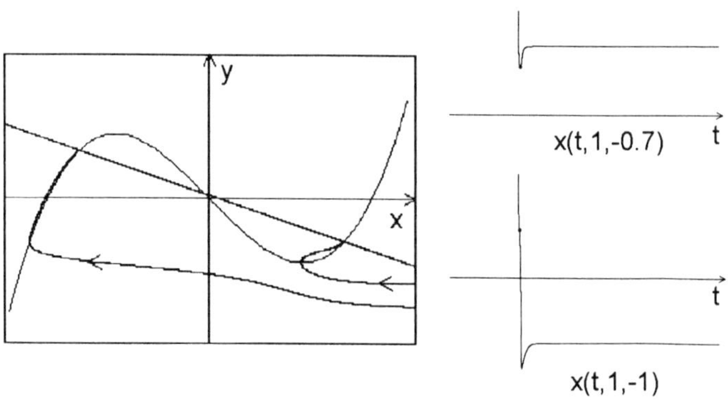

Fig. 5.3.2. Excitation oscillations for $a = 0.1$, $b = 3$, $c = 5$.

The physical system finds some 'energy' to increase its state functions. Such a medium is referred to as excitable. This is a degenerate case, because not all trajectories present such a phenomenon.

A similar phenomenon occurs in the case of the transient trajectories possesing a single maximum.

In the absence of limit cycles, oscillations of $x(t)$ (if they exist) either stabilize themselves at a constant value (if the transient trajectories tends to an attractive point or if they are stable manifolds of a saddle) or become unbounded (if they are attracted by the point at infinity).

The trajectories have complicate shapes only near the equilibrium points, because, due to Theorem 1.2.5, far from these points the phase portrait is homeomorphic to the phase portrait of the product system, whose trajectories are parallel straight lines. Of course this is true irrespective of the presence or absence of the limit cycles.

As an example, the phase portrait for $(b, a) = (10, 0)$ (that is a point from domain 18 of the parametric portrait) is given in Figure 5.3.3a near the equilibrium points and in Figure 5.3.3b far from these points.

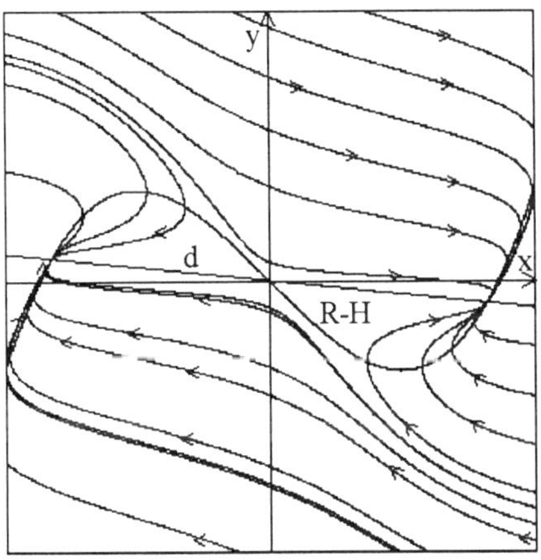

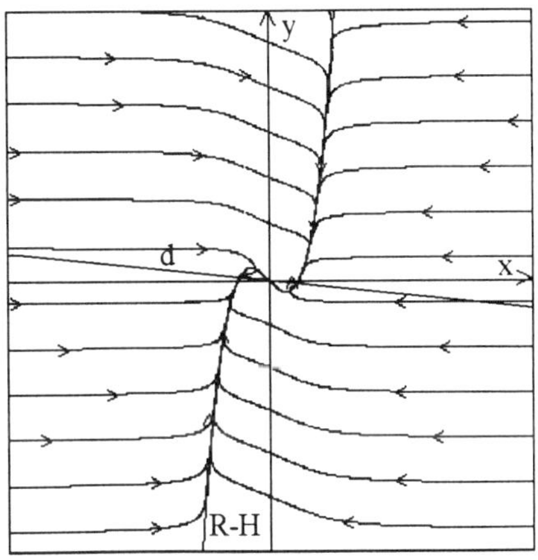

Fig. 5.3.3. Phase portraits for $(b, a) = (10, 0)$, $c = 5$ and a) $x, y \in (-2, 2)$; b) $x, y \in (-15, 15)$.

5.3.2. TYPES OF TRANSIENT REGIMES IN THE PRESENCE OF LIMIT CYCLES

In the presence of limit cycles, the dependence of x on t when $(x(t), y(t))$ is situated on a transient trajectory is more complicated. The graph of x versus t can present regions of damped out oscillations (if the current point is in the neighborhood of an attractive or repulsive equilibrium point) and regions of almost periodic oscillations (if the point moves around a limit cycle).

If a and b vary, the form and the size of the limit cycles vary too. As a consequence the corresponding periodic oscillations are varying as well. Moreover, to the transient trajectories having limit cycles as α- or/and ω-limit set, non-periodic oscillations correspond. However, as the time is increased these oscillations will be closer and closer to periodic oscillations. These properties are put into evidence in Figures 5.3.4–5.3.11, where we plotted the limit cycles, the equilibria, as well as the oscillations of $x\left(t,x^0,y^0\right)$ and $y\left(t,x^0,y^0\right)$ for $a=10^{-10}$, $c=5$ and variable b. The values of b were chosen [122] such that (b,a) be successively inside the domains 2, 5, 8, 9, 11, 12, 13 and 16 of the parametric portrait, given in Figure 5.1.1.

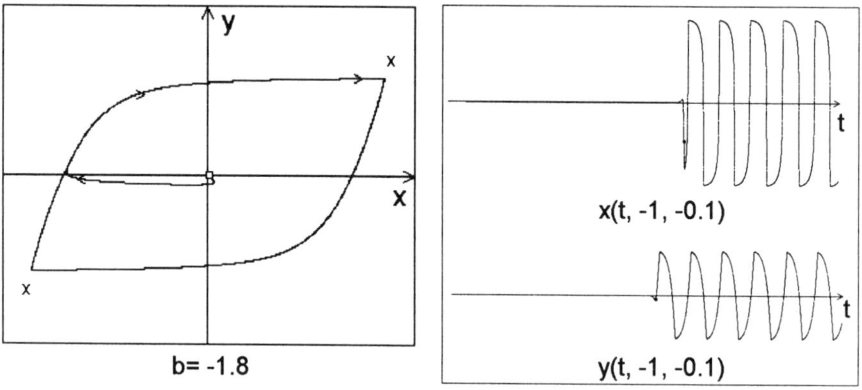

Fig. 5.3.4. Characteristic phase portrait and oscillations for (b,a) in domain 2 of the parametric portrait.

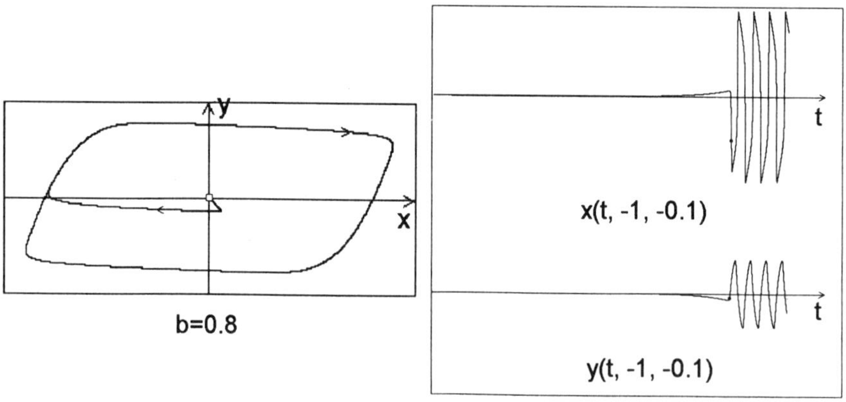

Fig. 5.3.5. Characteristic phase portrait and oscillations for (b,a) in domain 5 of the parametric portrait.

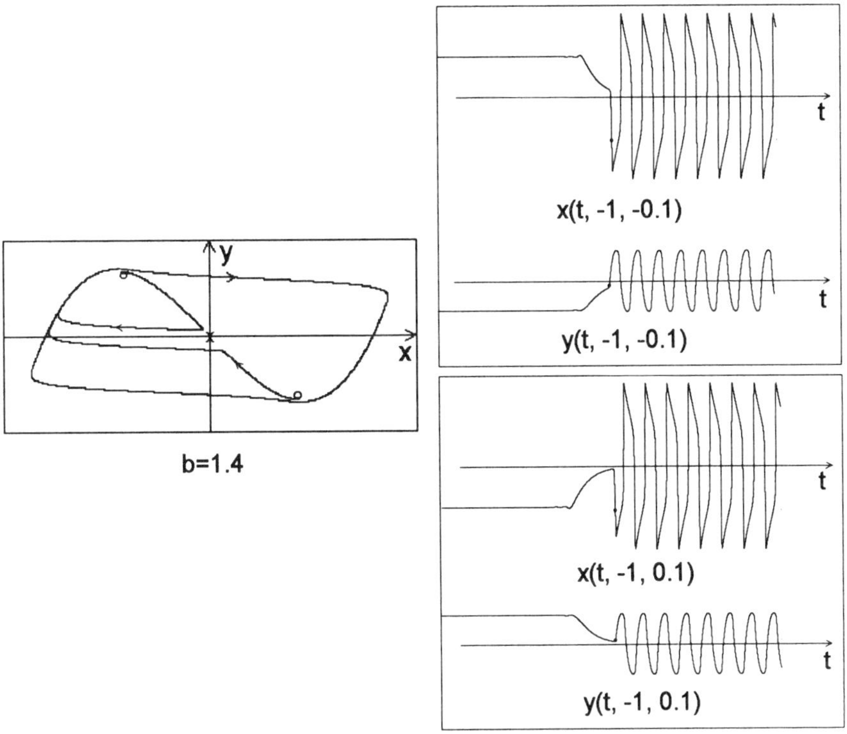

Fig. 5.3.6. Characteristic phase portrait and oscillations for (b, a) in domain 8 of the parametric portrait.

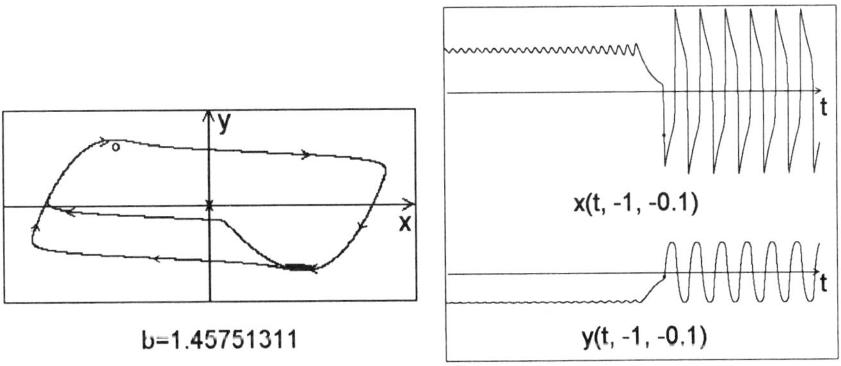

Fig. 5.3.7. Characteristic phase portrait and oscillations for (b, a) in domain 9 of the parametric portrait.

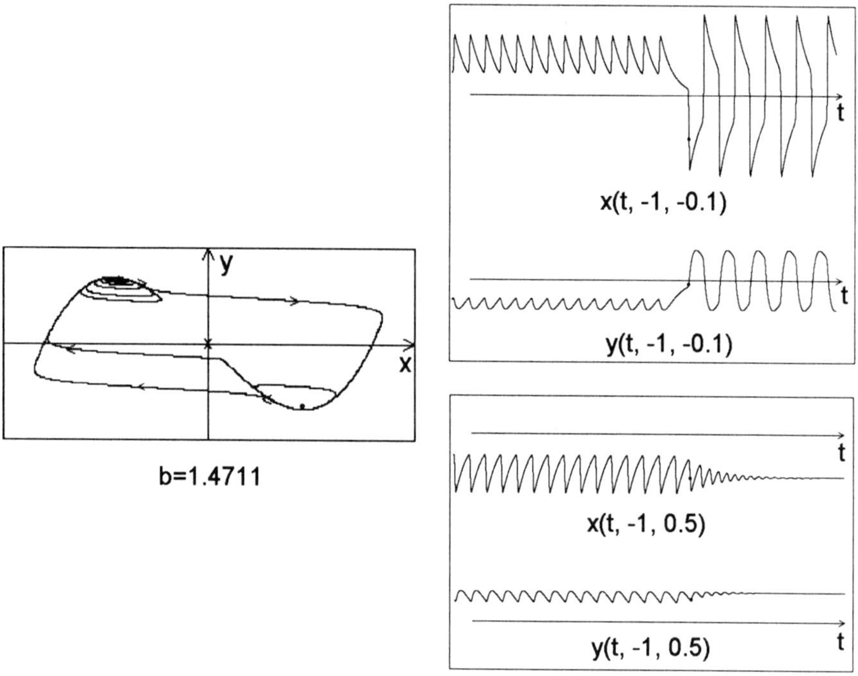

Fig. 5.3.8. Characteristic phase portrait and oscillations for (b, a) in domain 11 of the parametric portrait.

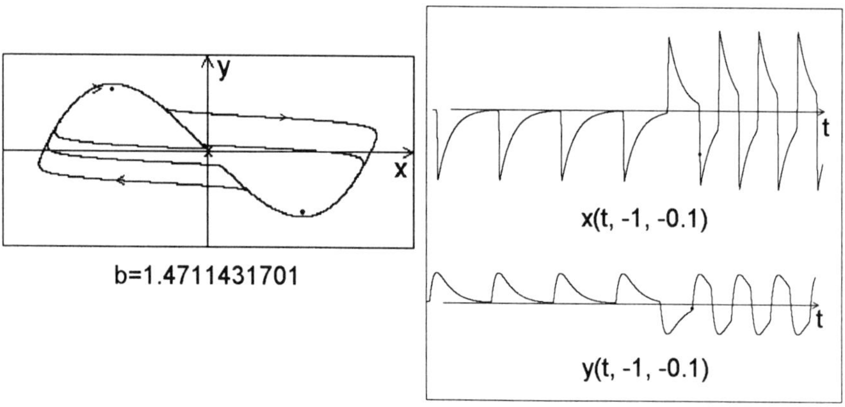

Fig. 5.3.9. Characteristic phase portrait and oscillations for (b, a) in domain 12 of the parametric portrait.

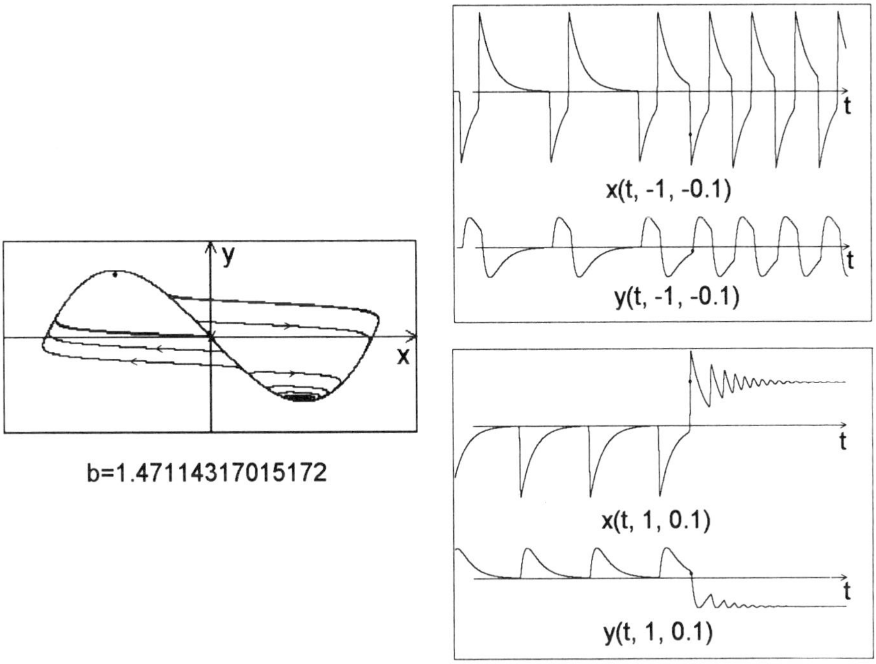

Fig. 5.3.10. Characteristic phase portrait and oscillations for (b, a) in domain 13 of the parametric portrait.

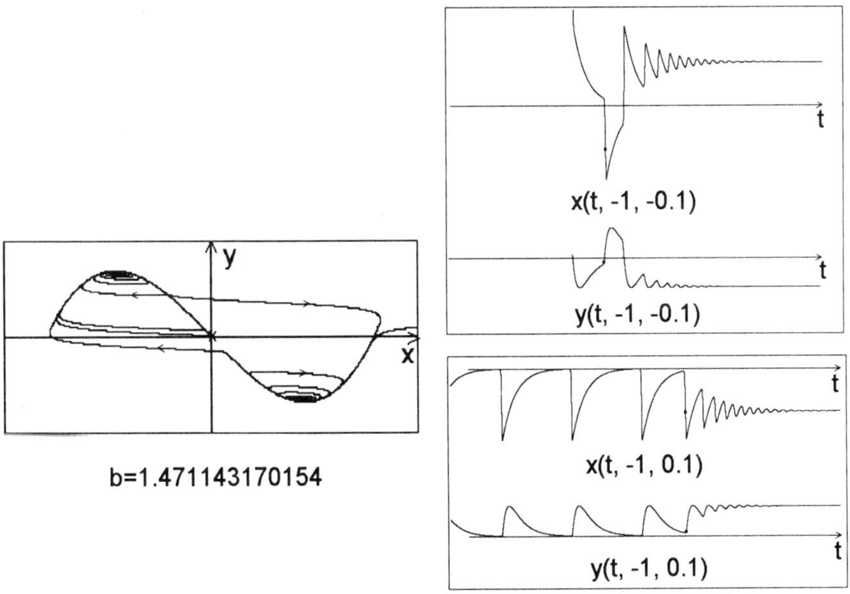

Fig. 5.3.11. Characteristic phase portrait and oscillations for (b, a) in domain 16 of the parametric portrait.

5.4. LIMIT CYCLES AND PERIODIC OSCILLATIONS

5.4.1. THE NUMBER OF THE LIMIT CYCLES FOR THE F–N MODEL

Given a dynamical system, the problem of how many limit cycles it possesses is known as the 16th Hilbert problem. In the case of the F–N model (1.1.17) we found that the maximum number of coexisting limit cycles is three [44]. This situation takes place for the domain 11 and two domains 13 of the parametric portrait (Figure 5.1.1).

In addition, the coexistence of two limit cycles corrersponds to the domains 7, 9, 12, 14, 17 and to the open segments Q_0Q_{21}, $Q_{19}Q_{27}$, $Q_{29}Q_6$, Q_6Q_{21}, $Q_{19}Q_6$, $Q_{23}Q_7$, Q_9Q_{25}, $Q_{27}Q_{29}$, for $a \geq 0$ and to their symmetrics with respect to the Ob-axis for $a < 0$ (Figure 5.1.1).

Finally, a single limit cycle is present in the phase portraits for the dynamical systems corresponding to (b, a) in the domains 2A, 2B, 5, 8, 10, 16, to the open segments and their symmetrics with respect to Ob-axis $Q_{17}Q_{23}$, $Q_{23}Q_0$, $Q_{21}Q_{19}$, $Q_{21}Q_7$, Q_7Q_9, Q_9Q_{19}, QQ_{23}, $Q_{17}Q_{25}$, $Q_{25}Q_{27}$, $Q_{29}Q_{30}$, Q_5Q_{13}, the Oa-axis for $a \in [0, 1)$ and to the points Q_0, Q, Q_6, Q_7, Q_9, Q_{29}, Q_{27}, Q_{25}, Q_{19}, Q_{21} and Q_{23} and their symmetrics with respect to the Ob-axis (Figure 5.1.1).

The regions of the (b, a)-plane corresponding to oscillatory regimes were hatched in Figures 5.1.1a, b.

In the following we are concerned with the periodic-oscillations coresponding to limit cycles. There exist two main types of such oscillations: sinusoidal and relaxation oscillations and a big variety of intermediate oscillations. The parameter determining the type of oscillation is c. For small c we have sinusoidal oscillations and, as $c \to \infty$, relaxation oscillations.

Indeed, considering the parameter $\varepsilon = c^{-2}$, as $c \to \infty$ we have $\varepsilon \to 0$ and an asymptotic analysis must be done (Chapter 4).

The transformation of oscillations into relaxation ones is shown in Figure 5.4.1. for $a = b = 0.5$ (domain 5 of the parametric portrait) and different larger and larger values of c.

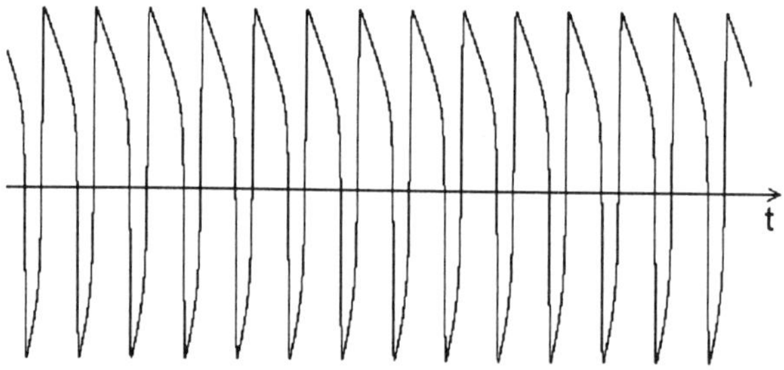

a) $c = 5$

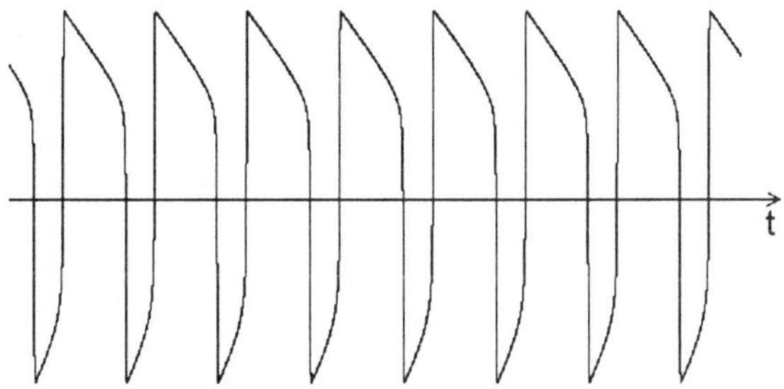

b) $c = 10$

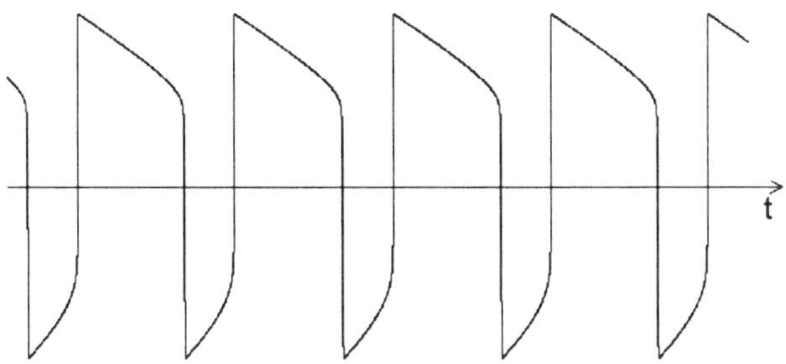

c) $c = 20$

Fig. 5.4.1. Oscillations for $a = b = 0.5$ and various values of c.

5.4.2. CONCAVE LIMIT CYCLES VERSUS INFLECTION POINTS

Inflection point arguments to explain the formation and destruction of concave limit cycles.
In the following, by **concave (convex) limit cycle** is meant a limit cycle surrounding a concave (convex) domain of the phase plane.

The form of limit cycles and the velocity of the current point on them, in the phase plane, determines the form of the corresponding periodic oscillations. Of primary interest in the heart functioning are the relaxation oscillations, characterized by four 'steps', namely two short and two long. This is the case of (more or less) rectangular limit cycles. For triangular limit cycles (looking like a duck's body), the corresponding oscillations have no longer four 'steps'. In addition, the presence of some concavity of the limit cycles transforms significantly the number of 'steps' and their duration, with dramatic consequences for the heart beats (Figure 5.4.2). Whence our interest for studying the conditions sufficient to induce concavity [55].

Since the concavity of a curve is related to its points of inflection, in Section 2.1.3 we initiated the investigation of the set $\mathfrak{I}$ locus of the inflection points of all phase plane trajectories.

There exist very few works devoted to these topics, partly because of the complex structure of $\mathfrak{I}$ and, correspondingly, to the equation of high degree defining the branches of $\mathfrak{I}$. Much more attention was paid to this question in earlier studies of nonlinear oscillations [30], rather than in contemporary researches.

In the following we derive sufficient conditions for the nonexistence of $\mathfrak{I}$, explain the formation of concave limit cycles and locate the regions in the parameter plane where they exist.

In particular, (2.1.9) shows that the points of the R–H nullcline defined by $\dot{x}= 0$, are not inflection points except for the case of points satisfying in addition the equation $\dot{y}= 0$. But $\dot{x}= 0,\ \dot{y}= 0$, defines the equilibria, hence R–H$\cap\ \mathfrak{I} = \Phi$. If in (2.1.9) we take $\dot{y}\ /\ \dot{x}$ as the unknown it follows that

$$\frac{\dot{y}}{\dot{x}} = \frac{-\left[b + c^2\left(1 - x^2\right)\right] \pm \sqrt{\left[b + c^2\left(1 - x^2\right)\right]^2 - 4c^2}}{2c^2}, \tag{5.4.1}$$

whence the sufficient conditions for the nonexistence of real solutions $\dot{y}\ /\ \dot{x}$ of (2.1.9)

$$x \in J_1 \equiv \left(-\frac{\sqrt{c^2 + 2c + b}}{c}, \frac{\sqrt{c^2 + 2c + b}}{c}\right) \tag{5.4.2}$$

for $b \in L_1 \equiv \left[-c^2 - 2c, -c^2 + 2c\right]$, and

$$x \in J_2 \equiv \left(-\frac{\sqrt{c^2 + 2c + b}}{c}, -\frac{\sqrt{c^2 - 2c + b}}{c}\right) \cup \left(\frac{\sqrt{c^2 - 2c + b}}{c}, \frac{\sqrt{c^2 + 2c + b}}{c}\right) \tag{5.4.3}$$

for $b \in L_2 \equiv \left(-c^2 + 2c, \infty\right)$.

Correspondingly, as b is increased from $-\infty$ to $+\infty$, inflection points exist up to $b = -c^2 - 2c$, then an interval for x appears, where no inflection points exist. This interval increases up to $(-2/\sqrt{c}, 2/\sqrt{c})$ for $b = -c^2 + 2c$, then it splits into two other intervals advancing towards $-\infty$ and $+\infty$ on the x-axis and they vanishe as $b \to \infty$. Substituting (1.1.17) in (5.4.1) we obtain the explicit form of the functions whose graphs are the two branches of $\mathfrak{I}$

$$\begin{aligned} y_{\pm} &= \frac{-2x + 2a}{b - c^2\left(1 - x^2\right) \pm \sqrt{\left[b + c^2\left(1 - x^2\right)\right]^2 - 4c^2}} \\ &- \frac{\left(x - \frac{x^3}{3}\right)\left[-b - c^2\left(1 - x^2\right) \pm \sqrt{\left[b + c^2\left(1 - x^2\right)\right]^2 - 4c^2}\right]}{b - c^2\left(1 - x^2\right) \pm \sqrt{\left[b + c^2\left(1 - x^2\right)\right]^2 - 4c^2}}. \end{aligned} \tag{5.4.4}$$

They are defined for $x \in \mathbf{R} - J_1$, for $b \in L_1$ and $x \in \mathbf{R} - J_2$, for $b \in L_2$. In addition, at $x = \pm\sqrt{1-\frac{1}{b}}$, y_+ is not defined if $b \in (-\infty, -c) \cup (1, c)$ and has the asymptotes $x = \pm\sqrt{1-\frac{1}{b}}$. Similarly, at $x = \pm\sqrt{1-\frac{1}{b}}$, y_- is not defined if $b \in (-c, 0) \cup (c, \infty)$ and has the asymptotes $x = \pm\sqrt{1-\frac{1}{b}}$. At $x = \pm\sqrt{1-\frac{1}{b}}$ (on the branch which is defined) there exist the points of inflection

$$(x, y) = \left(\pm\sqrt{1-\frac{1}{b}}, \pm\sqrt{1-\frac{1}{b}} \frac{c^2\left(\frac{2}{3}+\frac{1}{3b}\right)-b}{b^2-c^2} \right).$$

At $x = \pm\sqrt{1-\frac{1}{b}}$, if $b = \pm c$, neither y_+ nor y_- exists: they are asymptotic to $x = \pm\sqrt{1-\frac{1}{b}}$ and so, no inflection point exist. Consequently, for a given x, in the phase plane of (1.1.17) two, one or none inflection points can exist. For large c, the intervals for x where no inflection point exist become narrower and narrower. Therefore the possibility for $\mathfrak{I}$ to cross the limit cycles is more probable. This probability increases if we account for the property that for a fixed y, six inflection points can exist, as shown in Section 2.1.3. Thus $\mathfrak{I}$ consists at most of six branches coalescing at the ends of J_1 and J_2, some of them tending asymptotically to the straight lines $x = \pm\sqrt{1-\frac{1}{b}}$.

The above discussion has involved only two parameters: b and c.

From (5.4.1) it can be seen directly that at the points of inflection the slope at the trajectories is not equal to zero. Hence let us denote by $N_\pm$ the right hand side of (5.4.1). Then (5.4.4) simply reads

$$y_\pm = \frac{-x + a - c^2\left(x - \frac{x^3}{3}\right)N_\pm}{b + c^2 N_\pm}, \tag{5.4.5}$$

where $N_- \approx c^{-2}x^{-2}$ and $N_+ \approx x^2$ as $|x| \to \infty$. In addition N_- is always negative and N_+ always positive. Denoting by $\Delta_\pm$ the difference of the ordinates of the R–H curve and $y_\pm$ we have

$$\Delta_\pm = \frac{b\left(\frac{x^3}{3} - x\right) + x - a}{b + c^2 N_\pm}, \tag{5.4.6}$$

whence the interesting result that $\Delta_{\pm} \to 0$ as $x \to x^*$. Here x^* is an equilibrium point, i.e., satisfying the equation

$$x^3 - 3x\left(1 - \frac{1}{b}\right) - \frac{3a}{b} = 0.$$

Therefore R–H and $\mathfrak{I}$ do not intersect each other, but the limit points of $\mathfrak{I}$ (not belonging to $\mathfrak{I}$) are the equilibria. This result explains the difficulty in locating the limit cycles which have some portions situated between R–H and y_-. Indeed, the proof of the existence of the limit cycle and its localization is based on the construction of an absorbing domain which does not contain any equilibrium and part of whose boundaries belong to R–H and y_-. This difficulty is increased further by the fact that y_- has the same asymptotic behaviour as R–H (Section 2.1.3).

Of course, owing to the presence of the 'windows' within which $\mathfrak{I}$ has no branch, it is to be expected that not all equilibria are limit points for $\mathfrak{I}$. However, for those equilibria situated near $\mathfrak{I}$, the presence of $\mathfrak{I}$ is crucial. In particular, this is the case of equilibria which are Hopf bifurcation points. It is in this case that the duck-like nonconvex limit cycles emerge in the F–N system.

For a simpler explanation assume that $b \in (-c, 0)$ is kept fixed and a is decreased from its Hopf bifurcation value a_H (Figure 5.4.2). Of course, we assume that this range for b does not contain a curve of homoclinic bifurcation values whose effect would be to destroy the limit cycle very soon after its emergence.

The corresponding equilibrium (x^*, y^*) will move along R–H and will be surrounded by a limit cycle. Owing to the bending effect of R–H, this limit cycle, at least in its incipient stages (i. e. for a near a_H), remains near R–H, and its form will be more or less like a small duck's body. It surrounds a convex domain (Figure 5.4.2a).

As a goes away from a_H the size of the limit cycle enlarges up to some a_t when it becomes tangent to $\mathfrak{I}$ (Figure 5.4.2b). From now on, i.e., for $a < a_t$, the limit cycle will intersect $\mathfrak{I}$ at two neighbouring points favouring the formation of the 'head' (Figure 5.4.2c). The further decrease in a will have as an effect the increase of the distance between the inflection points situated on the limit cycle, favouring the formation of the large 'body'. At certain value of a, namely a_d, the limit cycle does not follow the unstable manifold of R–H any more, so the canard phenomenon ceases to exist. Decreasing further a, the inflection points become closer up to the situation when, for a_e, they coalesce, hence the limit cycle is again tangent to $\mathfrak{I}$ (Figure 5.4.2d). This ends the stage of concavity of the limit cycle. For still smaller a below a_e the limit cycle becomes convex and changes very slowly its from as a function of a (Figure 5.4.2e). Thus, the concave limit cycle exists for $a \in (a_e, a_t)$.

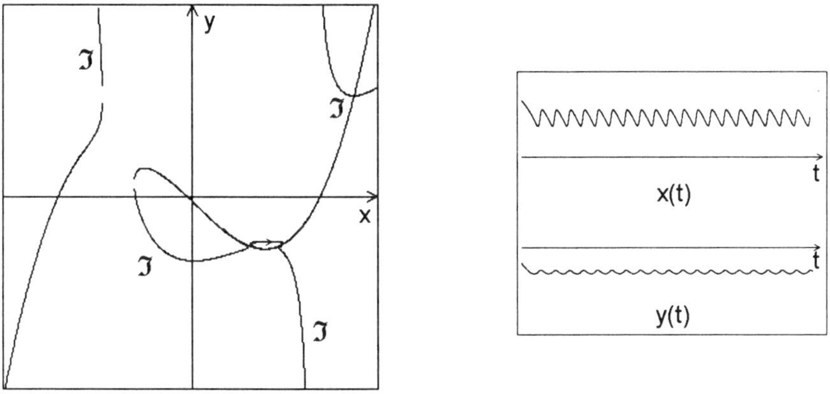

a) $a = 1.406$

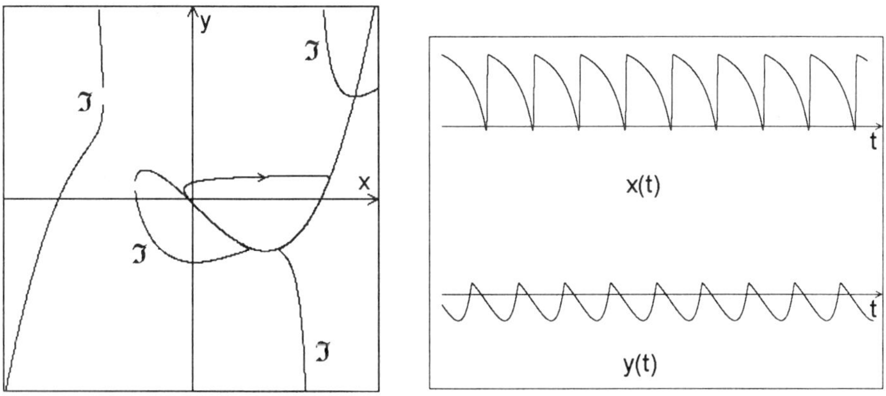

b) $a = 1.4008328398 \approx a_t$

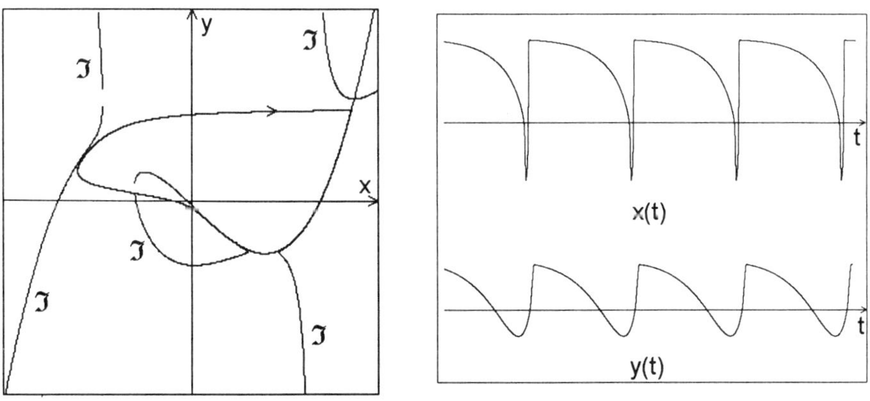

c) $a = 1.400832835$

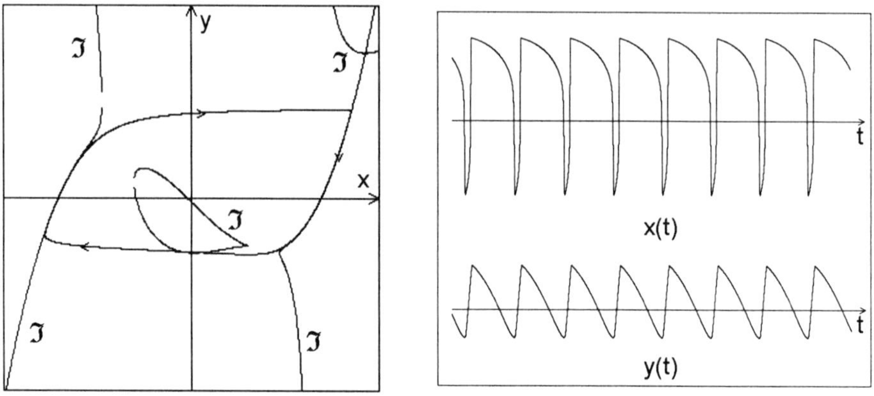

d) $a = 1.2 \approx a_e$

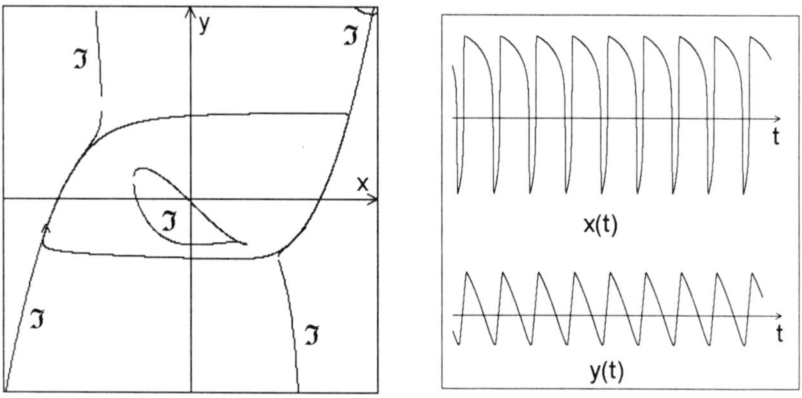

e) $a = 1$

Fig. 5.4.2. Limit cycles and inflection points for $b = -0.6$, $c = 5$, $x, y \in (-2.5, 2.5)$ and the corresponding oscillations.

The large variety of the intermediate forms will correspond to a large variety of the associated oscillations (Figure 5.4.2) and heart beats.

All these are sufficient arguments to notice the fundamental role of the presence of the curve of inflection points, in the neighborhood of the limit cycle to determine significant changes in the form of the limit cycle leading to the duck-like shape.

Regions of the parameter space corresponding to concave limit cycles. Numerical experiments showed that there exists a region near the curve H_1 of Hopf bifurcation values to which concave and attractive limit cycles correspond. However, since the limit cycles can be destroyed by a homoclinic bifurcation or by the formation of non-hyperbolic limit cycles, in these cases it is possible that the limit cycles disappear before the reaching of the value a_t, hence of the first tangency with $\mathfrak{I}$. In this case the limit cycles have only a small 'body'.

The range of the parameter plane to which concave attractive limit cycles correspond was drawn for $c = 5$ and $a \geq 0$ in Figure 5.4.3. It is bounded by segments of the curves BTn_1 and Ba_1, the segment of the Ob-axis with $b \in (1, b_T)$ and two curves A_e, A_t obtained numerically, corresponding to points (b, a_e) and (b, a_t). The representation is qualitative, owing to the small distance between H_1 and Ba_1 or between H_1 and A_t. Notice that for (b, a) situated in the narrow domain bounded by H_1, Ba_1, BTn_1 and A_t, convex small limit cycles and two steps oscillations correspond, whilst for (b, a) in the region delimited by BTn_1, A_e and the segment of Ob-axis with $b \in (b_{Q5}, 1)$, convex large limit cycles and four steps oscillations correspond.

The duration of existence of the concave limit cycles, i. e. $a_t - a_e$, strongly depends on the other two parameters.

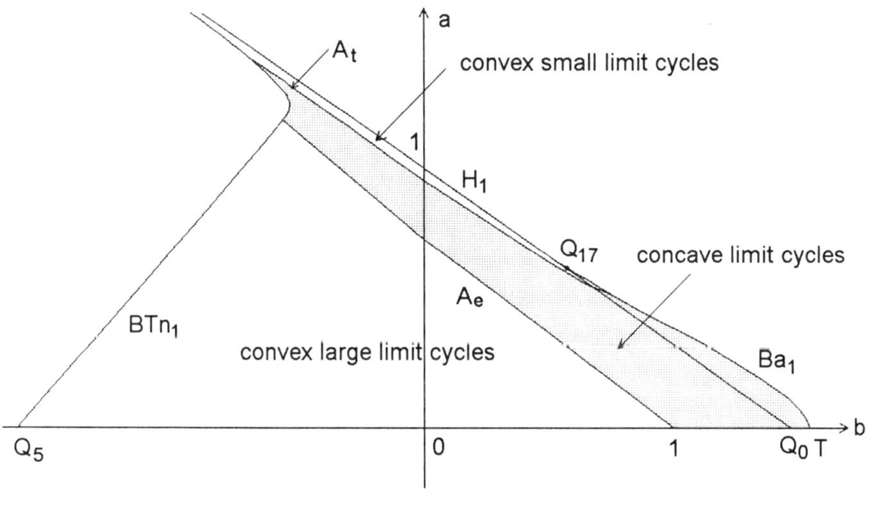

Fig. 5.4.3. Range corresponding to concave attractive limit cycles (hatched).

Numerical experiments have also shown the interesting point that the formation of the 'head' takes place in an extremely narrow range for a (for instance of order 10^{-14}), whilst the formation of the 'body' was quite slow, within a range of a of order 10^{-2}. This may be a source of errors in numerical solution for unwary persons. The fast formation of the 'head' accompanied by signifiant change in the oscillations is of great importance not only for the physiology, but also for enzymology and coding theory. It models *phenomena sensitive to a small change in parameters.*

Similarly, regions of the parameter plane where the repulsive limit cycles are concave can be found.

For $b > b_{17}$, where $Q_{17}(b_{17}, a_{17})$ is the Bautin bifurcation value, the duck-cycles were of supercritical type in the Eckhaus terminology (Section 4.5). In this case we distinguish a value a_e at which the outer limit cycle becomes concave and a value a_{nh} (corresponding to (b_{nh}, a_{nh}) on the curve Ba_1) at which the non-hyperbolic limit cycle obtained by the coalescence of the two duck-cycles disappears.

In [36] a_{nh} is denoted by $-\alpha_s$ and a_e by $-\alpha_c$. It is interesting to note that for $a > a_{nh}$, hence beyond the value of disappearance of the duck-cycles, there exist duck trajectories, but they are no longer cycles but unbounded.

The form of these trajectories, near the stable and unstable manifolds of R–H, have portions which look like the former non-hyperbolic limit cycle.

This phenomenon was noticed in [146] for a system with one parameter. Also, we encontered it near homoclinic bifurcations.

In [146] for the system (4.5.9), the Callot proof that the life of duck cycles is very short was reproduced. In other words the length of the interval of variation of the parameter a within which the duck-cycles exist is very small. In our case the range of (b, a)-plane where the canard phenomenon is present is formed by the region in Figure 5.4.3 corresponding to a convex small limit cycle and a narrow region situated along the curves A_t and Ba_1 and beneath them. To this narrow region correspond attractive concave limit cycles which are at the same time duck cycles.

5.5. THE INITIATION OF HEART BEATS

In order to explain the initiation of heart beats we shall use the form (1.1.11) of the F–N system. Here z stands for the injected current. The behaviour of the system (1.1.11) is given by its phase portrait. This can be obtained from the phase portrait of the system (1.1.17) written in the form (1.1.15) as follows.

The phase portrait for (1.1.11) and the values of the parameters c, z, b and a can be obtained from the phase portrait of (1.1.15) and the values of the parameters c, b, $\overline{a} = a + bz$ by the translation $y = \overline{y} - z$. In addition, if the trajectory through (x^0, y^0) is drawn for the system (1.1.11), then the trajectory through $(x^0, y^0 + z)$ must be drawn for the system (1.1.15).

As an example let us recover FitzHugh's results. If $c = 3$, $a = 0.7$ and $b = 0.8$ for a cathodic injected current $z = -0.128$, we must consider the system (1.1.15) for the following values of the parameters: $c = 3$, $b = 0.8$ and $\overline{a} = 0.5976$. Thus, this situation corresponds to a point from the domain 6 of the parametric portrait and no oscillatory regimes exist. The phase portraits and the nullclines of the systems (1.1.11) and (1.1.15) are given in Figures 1.1.6b and 5.5.1 and it is easy to see that Figure 1.1.6b is obtained from Figure 5.5.1 by the translation $y = \overline{y} + 0.128$.

If the injected current changes at $z = -0.4$, then the parameters for the system (1.1.15) are $c = 3$, $b = 0.8$, $\overline{a} = 0.38$ and they correspond to a point of the parametric portrait situated in the domain 5, where a stable limit cycle exists. The corresponding phase portraits and the nullclines of the systems (1.1.11) and (1.1.15) are given in Figures 1.1.6c and 5.5.2 and Figure 1.1.6c is obtained from Figure 5.5.2 by the translation $y = \overline{y} + 0.4$.

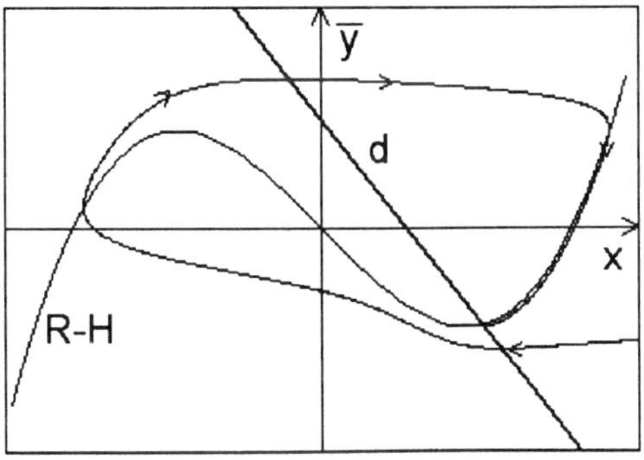

Fig. 5.5.1. Partial phase portrait of the system (1.1.15) for $c = 3$, $\overline{a} = 0.5976$, $b = 0.8$, $x \in (-2.2, 2.2)$ and $y \in (-1.5, 1.5)$.

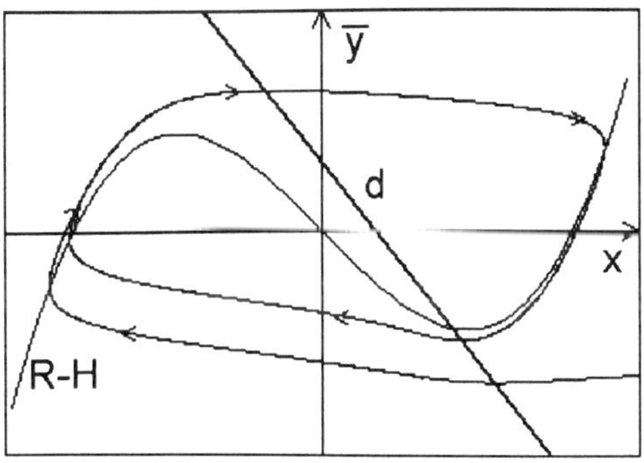

Fig. 5.5.2. Partial phase portrait of the system (1.1.15) for $c = 3$, $\overline{a} = 0.38$, $b = 0.8$ $x \in (-2.2, 2.2)$ and $y \in (-1.5, 1.5)$.

By this method any possible phase portrait of system (1.1.11) can be obtained from the corresponding phase portrait of system (1.1.15), and thus of system (1.1.17) dealt with in our work.

Consider the first quadrant of the parameter plane (Figure 5.1.1) which is the most important range of the parameters from physiological point of view. As z is varied, the transition from a regime without oscillations to an oscillatory regime for which an attractive limit cycle is present can proceed in two ways.

One is a result of the crossing, in the parameters plane, of the curve Ba_1 consisting of non-hyperbolic limit cycle bifurcation values. As a result two limit cycles (the inner one being repulsive and the outer one being attractive) emerge in the domains 7, 14, 17 or 13. The repulsive limit cycle disappears as H_1 or D_1 is crossed and the parameters are situated in the domains 5 or 8. This situation explains the well known biological law of *all or none responses* of the nodal system of the heart. This is associated with the sudden emergence of the attractive limit cycle at the points of Ba_1.

The second way of transition from non-oscillatory to oscillatory regimes takes place by crossing the curve H_1 and passing from domain 6 to domain 5 of the parametric portrait. Then a smooth Hopf bifurcation takes place and the canard phenomenon is present. The big attractive limit cycle develops after a fast (with respect to parameters) succession of transient limit cycles. That is, the heart beats do not appear at their maximum intensity as z reaches the threshold value, but a lot of transient beats appear first. This phenomenon, owing to its rapid development, is rarely detected in experiments.

5.6. CONCLUDING REMARKS OF INTEREST TO PHYSIOLOGISTS

Under the influence of Sparrow's famous monograph [132] and taking into account the recent advances in dynamical system theory, we have intended to provide the physiologists with the detailed and comprehensive study of the F–N model (1.1.17).

In fact, in applications it is highly desirable to possess a well defined scheme of all components of such a study and to have a more ot less algorithmic approaches to find the response of the model for all possible data (i.e., initial conditions and parameters).

Then, according to the qualitative response of some model, physiologists will decide if it suits their expectation and which are the data corresponding to realistic responses. In this way the model will be considered as valid only for those data. In mathematical terms, these data belong to the domain of the physiological validity of the model.

On the other hand, in life sciences the thermodynamic approach is still in its infancy, such that a lot of models must be investigated leading to the discovery of the main models. Thus the evolution of a model must proceed quickly, during a period of time reasonable for applications.

Keeping in mind all these ideas, in the present book we primarly give mathematical results revealed by the F–N model, and secondly accompany them by numerous figures, tables, comments, and remarks.

Our attention focusses on two questions of interest in applications: oscillations and bifurcations. We have found many types of oscillations, corresponding to very different heart functionings.

We also show numerous types of static and dynamic bifurcations. The related question of structurally stable dynamical systems is also treated.

In the parameter space the regions of structural stability or of bifurcations of codimension-one and -two are made evident in the form of the parametric portrait. This is supplemented by the phase portrait for each stratum of the parametric portrait. In this way we obtain an overall view — the global bifurcation diagram — of all phenomena possible for all real parameters a and b and for the third parameter $c > 1 + \sqrt{3}$.

It is just this bifurcation diagram which gives the synthetic view of the F–N model. In order to understand its message physiologists, biologists, and other specialists interested in dynamics are kindly invited to read the first chapter. Therein they can find the keys for decoding the information included in the bifurcation diagram.

For each triple of the parameters there exists a response of the F–N model, i.e., the corresponding heart beats for all initial data. All these beats are described by the phase trajectories. The set of all phase trajectories form the phase portrait. Hence the phase portrait gives the concise information of all types of beats possible at the concrete physiological situation characterized by the chosen values of the three parameters.

As the parameters are changed, the phase portraits change too. However, certain changes are not essential. In this case we say that the phase portraits keep being topologically equivalent. Therefore the topological properties of the phase trajectories remain unchanged. For instance, if the trajectory through some points is unbounded, this mean that the characteristics of the dynamics (i.e., the static functions) go to infinity as time tends to infinity. This situation is preserved by topological equivalence. Also, if a phase trajectory reduces to a point, i.e., it corresponds to a stationary functioning of heart, this trajectory will remain a point if the phase portraits are topologically equivalent.

The trajectories consisting of a single point are called the equilibrium points.

The most important heart action is periodic. It corresponds to closed phase trajectories, e.g. limit cycles. The topological equivalence preserves the property of these trajectories to remain closed. The corresponding heart beats will be periodic. However, an important remark for applications is that two phase portraits may be topologically equivalent but some of their limit cycles may correspond to different numbers of heart beats (in two 'steps', three 'steps' or four 'steps'). Hence some features of interest for physiologists are not distinguishable by the topological equivalence. For them refined types of mathematical equivalences are needed.

The non–periodic oscillatory beats correspond to phase trajectories which come closer and closer to some limit cycle as the time tends to $+\infty$ or $-\infty$. This behaviour is also preserved by topological equivalence. The trajectories corresponding to a non–periodic oscillatory heart action are referred to as transient phase trajectories.

Another type of transient trajectories are those ending at equilibria. They correspond to heart beats which stabilize to stationary functioning.

In the two-dimensional phase space (which is the case of the F–N model) the most important trajectories are equilibria and limit cycles. This is because they are the single trajectories with attractive properties. In other words, according to the F–N model the heart beats can stabilize themselves either to stationary action or to periodic oscillations. Whence the characterization of the phase portraits by equilibria and limit cycles. If two phase portraits are topologically equivalent, they will contain the same number of equilibria and limit cycles with the same attractivity properties. This is why, mathematically, they will be considered as being not distinguishable. Nevertheless, physiologically the value of the state functions at the same time will differ. It follows that the information yielded by us is qualitative, but it concerns fundamental features of the heart beats.

Of course, we have also given numerical results, i.e., quantitative features. However, they were only the representatives for a whole class of qualitatively similar results.

As far as the topologic feature of the phase portraits remain unchanged (i.e., the phase portraits are topologically equivalent) at the variation of parameters in an entire neighborhood of some triple $\left(\tilde{a}, \tilde{b}, \tilde{c}\right)$, we say that the F–N model is robust or structurally stable. If in every neighborhood of $\left(\tilde{a}, \tilde{b}, \tilde{c}\right)$ there exist points $\left(\tilde{\tilde{a}}, \tilde{\tilde{b}}, \tilde{\tilde{c}}\right)$ corresponding to phase portraits non-topologically equivalent to the phase portrait for $\left(\tilde{a}, \tilde{b}, \tilde{c}\right)$, we say that the dynamical system corresponding to $\left(\tilde{a}, \tilde{b}, \tilde{c}\right)$ is a bifurcation. This means that the heart beats can undergo dramatic changes at a certain small perturbation of the parameters. Hence this is a dangerous case. The danger is higher the higher is the codimension of the bifurcation. *Grosso modo*, the codimension is equal to the number of behaviours topologically non-equivalent to the beats corresponding to $(\tilde{a}, \tilde{b}, \tilde{c})$. This is why we investigated all points of codimension-one and -two, giving the local bifurcation diagrams around them.

A codimension-three bifurcation corresponds to a point of the parameter space situated at the intersection of some boundaries of the strata. In the (b, a) parameter plane these boundaries are points and curves. As c varies, in the (b, a, c) parameter space these boundaries become curves and surfaces respectively. Their intersections consist of codimension-three bifurcation values. Since some bifurcation curves were deduced only numerically, theoretically it is dificult to derive the codimension-three bifurcation values. The detection of codimension-three bifurcations is a very different problem and it will be treated elsewhere.

The variation of parameters revealed certain changes of the limit cycles corresponding to notable changes in the type of the corresponding periodic oscillations. Mostly the form of these limit cycles resembles a duck (canard in French), but their main characteristic is that they are concave. In addition, the formation of the concavities was very sudden, i.e., they were formed during an infinitesimal variation of the parameters. Private discussions with several biologists and physiologists have led us to the conclusion that the animal (or human) organism 'feels' these changes and that they can be very important in determined circumstances. This is why we have given numerical results for their formation. To our great surprise there exists a large band of variation of a and b corresponding to the presence of concave limit cycles. At the same time the extremely thin region of steep variation (with respect to parameters) of the concave limit cycles closely follows the Hopf bifurcation curve. Hence this steep variation takes place very close to the values of the parameters at which the stabilization of the heart beats turns from a stationary to a periodic regime.

Another important large time behaviour of the heart beats is related to the relaxation oscillations. They occur at large values of the parameter c and, in the phase space, are described by limit cycles along which the dynamics is fast on some parts and very slow of the remaining parts. The fast dynamics correspond to a steep variation (jump) between the relaxation regions. In our book the relaxation oscillations are studied by asymptotic methods.

Apart from these general results, our book contains many other more particular descriptions of dynamics, especially at codimension-one and -two bifurcation points and around them.

The rich dynamics generated by the F–N model, presented in a very compact form in the global bifurcation diagram, can be further revealed by the analysis of the large number of figures in this diagram and throughout the paper. We are sure that our results can be processed in some other ways, too, to provide information of interest for application. A lot of such information may be trivial from the mathematical viewpoint, but it can be proved to be critical for physiology or biology. This is why the authors acknowledge the suggestion of Prof. Jean-François Hervagault of drawing the graphs of y versus t, and we shall be grateful for any other suggestion which may improve the understanding of the response of the F–N model.

5.7. OPEN MATHEMATICAL PROBLEMS

This book is an application-oriented paper. Correspondingly, it contains all rigorous theoretical results which can be obtained by taking into account the current level of development in dynamical systems and bifurcation theory.

The use of the curves $\mathfrak{I}$, the loci of points of inflection of the phase trajectories, has proved to be crucial in studying the formation and destruction of concave limit cycles. In fact, we have recognized a direction successful in the early stages of development of dynamical systems theory and very rarely used nowadays. The results on concavity of limit cycles are seldom met in literature.

Our theoretical results on the concavity of limit cycles are supplemented with numerical investigation owing to the lack of an exact expression for the functions defining the limit cycles. One can try to perform an approximate study by using asymptotic expressions for the functions.

In addition, graphs similar to that in Figure 5.4.3, for the region in the parameter (b, a)-plane corresponding to concave limit cycles, must be derived for some other values of c, too.

Apart from topics of mathematical interest, we have added tables and graphs which are very suggestive for those interested in applications. For instance, we have shown the notable variation of the oscillations corresponding to different concave limit cycles. These oscillations determine various regimes in the heart beats. We have investigated the oscillations corresponding to the concave case much more than those corresponding to the convex case.

We have been forced to restrict our investigations to the situation $c \geq 1 + \sqrt{3}$. However, for $0 \leq c < 1 + \sqrt{3}$ we do not expect new phenomena to occur except for codimention-three bifurcations, which for the time being cannot be foreseen theoretically.

Numerically and partially theoretically we have found the existence of at most three coexisting limit cycles. At first glance all criteria existing in the papers available to us did not apply to our case. Hence the sixteenth Hilbert problem for the F–N system is one of the most important open problems of our paper.

There are studies concerning the size of the domain enclosed by the limit cycles, and, therefore, the amplitude of the corresponding oscillations. We shall apply them to the F–N model elsewhere.

We have in view a future detailed investigation of the relaxation oscillations by the method of boundary layer functions.

A further investigation of the connection with the assumptions involved into the method of curves of points of inflection of the phase trajectories (Section 5.4.2) and the constraints imposed by a thorough standard or nonstandard asymptotic study of the F–N model is still awaited.

Because of their complexity, the space graphs associated with the simultaneous presence of three limit cycles and the Bautin bifurcation phenomenon were not included. They will be drawn when we find a suggestive graphical three-dimensional representation which takes into account changes at a very small and a very large scale at the same time.

All the quoted open problems imply hard work which will last for years. The presence of (even) three parameters makes the question of the 'absolute' completeness of the study unreasonable. Indeeed, a lot of the main manifolds in the parametric portrait can be described only numerically.

Anyhow, we hope that the framework of a study of dynamics and bifurcation established by us in this book will inspire many specialists.

APPENDIX A

LIAPUNOV COEFFICIENTS

Rec1 = (-36* g1* g2* ImL^3 + 27* g1* ImL^4* r + 92* g1^2* ImL^2* ReL - 16* g2^2* ImL^2* ReL - 4* g1* g2* ImL* ReL^2 + 30 * g1* ImL^2* r* ReL^2 + 12* g1^2* ReL^3+ 3* g1* r* ReL^4)/((ImL^2 + ReL^2)* (9* ImL^2 + ReL^2))

Imc1 = (-24* g1^2* ImL^3 - 60* g2^2* ImL^3 + 27* g2* ImL^4* r + 108* g1* g2* ImL^2 * ReL - 8* g1^2* ImL* ReL^2 - 12* g2^2 * ImL* ReL^2 + 30* g2* ImL^2* r* ReL^2 + 12* g1* g2* ReL^3 + 3* g2* r* ReL^4)/ ((ImL^2 + ReL^2)* (9* ImL^2 + ReL^2))

Rec2 = (-69120* g1^3* g2* ImL^11 - 69120* g1* g2^3* ImL^11 + 51840* g1^3* ImL^12 * r + 119232* g1* g2^2* ImL^12* r - 11664* g1* g2* ImL^13* r^2 + 223296* g1^4* ImL^10* ReL + 441552* g1^2* g2^2* ImL^10* ReL - 175728* g2^4* ImL^10* ReL - 708264* g1^2* g2* ImL^11* r* ReL + 190296* g2^3* ImL^11* r* ReL + 29889* g1^2* ImL^12* r^2* ReL + 6561* g2^2* ImL^12* r^2* ReL - 707232* g1^3* g2* ImL^9* ReL^2 + 440160* g1* g2^3* ImL^9* ReL^2 + 1082952* g1^3* ImL^10* r* ReL^2 + 353880* g1* g2^2* ImL^10* r* ReL^2 - 133812* g1* g2* ImL^11* r^2* ReL^2 + 263824* g1^4* ImL^8* ReL^3 + 999984* g1^2* g2^2* ImL^8* ReL^3 - 21280* g2^4* ImL^8* ReL^3 - 3951288* g1^2* g2* ImL^9* r* ReL^3 + 382152* g2^3 * ImL^9* r* ReL^3 + 343440* g1^2* ImL^10* r^2* ReL^3 + 75816* g2^2* ImL^10* r^2* ReL^3 - 2478784* g1^3* g2* ImL^7* ReL^4 - 543808* g1* g2^3* ImL^7* ReL^4 + 5191656* g1^3* ImL^8* r* ReL^4 + 418296* g1* g2^2* ImL^8* r* ReL^4 - 282996* g1* g2* ImL^9* r^2* ReL^4 + 1352480* g1^4* ImL^6* ReL^5 + 1772256* g1^2* g2^2* ImL^6* ReL^5 + 27712* g2^4* ImL^6* ReL^5 - 4934608* g1^2* g2* ImL^7* r* ReL^5 + 201392* g2^3* ImL^7* r* ReL^5 + 731313* g1^2* ImL^8* r^2* ReL^5 + 165321* g2^2* ImL^8* r^2* ReL^5 - 1426560* g1^3* g2* ImL^5 * ReL^6 - 472320* g1* g2^3* ImL^5* ReL^6 + 6104528* g1^3* ImL^6* r* ReL^6 + 303344* g1* g2^2 * ImL^6* r* ReL^6 - 218376* g1* g2* ImL^7* r^2* ReL^6 + 680320* g1^4* ImL^4* ReL^7 + 717216* g1^2 * g2^2* ImL^4 * ReL^7 + 24608* g2^4* ImL^4* ReL^7 - 1996208* g1^2* g2* ImL^5* r* ReL^7 - 9008* g2^3* ImL^5* r* ReL^7 + 571320 * g1^2* ImL^6* r^2* ReL^7 + 134568* g2^2* ImL^6* r^2* ReL^7 - 292160* g1^3* g2* ImL^3* ReL^8 -115136* g1* g2^3 * ImL^3* ReL^8 + 2245456* g1^3* ImL^4* r* ReL^8 + 148528* g1* g2^2* ImL^4* r* ReL^8 - 65016* g1* g2* ImL^5 * r^2 * ReL^8 + 111264* g1^4* ImL^2* ReL^9 + 97104* g1^2* g2^2* ImL^2* ReL^9 + 3888* g2^4* ImL^2* ReL^9 - 321928 * g1^2* g2* ImL^3* r* ReL^9 - 20488* g2^3* ImL^3* r* ReL^9 + 173907* g1^2* ImL^4* r^2* ReL^9 + 43875* g2^2* ImL^4 * r^2* ReL^9 - 18144* g1^3* g2* ImL* ReL^10 - 7776* g1* g2^3* ImL* ReL^10 + 316520* g1^3*

ImL^2* r* ReL^10 + 30776 * g1* g2^2* ImL^2* r* ReL^10 - 7812* g1* g2* ImL^3* r^2* ReL^10 + 5616* g1^4* ReL^11 + 3888* g1^2* g2^2* ReL^11 -17304 * g1^2* g2* ImL* r* ReL^11 - 1944* g2^3* ImL* r* ReL^11 + 21240* g1^2* ImL^2* r^2* ReL^11 + 5616* g2^2 * ImL^2 * r^2 * ReL^11 + 15048* g1^3* r* ReL^12 + 1944* g1* g2^2* r* ReL^12 - 324* g1* g2* ImL* r^2* ReL^12 + 891* g1^2* r^2* ReL^13 + 243* g2^2* r^2* ReL^13)/ (6* (ImL^2 + ReL^2)^3* (4* ImL^2 + ReL^2)* (9* ImL^2 + ReL^2)^2* (ImL^2 + 9* ReL^2))

Imc2 = (4032* g1^4* ImL^11 - 109008* g1^2* g2^2* ImL^11 - 113040* g2^4* ImL^11 + 117288* g1^2* g2* ImL^12* r + 150984 * g2^3* ImL^12* r -6561* g1^2* ImL^13* r^2 - 12393* g2^2* ImL^13* r^2 + 355104* g1^3* g2* ImL^10* ReL + 552096 * g1* g2^3* ImL^10* ReL - 163944* g1^3* ImL^11* r* ReL - 613224* g1* g2^2* ImL^11* r* ReL + 11664 * g1 * g2 * ImL^12 * r^2* ReL - 91024* g1^4* ImL^9* ReL^2 - 1216880* g1^2* g2^2* ImL^9* ReL^2 -552160* g2^4* ImL^9* ReL^2 + 1432440 * g1^2* g2* ImL^10* r* ReL^2 + 1067904* g2^3* ImL^10* r* ReL^2 - 75816* g1^2* ImL^11* r^2* ReL^2 -142722 * g2^2 * ImL^11* r^2* ReL^2 + 1601856* g1^3* g2* ImL^8* ReL^3 + 1980576 * g1* g2^3* ImL^8* ReL^3 - 1176984 * g1^3 * ImL^9 * r* ReL^3 - 3343704* g1* g2^2* ImL^9* r* ReL^3 + 133812* g1* g2* ImL^10* r^2* ReL^3 - 840928* g1^4 * ImL^7 * ReL^4 - 2310624* g1^2* g2^2* ImL^7* ReL^4 -502208* g2^4* ImL^7* ReL^4 + 3907400* g1^2* g2* ImL^8* r* ReL^4 + 1520720 * g2^3* ImL^8* r* ReL^4 - 165321* g1^2* ImL^9* r^2* ReL^4 -306819* g2^2* ImL^9* r^2* ReL^4 + 1192544* g1^3 * g2* ImL^6* ReL^5 + 1388576* g1* g2^3* ImL^6* ReL^5 - 1588336* g1^3* ImL^7* r* ReL^5 -4156336 * g1 * g2^2 * ImL^7 * r* ReL^5 + 282996* g1* g2* ImL^8* r^2* ReL^5 - 450368* g1^4* ImL^5* ReL^6 - 1077792 * g1^2 * g2^2 * ImL^5* ReL^6 - 150304* g2^4* ImL^5* ReL^6 + 3639856* g1^2* g2* ImL^6* r* ReL^6 + 739264* g2^3* ImL^6* r* ReL^6 - 134568 * g1^2 * ImL^7* r^2* ReL^6 -243756* g2^2* ImL^7* r^2* ReL^6 + 319712* g1^3* g2* ImL^4* ReL^7 + 325856 * g1 * g2^3 * ImL^4 * ReL^7 - 682448* g1^3* ImL^5* r* ReL^7 -1676048* g1* g2^2* ImL^5* r* ReL^7 + 218376 * g1* g2* ImL^6 * r^2 * ReL^7 - 77024* g1^4* ImL^3* ReL^8 - 179024* g1^2* g2^2* ImL^3* ReL^8 - 13488* g2^4* ImL^3* ReL^8 + 1192760 * g1^2 * g2* ImL^4* r* ReL^8 + 144296* g2^3* ImL^4* r* ReL^8 - 43875* g1^2* ImL^5* r^2* ReL^8 -76383* g2^2 * ImL^5 * r^2 * ReL^8 + 30720* g1^3* g2* ImL^2* ReL^9 + 21696* g1* g2^3* ImL^2* ReL^9 - 113320* g1^3* ImL^3* r* ReL^9 - 264040 * g1* g2^2* ImL^3* r* ReL^9 + 65016* g1* g2* ImL^4* r^2* ReL^9 - 3888* g1^4* ImL* ReL^10 - 9072* g1^2* g2^2 * ImL * ReL^10 + 151704* g1^2* g2* ImL^2* r* ReL^10 + 8832* g2^3* ImL^2* r* ReL^10 - 5616* g1^2* ImL^3* r^2* ReL^10 - 9522* g2^2* ImL^3* r^2* ReL^10 + 864* g1^3* g2* ReL^11-6168* g1^3* ImL* r* ReL^11-13848* g1* g2^2* ImL* r* Re L^11 + 7812* g1* g2* ImL^2* r^2* ReL^11 + 6552* g1^2* g2* r* ReL^12 -243* g1^2* ImL* r^2* ReL^12 -405* g2^2 * ImL * r^2* ReL^12 + 324* g1* g2* r^2* ReL^13)/ (3* (ImL^2 + ReL^2)^3* (4* ImL^2 + ReL^2)* (9* ImL^2 + ReL^2)^2* (ImL^2 + 9* ReL^2))

BRIEF DESCRIPTION OF THE SOFT DIECBI

DIECBI is a package of programs devoted to the study of continuous two-dimensional dynamical systems. It can be used for systems with at most 12 parameters.

Among the capabilities of DIECBI we quote the following:

— solves ODE's, using an improved Runge–Kutta method, with fixed and variable steps;

— determines the coordinates of the points of the phase trajectories with 18 digits;

— plots the phase trajectories. The initial conditions may be chosen by the user or are automatically established;

— plots the oscillations as functions of time;

— determines the equilibria by using the Newton–Raphson method;

— determines the eigenvalues and the eigenvectors of the system linearized around an equilibrium point;

— plots the variation of the real or imaginary part of an eigenvalue with respect with one parameter;

— plots the stable and unstable manifolds of saddles;

— plots the basins of attraction;

— plots various functions such as the nullclines or the inflection points of the phase trajectories;

— the user can change parameters, initial condition, the integration errors, the number of integration steps and the scale of the screen.

The phase portrait is determined for a domain of the phase plane which is chosen by the user (domain of analysis). However the computation of the coordinates of the points of every trajectory may continue in a bigger domain (domain of computation) when the trajectory get out of the domain of analysis. Of course, the possibility that the trajectory can return in the domain of analysis is taken into account. The computation continues until either the domain of existence is crossed or the number of integration steps is reached.

The most important numerical task in simulating continuous dynamical systems is the computation of trajectories. In literature there are many integration algorithms that solve the Cauchy problem for a system of ODEs.

Here we present only the improvements that we made for an integration algorithm, namely Runge–Kutta of four order.

All integration algorithms attempt to approximate the solution of the continuous-time system

$$\begin{aligned} \dot{\mathbf{x}} &= \mathbf{f}(\mathbf{x}), \\ \mathbf{x}(t_0) &= \mathbf{x}^{(0)}, \end{aligned} \tag{B0.1}$$

by generating a sequence of points $\mathbf{x}^{(0)}, \mathbf{x}^{(1)}, \ldots$ at times $t_0, t_1, \ldots$ with $\left\|\mathbf{x}^{(k)} - \mathbf{x}(t_k)\right\| < \varepsilon$. Recall that a differentiable function $g : \mathbf{R} \to \mathbf{R}$ is increasing on the interval (t_1, t_2) iff $g'(t) > 0$ on this interval and is decreasing iff $g'(t) < 0$.

Let $\mathbf{x}^{(k)}$ and $\mathbf{x}^{(*)}$ be two successive approximations of the solution $\mathbf{x}$ at two times t_k and t_* and let $Tol1/2$ be the admitted error, i.e.,

$$d_k = \left\|\mathbf{x}^{(k)} - \mathbf{x}^{(*)}\right\| \leq \frac{Tol1}{2}. \tag{B0.2}$$

If $d_k > Tol1/2$ then the integration step is halphen by the computer and the computation goes on. This proccess is repeated until we obtain a point $\mathbf{x}^{(*)}$ such that $d_k \leq Tol1/2$.

At the point $\mathbf{x}^{(k)}$ we can have the following situations:

1. $\dot{x}_i = f_i\left(\mathbf{x}^{(k)}\right) \equiv \widetilde{x}_i(t_k) \geq 0$, for every $i \in \mathcal{I}$, where $\mathcal{I} = \{1, 2, ..., n\}$;
2. $\dot{x}_i = f_i\left(\mathbf{x}^{(k)}\right) \equiv \widetilde{x}_i(t_k) \leq 0$, for every $i \in \mathcal{I}$;
3. There exist I and J with $I, J \subset \mathcal{I}$, $I \cup J = \mathcal{I}$ such that $\widetilde{x}_i(t_k) \geq 0$, for every $i \in I$ and $\widetilde{x}_j(t_k) < 0$, for every $j \in J$.

After the computation of the point $\mathbf{x}^{(*)}$ we can have

1.1. $\widetilde{x}_i(t_*) \geq 0$, for every $i \in \mathcal{I}$;

1.2. There exists $I_1 \subset \mathcal{I}$ such that $\widetilde{x}_i(t_*) < 0$, for every $i \in I_1$ and $\widetilde{x}_i(t_*) \geq 0$, for every $i \in \mathcal{I} - I_1$.

2.1. $\widetilde{x}_i(t_*) < 0$, for every $i \in \mathcal{I}$;

2.2. There exists $J_1 \subset \mathcal{I}$ such that $\widetilde{x}_j(t_*) \leq 0$, for every $j \in J_1$and $\widetilde{x}_i(t_*) < 0$, for every $i \in \mathcal{I} - J_1$.

3.1. $\widetilde{x}_i(t_*) \geq 0$, for every $i \in I$ and $\widetilde{x}_j(t_*) < 0$, for every $j \in J$,

3.2. There exist I_1 and J_1 $(I_1 \subset I, J_1 \subset J)$ such that $\widetilde{x}_i(t_*) < 0$, for every $i \in I_1 \cup (J - J_1)$ and $\widetilde{x}_j(t_*) \geq 0$, for every $j \in J_1 \cup (I - I_1)$.

The point $\mathbf{x}^{(*)}$ for which the cases 1.1, 2.1 or 3.1 hold is referred to as a **consistency point.** In this case all the state variables are assumed to be monotonous functions of time on the interval (t_k, t_*). If one of the cases 1.2, 2.2 or 3.2 hold, the point $\mathbf{x}^{(*)}$ is called an **inconsistency point**. In this situation at least one of the state variables is not a monotonous function of time on the interval (t_k, t_*).

If by applying a numerical method of integration an inconsistency point $\mathbf{x}^{(*)}$ is obtained, then the integration step is halphen and a new $\mathbf{x}^{(*)}$ is computed. The computation proceeds until a consistency point is obtained or until the obtained inconsistency point $\mathbf{x}^{(*)}$ satisfies the inequality

$$\left\| \mathbf{x}^{(k)} - \mathbf{x}^{(*)} \right\| \leq \frac{Tol2}{2},$$

where $Tol2$ is an error established by the user. Of course we have $Tol2 < Tol1$.

Then $x^{(k+1)} = x^{(*)}$ i.e., the point $\mathbf{x}^{(*)}$ is accepted in the sequence for approximation the $\mathbf{x}$ solution. The effect of imposing the error $Tol2$ is to improve the resolution during the computation of the phase trajectory.

References

1. Agladze, K., Keener, J., Müller, S.C., Panfilov, A.V. (1994) Rotating spiral waves created by geometry, *Science*, **264**, pp. 1746–1748.
2. Aliev, R., Panfilov, A. (1996) A Simple Two - Variable model of cardiac excitation, *Chaos, Solitons & Fractals*, **7** (3), pp. 293–301.
3. Andronov, A. A., Leontovich, E. A., Gordon, I., Maier, A. G. (1967) *Theory of bifurcations of dynamical systems on a plane*, Halsted, New York.
4. Andronov, A., Vitt, A., Chaikin, C. E. (1966) *Theory of oscillations*, Pergamon, New York.
5. Arnold, V. I. (1983) *Geometrical methods in the theory of ordinary differential equations*, Springer, New York.
6. Arnold, V. I. (1974) *Equations diferentielles ordinaires*, Mir, Moscow.
7. Arrowsmith, D. K., Cartwright, J., Lansbury, A., Place, C. M. (1993) The Bogdanov map: bifurcations, mode locking and chaos in a dissipative system, *Int. J. Bif. Chaos*, **3**, pp. 803–842.
8. Arrowsmith, D. K., Place, C. M. (1990) *An introduction to dynamical systems*, Cambrige University Press, Cambrige.
9. Arrowsmith, D. K., Place, C. M. (1982) *Ordinary differential equations. A qualitative approach with applications*, Chapman and Hall, London.
10. Aulbach, B. (1984) Continuous and discrete dynamical systems near manifolds of equilibria, *LNM*, **1094**, Springer, Berlin.
11. Bautin, N., Leontovich, E. (1976) *Methods and tricks for qualitative study of dynamical systems on the plane*, Nauka, Moscow (Russian).
12. Bautin, N. (1949) *Behavior of dynamical systems near the boundaries of stability regions*, Ogiz Gostexizdat, Leningrad-Moscow (Russian).
13. Bazykin, A., Kuznetsov, Y., Khibnik, A. (1989) *Portraits of bifurcations: bifurcation diagrams of planar dynamical systems*, Znanie, Moscow (Russian).
14. Benoit, E. (1990) Canards et enlacements, *Publ. Math. IHES*, **72**, pp. 63–91.
15. Benoit, E. (1983) Systemes lents–rapides dans $\mathbf{R}^3$ et leurs canards, *Asterisques*, pp. 109–110, pp. 159–191.
16. Benoit, E. (1979) *Equation de Van der Pol avec terme forçant*, Thése 3ème cycle, Université Paris VII.
17. Bogdanov, R.I. (1975) Versal deformations of a singular point on the plane in the case of zero eigenvalues, *Func.Annal. Appl.*, **9**, pp. 144–145.
18. Callot, J-L, Diener, F., Diener, M. (1978) Le probleme de la chasse au canard, *C. R. Acad. Sc. Paris*, **A, 286**, pp. 1059–1061.
19. Carrier, G. F., Lewis, J. A. (1950) The relaxation oscillations of the Van der Pol oscillator, *Adv. in Appl. Mech.*, **3**, pp. 12–16.
20. Cartwright, M.L. (1952) Van der Pol's equation for relaxation oscillation, in *Contrib. to the theory of nonlinear oscillations*, **II**, Lefschetz S. (ed.), *Annals of Math. Studies*, **29**, pp. 3–18.
21. Cartwright, M.L., Litllewood, J.E. (1945) On nonlinear differential equations of the second order I. The equation $\ddot{y} - k(1-y)\dot{y} + y = b\lambda k \cos(\lambda t + \alpha)$, k large, *J. London Math.Soc.*, **20**, pp. 180–189.
22. Chance, B., Pye, E.K., Ghosh, A.K., Hess, B. (eds.) (1978) *Biological and biochem-*

ical oscillators, Academic, New York.

23. Chauvet, G. (1986) *Traité de physiologie theorique*, Mason, Paris.
24. Chow, S. N., Li, C., Wang, D. (1994) *Normal forms and bifurcations of planar vector fields*, Cambridge University, Cambridge.
25. Chow, S. N., Hale, J. (1982) *Methods of bifurcation theory*, Springer, New York.
26. Chu, K. W. E., Govaerts, W., Spence, A. (1993) Matrices with rank deficiency two in eigenvalue problems and dynamical systems, *SIAM J. Numer. Annal.*, **30**, pp. 1–21.
27. Cronin, J. (1987) *Mathematics aspects of Hodgkin–Huxley neural theory*, Cambridge Studies in Mathematical Biology **7** , Cambridge University Press.
28. Cronin, J. (1977) Some mathematics of biological oscillations, *SIAM Reviews*, **19**, pp. 100–138.
29. Cross, M. C., Hohenberg, P. C. (1993) Pattern formation outside of equilibrium, *Reviews of Modern Physics*, **65, 31**, pp. 854–1086.
30. Davies, H.T., (1960) Introduction to nonlinear differential and integral equations, *U.S. Government Printing Office*, Washington.
31. Davis, M. (1997) *Applied nonstandard analysis*, Wiley, New York.
32. Diener, F. (1980) Les canards de l'équation $\ddot{y} + (y+a)^2 + y = 0$, *Report Ser. Math. Pures et Appl.*, I.R.M.A., Strasbourg.
33. Diener, F. (1979) *Quelques examples de bifurcations et leurs canards*, IRMA, 75/P37.
34. Dumortier, F., Roussarie, R., Sotomayor, J., Zoladek, H. (1991) *Bifurcations of planar vector fields, nilpotent singularities and abelian integrals*, Springer, Berlin.
35. Ebeling, W., Herzel, H., Richert, W., Schimansky–Geier, L. (1986) Influence of noise on Duffing-Van der Pol oscillators, *ZAMM*, **66**, pp. 141–146.
36. Eckhaus, W. (1983) Relaxation oscillations including a standard chase on French ducks, in *Asymptotic analysis II.,LNM*, **985**, F. Verhulst (ed), Springer, Berlin, pp. 449–497.
37. Eckhaus, W. (1979) *Asymptotic analysis of singular perturbations*, North–Holland, Amsterdam.
38. Favello, L., Ballossino, N., Reineri, M. T. (1988) Electrical excitation propagation front in the cardiac muscle, in *Biomathematics and Related Computational Problems, (ed. Ricciardi)*, Kluwer, Dordrecht, pp. 649–658.
39. FitzHugh, R. (1969) Mathematical models of excitation and propagation in nerve, in *Biological engineering*, H. P. Schwann (ed.), McGraw–Hill, New York, pp. 1–85.
40. FitzHugh, R. (1961) Impulses and physiological states in theoretical models of nerve membrane, *Biophysical J*, **1**, pp. 445–466.
41. FitzHugh, R. (1955) Mathematical models of threshold phenomena in the nerve membrane, *Bull. Math. Biophysics*, **17**, pp. 257–269.
42. Frank, L. S. (1990) Singular perturbations. Spaces and singular perturbations on manifolds without boundary, *Studies in Math and its Appl.*, **23**, Elsevier, Amsterdam.
43. Friedman, M., Goshen, S., Rabinovich, A., Thierberger, R., (1993) Phase locking of Bonhoeffer – Van der Pol Model, in *Future Direction of nonlinear dynamics in physical and biological systems* (ed. P. L. Christiensen et al.), Plenum, New York, 1993.
44. Georgescu, A., Rocşoreanu, C., Giurgiţeanu, N. (1999) *Regimes with two or three limit cycles in the F–N system*, ZAMM, **79**, S293–S294.
45. Georgescu, A., Moroianu, M., Oprea, I. (1999) *Bifurcation theory. Principles and applications*, Univ. of Piteşti Press (Romanian).
46. Georgescu, A., Rocşoreanu, C., Giurgiţeanu, N. (1998) Equilibria and relaxation oscillations of the nodal system of the heart 2. Hopf bifurcation, *Rev. Roum. Sci. Tech. Mec. Appl.*, **43**,(3) (in press).
47. Georgescu, A., Rocşoreanu, C., Giurgiţeanu, N. (1997) Two types of dynamic bi-

furcation in a biodynamical system, *Proc. 4th Conf. on Appl. and Ind. Math., ROMAI*, pp. 53–64.
48. Georgescu, A., Giurgiţeanu, N. (1997) Study of the phase portrait. III. Influence of the linearization to the nonlinear dynamical system, *St.cerc.mec.apl.*, **56**,(3–4), pp. 141–153 (Romanian).
49. Georgescu, A., Giurgiţeanu, N. (1997) Study of the phase portrait. II. Inflection points of the phase trajectories of the Van der Pol dynamical system, *St.cerc.mec.apl.*, **56**,(1–2), pp. 15–31 (Romanian).
50. Georgescu, A., Rocşoreanu, C., Giurgiţeanu, N. (1997) Bifurcation and oscillatory regimes for the FitzHugh–Nagumo system, *Proc. 8th Int. Coll. on Diff. Eq., Plovdiv*, (Angel Dishliev, ed), **III**, pp. 33–41.
51. Georgescu, A. (1995) *Asymptotic treatment of differential equations*, Chapman and Hall, London.
52. Georgescu, A., Oprea, I. (1994) *Bifurcation theory from the applications point of view*, Univ. of Timişoara Press (Romanian).
53. Georgescu, A. (1992) *Synergetics. Solitons. Fractals, determinism, turboulence.* Univ. of Timişoara Press, (Romanian).
54. Giannakopoulos, F., Oster, O. (1997) Bifurcation properties of a planar system modelling neural activity, *Diff. Eqs. Dynamical Systems*, **5**, (3/4), pp. 229–242.
55. Giurgiţeanu, N., Georgescu, A., Rocşoreanu, C. (1997) Study of the phase portrait. IV. Absence of the canard bifurcation, *St.cerc.mec.apl.*, **56**,(5–6), pp. 297–305 (Romanian).
56. Giurgiţeanu, N. (1997) *Computational economical and biological dynamics – DIECBI*, Europa, Craiova, (Romanian).
57. Giurgiţeanu, N. (1997) *Contributions to the study of systems of differential equations by numerical methods. Applications to biology*, Romanian Academy, Bucharest, PhD Thesis.
58. Giurgiţeanu, N. (1996) Study of the phase portrait. I. Improvements of some numerical methods of integration, *St.cerc.mec.apl.*, **55**, pp. 1–2, (Romanian).
59. Glass, L. (1996) Dynamics of cardiac arrhythmias, *Physics Today*, 40–45.
60. Glass, L., Mackey, M.C. (1988) *From clocks to chaos: the rhythms of life*, Princeton University Press.
61. Grasman, J., (1987) *Asymptotic methods for relaxation oscillations and applications*, Springer, New York.
62. Gray, R., et al. (1995) Mechanisms of cardiac fibrillation, *Science*, **270**, pp. 1222–1225.
63. Griewank, A., Reddien, G. (1984) Characterization and computation of generalized turning points, *SIAM J. Num. Annal.*, **21**, pp. 176–185.
64. Guckenheimer, J. (1986) Multiple bifurcation problems for chemical reactors, *Physica*, **20D**, pp. 1–20.
65. Guckenheimer, J., Holmes, P. (1983) *Nonlinear oscillations, dynamical systems and bifurcations of vector fields*, Springer, New York.
66. Guckenheimer, J. (1981) On a codimension two bifurcation, *LMN*, **898**, Springer, Berlin, pp. 99–142.
67. Hale, J. K., Kocak, H. (1991) *Dynamics and bifucations*, Springer, New York.
68. Hale, J. K. (1984) Introduction to dynamic bifucation, *LNM*, **1057**, Springer, Berlin, pp. 106–151.
69. Hartzman, C.S., Naugler, D.R. (1980) Separatrices, non-isolated invariant sets and the Seifert conjecture, in *Global theory of dynamical systems*, (Z. Nitecki, C. Robinson, eds), *LMN*, **819**, Springer, Berlin, pp. 224–245.
70. Hastings, S. P. (1976) On the homoclinic and periodic orbits of the FitzHugh–Nagumo equations, *Quart. J. Math., Oxford Series*, **2** (27), pp. 123–134.
71. Hirsch, M., Smale, S. (1974) *Differential equations, dynamical systems and linear algebra*, Academic, New York.
72. Hodgkin, A. L., Huxley, A. F. (1952) A quantitative description of membrane

current and its application to conduction and excitation in nerve, *J. Physiol.*, **117**, pp. 500–505.

73. Holden, A. (1993) Continuous or discrete state dynamical systems as models for computation in neural systems, in *Future directions of nonlinear dynamics in physical and biogical systems*, (P. L. Cristiansen et al. eds), Plenum, New York, pp. 513–516.
74. Honerkamp, J. (1983) The heart as a system of coupled nonlinear oscillators, *J. Math. Biology*, **18**, pp. 69–88.
75. Hoppenstead, F. C., Peskin, C. S. (1992) *Mathematics in medicine and life sciences*, Springer, New York.
76. Hoppenstead, F. C. (ed) (1979) Nonlinear oscillations in biology, *Lectures in Applied Mathematics*, **17**, AMS, Providence.
77. Ideker, E., et al. (1991) Basic mechanisms of ventricular defibrillation, in *Theory of heart - biomechanics, biophysics and nonlinear dynamics of cardiac function*, (Glass, L., Hunter, P., McCulloch, A., eds), Springer , New York, pp. 533–561.
78. Jackson, A., Grossu, I. (1995) An open plus-closed-loop (OCLP) control of complex dynamical systems, *Physica*, **D**, (85), pp. 1–9.
79. Jalife, J. (ed.) (1990) *Mathematical approaches to cardiac arrhythmias*, Ann. NY Acad. Sci.
80. Jordan, D. W., Smith, P. (1989) *Nonlinear ordinary differential equations*, Clarendon, Oxford, 2nd ed. Oxford Univ. Press.
81. Joseph, D.D. (1981) Lectures on bifurcation from periodic orbits, *LNM*, **898**, Springer, Berlin, pp. 1–12.
82. Kakiuchi, N., Tchizawa, K. (1998) On another type of ducks in the FitzHugh–Nagumo equation, *Sci. Bull. Univ. Piteşti*, **2**, pp. 141–146.
83. Kakiuchi, N., Tchizawa, K. (1997) On an explicit duck solution and delay in the FitzHugh–Nagumo equation, *J. of Diff. Eq.*, **141**, (2), pp. 327–339.
84. Keener, J. P. (1991) Wave propagation in myocardium, in *Theory of heart - biomechanics, biophysics and nonlinear dynamics of cardiac function*, (Glass, L., Hunter, P., McCulloch, A., eds), Springer, New York, pp. 405–437.
85. Keener, J. P. (1987) Causes of propagation failure in excitable media, in *Temporal disorder in human oscillatory systems, Springer Series in Synergetics* **36**, Springer, Berlin, pp. 134–140.
86. Keener, J. P. (1986) Geometrical theory for spirals in excitable media, *SIAM J. Appl. Math.*, **46**, pp. 1039–1056.
87. Keener, J. P. (1980) Waves in excitable media, *SIAM J. Appl. Math.*, **39**, pp. 528–547.
88. Kootsey, M. (1991) Electrical propagation in distributed cardiac tissue, in *Theory of heart - biomechanics, biophysics and nonlinear dynamics of cardiac function*, (Glass, L., Hunter, P., McCulloch, A., eds), Springer, New York, pp. 391–405.
89. Kovacs, S. (1991) A clinical perspective on theory of heart, in *Theory of heart - biomechanics, biophysics and nonlinear dynamics of cardiac function*, (Glass, L., Hunter, P., McCulloch, A., eds), Springer, New York, pp. 601–611.
90. Krinsky, V.I. (ed.) (1984) *Autowaves: results, problems, and outlooks*, Springer, New York.
91. Krinsky, V. I. (1973) Excitation wave propagation during heart fibrillation, in *Biological and biochimical oscillators*, B. Chance, K. Pye (eds), Academic, New York.
92. Kuznetsov, Yu. (1995) *Elements of applied bifurcation theory*, Springer, New York.
93. LaSalle, J. (1949) Relaxation oscillations, *Quart. Appl. Math.*, **7**, pp. 1–19.
94. Levi, M. (1981) Qualitative analysis of the periodically forced relaxation oscillations, Memoirs of the AMS, **32** (244) , first of 3 numbers, pp. 1–147.
95. Levi, M. (1980) Periodically forced relaxation oscillations, *LNM*, **819**, Springer, Berlin, pp. 300–317.
96. Liénard, A. (1928) Etude des oscillations entretenues, *Revue Generale de*

l'Electricite, **23**, pp. 901–912, pp. 946–954.

97. Lloyd, N. G. (1972) On the non-autonomous Van der Pol equation with large parameter, *Proc. Camb. Phil. Soc.*, **72**, pp. 213–227.
98. Lutz, R., Goze, M. (1981) Nonstandard analysis. A practical guide with applications, *LMN* **881**, Springer, Berlin.
99. Meron, E. (1992) Pattern formation in excitable media, *Physics Reports*, **218A**, pp. 1–66.
100. Meyer, D. (1973) *Two dimensional analysis of chemical oscillators*, in Biological and biochemical oscillators, (B. Chance, K. Pye, eds), Academic, New York.
101. Murray, J. D. (1993) *Mathematical biology*, Springer, Berlin.
102. Nagumo, J., Arimoto, S., Yoshizawa, S. (1962) An active pulse transmission line simulating nerve axon, *Proc. IRE*, **50**, pp. 2061–2070.
103. Nekorkin, V. I., Chua, L.O. (1993) Spatial disorder and wave fronts in a chain of coupled Chua's circuits, *Int. J. of Bif and Chaos*, **3**, **5**, pp. 1281–1291.
104. Nitecki, Z. (1971) *Differentiable dynamics. An Introduction to the orbit structure of diffeomorfisms*, M.I.T., Cambridge, Mass.
105. Noble, D. (1979) *The initiation of the heartbeat*, Oxford, Clarendon.
106. Obădeanu, V. (1992) *Introduction to analytic biodynamics*, Univ. of Timişoara Press, (Romanian).
107. Ohta, T., Mimura, M., Koloyashi, R. (1989) Higher dimensional localized patterns in excitable media, *Physica*, **D**, (34), pp. 115–144.
108. Othmer, H.G. (ed.) (1986) *Spiral waves in excitable media*, Springer, New York.
109. Panfilov, A. V., Holden, A. V. (eds) (1996) *Computational biology of the heart*, Wiley, New York.
110. Panfilov, A. V., Hogeweg. P. (1993) Spiral break-up in a modified FitzHugh–Nagumo model, *Phys. Lett.*, A, **176**, pp. 295–299.
111. Pavlidis, T. (1973) *Biological oscillators: their mathematical analysis*, Academic, New York.
112. Peixoto, M.M. (1962) Structural stability on 2-dimensional manifolds, *Topology*, **1**, pp. 101–120.
113. Peixoto, M.M. (1959) On structural stability, *Ann.of Math.*, **69**, pp. 199–222.
114. Petrişor, E. (1992) *Chaotic dynamical systems*, Univ. of Timişoara Press, (Romanian).
115. Pumir, A., Plaza, F., Krinsky, V (1994) Efect of an externally applied electric field on excitation propagation in the cardiac muscle, *Chaos*, **4**, pp. 547–555.
116. Pumir, A., Plaza, F., Krinsky, V. (1994) Control of rotation waves in cardiac muscle: analysis of the effect of an electric field, *Proc. R. Soc. London*, **B**, pp. 129–134.
117. Rensing, R. (ed.) (1987) *Temporal disorder in human oscillatory systems*, Springer, Berlin.
118. Rinzel, J. (1981) Models in neurobiology, in *Nonlinear phenomena in physics and biology*, (Enns, R., Jones, B., Miura, R., Rangneker, S. eds), Plenum, New York.
119. Rinzel, J. (1980) Impulse propagation in excitable systems, in *Dynamycs and modelling of reactive systems*, (Stewart, J., Ray, Conley, eds) Academic, New York.
120. Rocşoreanu, C., Giurgiţeanu, N., Georgescu, A. (1999) New types of codimension-one and -two bifurcation in the plane, *Preprint* **12**, Romanian Academy, Bucharest.
121. Rocşoreanu, C. (1998) Equilibria and relaxation oscillations of the nodal system of the heart 1. Bifurcation diagrams of equilibria, *Rev. Roum. Sci. Tech. Mec. Appl.*, **43**, (2) (in press).
122. Rocşoreanu, C. (1998) Equilibria and relaxation oscillations of the nodal system of the heart 3. Oscillatory and transient regimes, *Rev. Roum. Sci. Tech. Mec. Appl.*, **43**, (4) (in press).
123. Rocşoreanu, C. (1997) *Dynamics and bifurcation in the FitzHugh–Nagumo equation*, Romanian Academy, Bucharest, PhD Thesis (Romanian).
124. Rocşoreanu, C., Giurgiţeanu, N., Georgescu, A. (1997) Degenerated Hopf bifur-

cation in the FitzHugh–Nagumo system 1. Bogdanov–Takens bifurcation, *Ann. of Univ. of Timişoara*, **XXXV**, (2), pp. 285–298.
125. Rocşoreanu, C. (1996) A mathematical model for obtaining the arterial pressure, *Proc. of the Third Conf. on Appl. and Ind. Math.*, ROMAI, Oradea, pp. 98–102.
126. Rocşoreanu, C., Giurgiţeanu, N., Georgescu, A, Degenerated Hopf bifurcation in the FitzHugh-Nagumo system 2. Bautin bifurcation, *Mathematica*, Cluj-Napoca, (in press).
127. Ruwisch, D., Bode, M., Scutz, P. (1994) Parallel analog computation of coupled cell cycles with electrical oscillatory , *Physics Letters A*, , **186**, pp. 137–144.
128. Sanders, J. A., Verhulst, F. (1985) *Averaging methods in nonlinear dynamical systems*, Springer, New York.
129. Schaeffer, D. (1981) General introduction to steady state bifurcation, *LNM*, **898**, Springer, Berlin, pp. 13–47.
130. Schecter, S. (1987) The saddle - node separatrix - loop bifurcation, *SIAM J. Math. Annals*, **18**, pp. 1142–1156.
131. Schecter, S., Singer, M. (1980) Singular points of planar vector fields, in *Global theory of dynamical systems*, (Z. Nitecki, C. Robinson, eds), *LNM*, **819**, Springer, Berlin, pp. 393–409.
132. Sparrow, C. (1982) *The Lorenz equations. Bifurcations, chaos and strange attractors*, Springer, New York.
133. Starmer, F., Biktashev, C., Romashko, V.N., Stepanov, M.R., Makarova, O.N. (1993) Vulnerability in an excitable medium. Analytical and numerical studies of initiating unidirectional propagation, *Biophysical J*, **65**, pp. 1775–1787.
134. Takens, F. (1973) Unfoldings of certain singularities of vector fields: generalized Hopf bifurcations, *J. Diff. Eqs.*, **14**, pp. 476–493.
135. Tchizawa, K. (1997) On a coordinate transformation to express an explicit duck solution in the FitzHugh–Nagumo equation, *Proc. 7th Int. Coll. on Diff. Eq.*, (Bainov, D., ed), VSP, pp. 405–407.
136. Tu, P. (1994) *Dynamical systems. An introduction with applications in economics and biology*, Springer, Berlin.
137. Tyson, J. J., Keener, J. P. (1988) Singular perturbation theory travelling waves in excitable media (a review), *Physica D*, **32**, pp. 327–361.
138. Van der Berg, I. (1987) Nonstandard asymptotic analysis, *LNM*, **1249**, Springer, Berlin.
139. Van der Pol, (1926) On relaxation oscillations, *Phil. Mag.*, **2**, pp. 978–992.
140. Vasilieva, A.B., Butuzov, V. F., Kalachev, L. V. (1995) The boundary function method for singular perturbation problems, *Studies in Appl. Math.*, **14**, SIAM, Philadelphia.
141. Verhulst, F. (1990) *Nonlinear differential equations and dynamical systems*, Springer, Berlin.
142. Winfree, A. T. (1991) Estimating the ventricular fibrillation threshold, in *Theory of heart - biomechanics, biophysics and nonlinear dynamics of cardiac function*, (Glass, l., Hunter, P., McCulloch, A., eds), Springer, New York, pp. 477–533.
143. Winfree, A.T. (1991) Alternative stable rotors in an excitable medium, *Physica D*, **49**, pp. 125–139.
144. Zeeman, E. C. (1972) Differential equations for the heart beat and nerve impulse, in *Towards a theoretical biology*, (C. H. Waddington, ed), Aldine, Cambridge , pp. 8–67.
145. Zipes, D.P., Jalife J. (eds.) (1995) *Cardiac electrophysiology: from cell to bedside.* 2nd ed., W.B. Saunders, Philadelphia.
146. Zvonkin, A. K., Shubin, M. A. (1984) Nonstandard analysis and singular preturbations of ordinary differential equations, *Russian Math. Surveys*, **39**, (2), pp. 69–131.
147. *** (1994) MATHEMATICA. A system for doing mathematics by computer.

Index

MATHEMATICAL MODELLING:
Theory and Applications

1. M. Křížek and P. Neittaanmäki: *Mathematical and Numerical Modelling in Electrical Engineering*. Theory and Applications. 1996
ISBN 0-7923-4249-6
2. M.A. van Wyk and W.-H. Steeb: *Chaos in Electronics*. 1997
ISBN 0-7923-4576-2
3. A. Halanay and J. Samuel: *Differential Equations, Discrete Systems and Control*. Economic Models. 1997 ISBN 0-7923-4675-0
4. N. Meskens and M. Roubens (eds.): *Advances in Decision Analysis*. 1999
ISBN 0-7923-5563-6
5. R.J.M.M. Does, K.C.B. Roes and A. Trip: *Statistical Process Control in Industry*. Implementation and Assurance of SPC. 1999
ISBN 0-7923-5570-9
6. J. Caldwell and Y.M. Ram: *Mathematical Modelling*. Concepts and Case Studies. 1999 ISBN 0-7923-5820-1
7. 1. R. Haber and L. Keviczky: *Nonlinear System Identification - Input-Output Modeling Approach*. Volume 1: Nonlinear System Parameter Identification. 1999 ISBN 0-7923-5856-2; ISBN 0-7923-5858-9 Set

   2. R. Haber and L. Keviczky: *Nonlinear System Identification - Input-Output Modeling Approach*. Volume 2: Nonlinear System Structure Identification. 1999 ISBN 0-7923-5857-0; ISBN 0-7923-5858-9 Set
8. M.C. Bustos, F. Concha, R. Bürger and E.M. Tory: *Sedimentation and Thickening*. Phenomenological Foundation and Mathematical Theory. 1999
ISBN 0-7923-5960-7
9. A.P. Wierzbicki, M. Makowski and J. Wessels (eds.): *Model-Based Decision Support Methodology with Environmental Applications*. 2000
ISBN 0-7923-6327-2
10. C. Rocşoreanu, A. Georgescu and N. Giurgiţeanu: *The FitzHugh-Nagumo Model*. Bifurcation and Dynamics. 2000 ISBN 0-7923-6427-9

KLUWER ACADEMIC PUBLISHERS – DORDRECHT / BOSTON / LONDON